graphs, models, and finite mathematics

Joseph Malkevitch
York College (CCNY)

Walter Meyer
Adelphi University

PRENTICE-HALL, INC., Englewood Cliffs, New Jersey

49396

Library of Congress Cataloging in Publication Data

MALKEVITCH, JOSEPH,
    Graphs, models, and finite mathematics.

    Includes bibliographies.
      1. Mathematics—1961– 2. Graph theory.
3. Mathematical models. I. Meyer, Walter,
joint author. II. Title
QA39.2.M335 1973   510   73-7580
ISBN 0-13-363465-5

10   9   8   7   6   5   4   3   2   1

Printed in the United States of America

PRENTICE-HALL INTERNATIONAL, INC., LONDON
PRENTICE-HALL OF AUSTRALIA, PTY. LTD., SYDNEY
PRENTICE-HALL OF CANADA, LTD., TORONTO
PRENTICE-HALL OF INDIA PRIVTAE LIMITED, NEW DELHI
PRENTICE-HALL OF JAPAN, INC., TOKYO

To Nina and Rochelle

contents

4
paths and connectivity

5
computers and related topics

6
functional models

7
facts and figures

8
probability

9
games and decisions

preface

This text is designed for courses in finite mathematics and mathematics for the liberal arts student. It offers two main novelities: an early and substantial treatment of graphs (networks) and an explicit treatment of the process of mathematical modeling.

Beginning the book with graph theory seems to us advantageous because graph theory not only has many useful applications, but also is so conceptually simple that students find it fun. We believe it is a more exciting way to start the course than the more traditional set theory and logic.

Our treatment of modeling brings to the surface what is implicit in any book which discusses mathematical applications. Although it is a small step, it seems to us desirable and potentially very useful for mathematicians to comment on the interface between mathematics and the subjects to which it is applied. The concept of a model also provides some unity for the diverse subjects discussed in this book.

The philosophy which guides our choice of subject matter and approach is this: The best way to teach mathematics to the nonmathematics major is to present him with mathematics which is interesting, useful, and accessible. This means, for example, that we avoid axiomatics, formality, and excessive rigor.

Most standard topics for a finite mathematics course are included: probability, matrices, game theory, and linear programming. Although there is no separate chapter on matrices, they appear in connection with graphs in the chapter on computers and later in the chapter on probability. Set theory has been placed in an appendix so that the instructor can insert it where he likes. It is used in the text so sparingly that in the classroom we usually introduce set concepts only when needed. Chapter 5 on computers is designed to give students a general understanding and we avoid the intricacies of a programming language. The concepts of algorithms, flow charts, and machine language are covered. In addition there are short discussions on binary numbers and circuit theory.

Chapter 11 on difference equations and the limits to growth is included because we feel that a strong case can be made for the inclusion of this material in finite mathematics courses. Difference equations, being discrete analogues of differential equations, are extremely useful in modeling many dynamic processes of ever increasing concern to us, such as ecological balances and the exponential growth tendencies in population and in pollution. Some of these difference equations are more realistic models than differential equations and have the advantage of not requiring knowledge of calculus.

Chapters 1 through 5 should be taught in sequential order. The subsequent chapters are essentially independent of one another. Chapter 7 (statistics) has some material helpful for Chapter 8 (Probability), and Chapter 11 (Difference Equations) may require Sections 1 through 4 of Chapter 6 (Functional Models). Although Chapter 9 (Game Theory) can be best appreciated with some understanding of elementary probability (Chapter 8, Sections 8.1, 8.2, and 8.6), we have taught Chapter 9 without first covering Chapter 8 with considerable success.

We have starred sections which we feel many instructors may wish to skip either because they are more difficult than the other sections or because they are not in the main stream of the content of the chapter.

We have used examples from the social sciences and life sciences whenever possible since we feel that these areas have sometimes been slighted. With this in mind, we would appreciate receiving examples of interesting but simple mathematical models from the social sciences and life sciences. Please send such examples to J. Malkevitch, Department of Mathematics, York College, Jamaica, New York 11432.

Many persons have contributed to this book by making suggestions and corrections and by assisting in the preparation of the manuscript. We wish to extend our sincere thanks to all of them. Particular thanks are due to Professor Leroy Dickey and Professor Kenneth Hoffman for reading the manuscript and making helpful suggestions; to Reatha C. King, Dean for the Sciences and Mathematics at York College (CUNY) for her interest and unfailing support; and to Arthur Wester, our editor.

Joseph Malkevitch
Walter Meyer

graphs, models,
and
finite mathematics

1
graphs in disguise

Here is a collection of problems. Try your hand at solving some of them. These problems are chosen from a wide variety of fields, including nutrition, business, and urban planning. The mathematics we develop in this book will enable us to solve these and other practical problems arising in many different areas. In addition to developing the mathematics necessary to answer specific questions, we shall be concerned with the process by which practical questions are converted into mathematical ones. This process, called *modeling*, is the subject of Chapter 2.

MAILMAN PROBLEM

A mailman must distribute mail to all the houses along the boldly lined sides of the streets indicated in the map shown in Figure 1.1. He needs to begin and end his route at the Post Office, marked *A* on the map, and he would like to make his route as efficient as possible. The mailman takes his problem to MAIL (Mathematical Aid to Inefficient Lettercarriers), which designs a route for him according to the following principle: Each boldly lined sidewalk should be traversed only once. In other words, it should never be necessary to travel down a sidewalk where deliveries have already been made just for the purpose of reaching another part of the route.

QUESTION 1
MAIL found such a route. Can you?

1

Figure 1.1

MAIL DISTRIBUTOR PROBLEM

Suppose the Post Office Department wishes to assist the mailman of the previous problem by hiring a mail distributor to deposit bundles of mail in special boxes, one of which is located at some corner of *each* intersection of the mailman's route (see Figure 1.1). The mail distributor also needs to begin and end at the Post Office, marked A in Figure 1.1, and wants to have a route that is as efficient as possible. Since the distributor is using a truck, however, his idea of efficiency is different from that of the mailman. He decides that no intersection, other than A, should be passed through more than once.

The distributor takes his problem to MAID (Mathematical Aid to Inefficient Distributors).

QUESTION 2

What did MAID tell him?

SCHEDULING PROBLEM

The student government of Utopia State College is holding a symposium on urban problems featuring six speakers on the opening day. Each speaker plans an hour lecture. If the speakers are scheduled in six different time slots, the day's activities will be a bit lengthy. On the other hand, since several speakers are especially popular, it is undesirable that they speak at the same hour because this would require a person to choose which of two or more especially popular speakers he wants to hear. After some reflection, it is decided that there should be no more than four time slots. The following table, constructed on the basis of a student poll, indicates which speakers may or may not speak simultaneously. One reads the table in the following way: the X in the column labeled *A* and the row labeled *E* indicates that speakers labeled *A* and *E* should not speak simultaneously. If there is no tabular entry in a given row and column, this indicates that the corresponding speakers can speak simultaneously.

	A	*B*	*C*	*D*	*E*	*F*
A					X	
B						
C				X	X	X
D			X		X	X
E	X		X	X		X
F			X	X	X	

QUESTION 3

Can this symposium be scheduled in four time slots, subject to the restrictions given in the table?

PAPER RECYCLING PROBLEM

A paper recycling company uses two materials, scrap paper and scrap cloth, to make two different grades of recycled paper. A single batch of grade *A* recycled paper is made from 4 tons of cloth and 18 tons of paper, while one batch of grade *B* needs 1 ton of cloth and 15 tons of paper. The company has on hand 10 tons of scrap cloth and 66 tons of scrap paper. There is a $1000 profit on each batch of grade *A* paper and a $500 profit on each batch of grade *B* paper. It is not absolutely necessary to use up all the paper and cloth.

QUESTION 4

If the company produces 1 batch of each type, it will use 5 tons of scrap cloth and 33 tons of scrap paper, which is less than the available supply. This produces a profit of $1500. Can you do better without exceeding available supplies?

QUESTION 5

How many batches of each type should the company make to maximize its profit?

THE MODEL HOME PROBLEM

A contractor who plans a housing development wishes to construct some model homes to advertise and to encourage advance sales. Since he is anxious to have them done as soon as possible, he wishes to minimize the time required for their construction. He divides the job into a number of component activities as listed in the table below. The second column of the table lists the time required for the various activities. To save time the contractor decides to perform activities simultaneously whenever possible since he has sufficient manpower available to do this. Of course, it is not possible to perform certain activities simultaneously. For example, the main structure cannot be started until after the foundation has been poured. The last column of the table indicates which activities must immediately precede certain others. For example, line 3 is interpreted to mean that the activity of building the main structure takes 14 days and that this activity can be started immediately after the foundation has been completed. Note that since site preparation immediately precedes the building of the foundation, building the main structure is also preceded, though not immediately, by site preparation.

	Activity	Time (days)	Immediate Predecessors
(1)	Site preparation	2	None
(2)	Build foundation	3	Activity 1
(3)	Build main structure	14	Activity 2
(4)	Electric wiring	5	Activity 3
(5)	Plumbing	4	Activity 3
(6)	Interior finishing	7	Activities 4 and 5
(7)	Exterior finishing	11	Activity 3
(8)	Furnishing	2	Activities 6 and 7
(9)	Landscaping	4	Activity 7

QUESTION 6

What is the minimum amount of time required to complete the whole job?

QUESTION 7

If the job begins on day 1, on which days should the various component activities begin to achieve this minimum amount of time?

2
modeling

2.1   introduction

In order to solve the problems in Chapter 1, we shall discuss in this chapter some mathematical models of a special type. To help understand what a mathematical model is, let us first discuss our commonsense understanding of the word *model*.

A good place to start might be with model airplanes. It is interesting to note how far model airplanes are from being true airplanes. For example, they are usually made of balsa wood or plastic, materials that are absurdly inappropriate for construction of real airplanes. The parts are put together with glue instead of rivets or welded joints, the size is way off, and it is common for model planes not to have any moving parts. Nevertheless, everyone recognizes a model airplane for what it is, a *representation* of an airplane.

Similarly, when a chemist makes a model of a complicated molecule, he uses round balls to represent atoms and connects these balls with rods, which represent the chemical bonds. Round balls aren't much like atoms nor are rods much like chemical bonds, but the resulting assemblage (see, for instance, Figure 2.1) is a tolerably good model for a molecule.

These examples illustrate some important points about models:

(1)   The model may differ substantially from the object or situation it represents.

(2)   The model is simpler than the object or situation it represents.

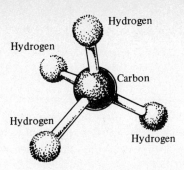

Figure 2.1   A methane molecule

Model airplanes are usually constructed just for the fun of it. Models often serve practical functions as well as aesthetic ones, however. A case in point would be the models or prototypes constructed by aircraft companies for the purpose of testing various design features for a plane that is just coming off the drawing boards. For example, the company may be concerned about the possibility of a wing snapping off an actual aircraft in flight, resulting in an air disaster. To test whether or not a new wing design will withstand the stresses it will encounter, a model may be constructed for testing in a wind tunnel. For the sake of simplicity and economy, design features such as landing gear and instrument panels that are not essential to the problem under consideration, namely, stresses on the wing, will not be included in the model.

For another insight into models, suppose for a moment you wished to take a cross-country drive from Miami to Seattle. How would you determine your route? One possibility would be to hop into your car and drive off, determining your route as you go along on the basis of road signs, trial and error, and your knowledge of geography. This seems so absurd and the alternative of using a road map is so obviously sensible that discussion may appear superfluous. Nevertheless, it is worth remarking that modeling the road system of the United States by a road map is a logically more sophisticated idea than simply getting in your car and driving off. The fact that map reading or map making is so simple and familiar to us in no way alters the fact that it is an important example of mathematical modeling. When man made his first map, he had made a tremendous conceptual stride.

Just as aircraft models are only approximations of real airplanes, road maps do not reflect every detail of the geography and cloverleafery they represent. Not only are the small twists, turns, and ups and downs straightened out, but certain more important details like toll rates and local speed limits are usually not included. Worst of all, small roads, such as the back streets of Topeka, may not even be on the map. Some of these details could be included but only at the cost of others because a map is limited in size and must be clear and readable. Precisely what information will be put into the map will depend on what the map is to be used for. A great deal is left to the judgment of the map maker.

The field of medicine is one in which the concept of a model is especially valuable. The value that we place on human life is so high that risking new drugs and speculative treatments on human patients is undesirable. Animals such as monkeys, dogs, rats, and mice are used as models for human beings. New drugs or surgical treatments are tried on these animals before testing them on human beings. For example, heart transplants were attempted on dogs much earlier than on human beings. Also, mechanical hearts have been transplanted into animals as a preparation for the development of a successful mechanical heart for humans.

In summary, then, a model is a *representation* of something else. It rarely reflects all the details of the reality it represents; for the sake of simplicity or economy, some things are left out. The simplest function of such a model is to give us an idea of what reality is like, as in the case of a hobbyist's model airplane. Often there is a more important function: Certain tests or investigations can be performed upon the model more simply or economically than upon whatever the model represents.

EXERCISES 2.1

1. List some other details about roads and terrain that are left off road maps. Can you illustrate how these details might be of interest to the motorist?

2. A filter is a mechanical object that is sometimes regarded as a crude model for the human kidney. Here are some other mechanisms that might be appropriate models for parts of the body. Can you match up each mechanism with a bodily organ?

(a)	Transport network	Blood system
(b)	Pump	Brain
(c)	Energy storage cell	Eye
(d)	Computer	Heart
(e)	Electrical wires	Female breast
(f)	Camera	Liver
(g)	Baby's bottle	Nerves

3. There are numerous contexts in which the word *model* occurs but which were not discussed in the text. An example might be *model home*. Can you give other examples? How do these examples fit the description in the last paragraph of this section?

4. When is a globe a better model for the earth than a flat map of the world? When is a flat map a better model than a globe?

5. A sphere is a mathematical idealization of certain objects that occur in the real world. Name four such objects and mention some ways in which they differ from a sphere.

6. Children's toys are often models of objects adults use. Give six examples.

7. Explain why rats are preferred to monkeys as models for human beings in certain medical and psychological experiments.

8. A psychologist is doing an experiment in an attempt to investigate learning behavior in humans. Explain why he might run mice rather than monkeys through a maze.

9.* Consult your local organic chemist to find out the uses of ball-and-rod models in chemistry.

10.* Investigate the history of the drug thalidomide. What danger does this spell for the transfer of drugs from experimental animals to human beings?

11.* How did the use of animals as models affect the marketing of cyclamates and DDT in the United States?

2.2   a model for the mailman

In this section we shall give an example of the special type of mathematical modeling we are interested in and we shall see how it sheds light on the mailman problem. Our models will not be physical ones such as the model aircraft. Rather, they will resemble road maps, in that they will be diagrams that can be drawn on a piece of paper. The sort of diagrams we have in mind are like those in Figure 2.2. When we get around to making formal definitions, we shall call such diagrams *graphs*.

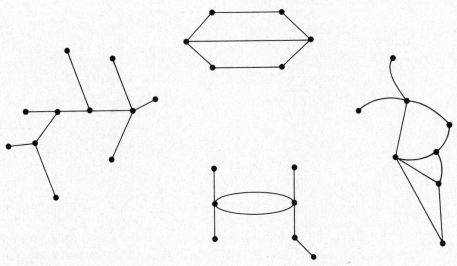

Figure 2.2

Our first step in modeling the mailman problem is to simplify the map of Figure 1.1 somewhat. We shall assume that the amount of walking the mailman does across intersections is small enough to be neglected and we shall replace each intersection with a dot, henceforth called a *vertex*. A curved or straight line segment, henceforth called an *edge*, drawn between two such dots will represent a sidewalk connecting the two intersections and along which the mailman needs to deliver mail. In this way we represent Figure 1.1 by the graph in Figure 2.3. Note that only the boldly lined sidewalks from Figure 1.1 are represented in Figure 2.3.

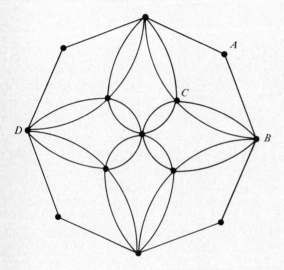

Figure 2.3

Recall that the mailman's job calls for him to travel down each edge of the graph. In some routes, however, it would be necessary for him to travel some sidewalk (edge) a second time, even after he delivered mail along that sidewalk (edge) once. This is precisely what MAIL wishes to avoid in designing a route for him. A route which starts at a specific vertex (corner), say, *A*, and ends at the same vertex and which uses each edge exactly once is called an *Euler circuit*. Such a route does exist in Figure 2.3, and you can probably find one by trial and error. Figure 2.4 shows an Euler circuit. (Just follow the numbers in the directions indicated by the arrows.) In Chapter 4 we shall find an elegant way to determine without trial and error whether a graph of this sort has an Euler circuit and, if it does, how to find one.

Before proceeding further, let us compare what we have done here to our earlier example of the model airplane in the windtunnel. In order to test

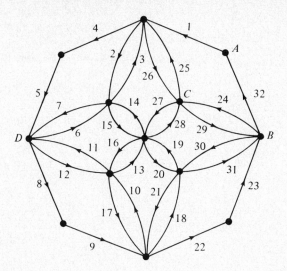

Figure 2.4

wing stress, the aircraft company builds a model of the proposed plane which is accurate only in those features which affect wing stress in a major way. In particular, landing gear, which has a minor and only indirect effect on wing stress, is left out. The mailman, in order to find the best route, draws a model of his route and plays pencil and paper games with his model instead of actually walking through the streets. This is analogous to the decision of the aircraft company to work in a wind tunnel rather than in the air. The model the mailman chooses represents each intersection by a single dot. Implicit in this is the assumption that he can disregard the time and effort spent in crossing intersections. This is analogous to the assumption that the landing apparatus doesn't affect wing stress a great deal.

After one has completed investigations in the model, one needs to consider what all this means for the original situation. For example, suppose the aircraft wing holds up in the tunnel; to what degree can one be confident that it will hold up when aloft? Is it conceivable that the retraction of the landing gear during takeoff may actually affect the wing stress enough to cause damage? Let us now interpret what the Euler circuit in Figure 2.4 means for our mailman. In Figure 2.5 we have transferred the corresponding circuit onto the original street map. When we examine this, some things become apparent that could not be so easily determined from the model in Figure 2.3. For instance, from Figure 2.5 we see that the mailman crosses the intersection at D three times. If D is the town's busiest intersection, it might have been better if MAIL's route had not required crossing this intersection so often. Furthermore, we see that so many streets must be crossed that the total distance covered in street crossings is considerable.

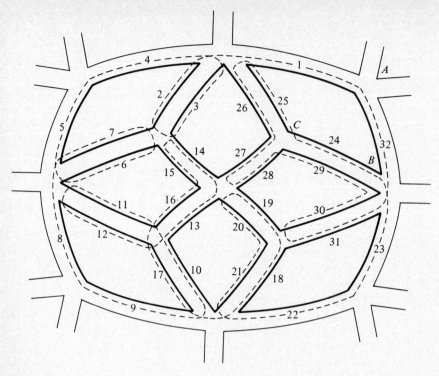

Figure 2.5

EXERCISES 2.2

1. Find an Euler circuit for the graph in Figure 2.3 different from the one shown in Figure 2.4.

2. Interpret the solution to Exercise 1 in the original street map of Figure 1.1. Is the number of crossings at intersection *D* in your solution fewer than the number in the solution given in the text?

2.3   *principles of modeling*

We have referred rather loosely to the concept of modeling and to the notion that a graph may be a model for a real situation. In the last section we actually carried out the process of modeling. Let us be a bit more explicit about what modeling means. There are four aspects to modeling with graphs:

CONVERSION: The real-world problem is simplified so that it can be converted to a problem about an appropriately constructed graph. Writing a list of ways in which the real-world problem has been simplified is often useful.

SOLUTION: The graph theory solution is obtained using either trial and error or mathematical techniques such as those we shall study in this book.

REINTERPRETATION: The graph theory solution is reinterpreted in terms of the real-world problem to obtain a solution in the real-world situation.

ESTIMATION OF FIT: Some estimate of how well the model fits reality is made in order to help determine how much faith we can have in the solution. The list of simplifying assumptions made in the conversion process will help in evaluating how good a fit we have.

The conversion step can be, in turn, broken down into three constituent parts:

(1) Decide what entities are to be represented as vertices of the graph.
(2) Decide which vertices are to be connected by edges. This is always done on the basis of some important relationships among the entities, which are represented as vertices.
(3) Restate the original problem in terms of the graph.

To illustrate these ideas, let us review the problem that we have modeled —the mailman problem.

THE MAILMAN PROBLEM REVISITED

CONVERSION:

(1) Vertices represent intersections of streets.
(2) Two vertices are connected when the intersections they represent are on the same street and one block apart down the street from one another.
(3) The problem of finding a route that covers each side of the street once and only once becomes the problem of placing a pencil at point A in Figure 2.3 and traversing each edge of the graph once and only once, ending at A.

SOLUTION: We already have given a solution to this problem (see Figure 2.4). You may have found a (different) solution of your own by trial and error. We shall postpone temporarily the question of how to do better than trial and error.

REINTERPRETATION: The trial and error solution in Figure 2.4 gives rise to the route shown in Figure 2.5.

ESTIMATION OF FIT: Because MAIL chose a very precise interpretation of how to provide an efficient route for the mailman—to traverse each bold-lined stretch of sidewalk once—the route that we found gives a perfect fit! But is the mailman's route really efficient in terms of the time necessary to

14 *modeling*

traverse his route or the total distance he must travel? Suppose, in Figure 2.3, that the intersection at *D* is extremely busy and without a traffic light. Our path requires the mailman to cross this intersection three times. There-fore, our solution may not really yield the best route for the mailman after all. Note also that by crossing streets the mailman must add distance to his route, and this is not reflected in the solution found in Figure 2.4 for the mathematical problem. This illustrates the fact that unless we put all the important data into the model, we may not get a sufficiently realistic answer.

Estimate of fit is an important aspect of the modeling process because we would like to be able to *use* the solution from our model. Often, however, it is the most difficult aspect of modeling to carry out. Usually so many simplifying assumptions are made during the conversion process that it is difficult to assess the adequacy of the solution.

The trial and error method we used to find an Euler circuit is obviously unsatisfactory from either a practical or theoretical point of view. It would be convenient if one could tell at a glance whether or not an Euler circuit existed in a given graph. For example, if the mailman is given a choice of 10 delivery districts and he wishes to determine whether any of them has an Euler circuit, he may do a lot of trying and a lot of erring before he can pick the best delivery district. Happily, there is a simple test, applicable to any graph, that tells us whether or not it has an Euler circuit. This test is:

For a connected graph to possess an Euler circuit, it is necessary that at each vertex an even number of edges meet. Conversely, if an even number of edges meet at each vertex of a given connected graph, then there will be an Euler circuit.

Notice that the test is very simple to apply and is not time-consuming. Applying it to the diagram of Figure 2.3, we see that at every vertex there are two, six, or eight edges meeting. Since two, six, and eight are all even numbers, the test assures us that there is an Euler circuit.

At this point, we shall not take time out to prove that this test really works, although we shall do so in Chapter 4. Note that the test above does not tell us how to find an Euler circuit even though we know one exists. In Chapter 4 we shall also consider methods of finding Euler circuits. We mention this test here mainly to illustrate how theoretical results concerning graphs can give a quick answer to a practical problem.

EXERCISES 2.3

1. State which of the graphs in Figure 2.6 have Euler circuits starting at the points marked *A*.

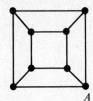 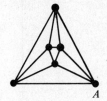 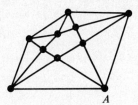

Figure 2.6   *A*   *A*   *A*

2. The mailman's supervisor offers to modify the mailman's territory by letting someone else deliver to one of the sidewalks (edges) connecting *B* and *C* (see Figure 1.1). The mailman declines the offer, commenting that it wouldn't save him any steps after all. After a moment's thought, however, the mailman tells his supervisor that if both the sidewalks (edges) between *B* and *C* are removed, he will agree to this modification. Can you explain this by reference to the Euler circuit test?

3. Make a list of the simplifying assumptions made in the conversion process for the mailman problem.

2.4 *modeling the mail distributor problem*

In this section we shall follow the modeling format of the last section in analyzing the mail distributor's attempts to deposit mail bundles at each intersection in Figure 1.1.

CONVERSION: Again we shall let vertices represent the intersections but our edges will now represent streets instead of sidewalks. Consequently, it will not be necessary to have two edges joining any pair of intersections but rather only one edge, as shown in Figure 2.7. Our problem now becomes that of tracing a path on this diagram, beginning and ending at *A* and passing through each *vertex* exactly once. Note the difference between this and the mailman's problem where we wished to pass along each *edge* exactly once.

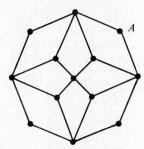

Figure 2.7

SOLUTION: Trial and error should convince you that no such route starting and ending at *A* and passing through each vertex—called a *Hamilton circuit*—exists for the graph in Figure 2.7. We shall see there is a simpler way of determining this, based on some theoretical considerations (see Exercise 2, below) rather than on trial and error.

REINTERPRETATION: The nonexistence of a Hamilton circuit in the graph of Figure 2.7 shows that there is no route for the mail distributor that goes through every intersection exactly once and begins and ends at *A*. The distributor will have to settle for something less efficient.

ESTIMATE OF FIT: The answer we have determined for the distributor's problem, while valuable and interesting, is by no means an end to the problem from a practical point of view. Although we have discovered that the ideal is unattainable, we have not given any indication of how close the distributor could come to this ideal. Can he arrange a route that duplicates only one intersection? What is the minimum number of duplications he can get away with? The fact that these questions can be posed and are unanswered indicates that the model in Figure 2.7 can be used as a jumping off point for further investigations.

EXERCISES 2.4

1. For each of the graphs indicated in Exercise 1, p. 15, determine by trial and error which ones have Hamilton circuits.

2. Can you prove (i.e., give ironclad reasons) why the graph in Figure 2.8 has no Hamilton circuits? (*Hint:* A circuit must alternate between the row of two vertices at the top and the column of three vertices below.)

Figure 2.8

3. Answer the questions raised in the discussion of estimate of fit by using trial and error in the graph in Figure 2.7.

4. Give an "exhaustive" list of the simplifying assumptions made in the conversion step of the distributor problem.

2.5 modeling the scheduling problem

In this section we shall follow the modeling format of Section 2.3 to analyze the scheduling problem. This may be more interesting than the last two problems because it is not so obvious what graph should be drawn.

CONVERSION: Each speaker will be represented by a vertex. Two vertices will be joined by an edge provided the speakers they represent should *not* speak during the same hour. If the speakers may speak simultaneously, they will not be connected by an edge. In this way we arrive at the graph of Figure 2.9(a).

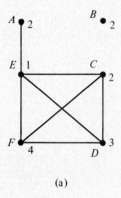
(a)
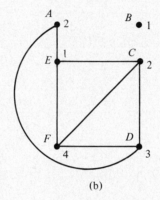
(b)

Figure 2.9

Notice that two of the edges in the graph cross. This could have been avoided by letting one of these edges curve around as in Figure 2.9(b). There is no harm in letting them cross, however, as long as we do not become confused and consider the intersection point a vertex representing some speaker. Our problem now is to provide each vertex (speaker) with one of the numbers 1, 2, 3, or 4 representing the four time slots so that two vertices joined by an edge never have the same number. For historical reasons, problems of assigning labels to the vertices of a graph so that vertices connected by an edge receive different labels are called *coloring* problems. One might think of coloring vertices so that vertices joined by an edge are colored differently.

SOLUTION: A solution is shown in Figure 2.9(a) itself. Another equally valid solution is shown in Figure 2.9(b).

INTERPRETATION: Interpreting the numbering of our first graph [Figure 2.9(a)] we can assert that four time slots are sufficient. One way to assign these times is to give the same time to speakers *A*, *B*, and *C* and then apportion the three other times among the other speakers, each speaker receiving a different time slot.

ESTIMATE OF FIT: According to the information given, we seem to have obtained a solution that fits the problem perfectly well. In the spirit of practical analysis, it may be interesting, however, to consider some changes in the data of the problem that would lead us to different solutions. The most natural change might be to assume that we are dealing with a larger number of speakers. This would not change the nature of the problem—it would still be a coloring problem—except to make the trial and error procedure for solving the problem more difficult. A second change might be that we wish to have about the same number of people speaking during each time slot. This might be desirable from the point of view of room availabilities and might ensure that no speaker would feel slighted by having to compete with a large number of other speakers while some speakers would have no competing speakers. In our example of six speakers, a step toward such equalization would be taken by using the second solution, Figure 2.9(b), in which the largest number of speakers in a time slot is two, compared with three in the first solution, which is shown in Figure 2.9(a). Is any further equalization possible with four time slots?

Our modeling of this problem illustrated a very significant fact. Often it is not obvious what graph to draw as a model for the situation. A certain amount of trial and error may be necessary. We might start out representing speakers as vertices but connecting them when they may speak simultaneously. Such a graph may be useful for other problems involving this symposium but it does not seem useful for the given problem. We can then try another criterion for connecting the vertices by edges. Here the only other criterion is the one we actually used. Usually there are not too many methods of connecting vertices that suggest themselves. After a bit of hunting around, one finds the useful method.

EXERCISES 2.5

1. Discuss some practical considerations that would make our modeling of the scheduling problem impractical.
2. For each of the graphs in Exercise 1 on p. 15, state the smallest number of colors needed to color them so that vertices joined by an edge receive different colors.
3. Color each of the graphs in Figure 2.10 using the minimum number of colors possible each time.

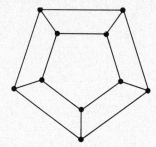

Figure 2.10

4. Color the vertices of the graph in Figure 2.11 with as few colors as possible.

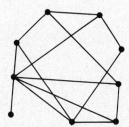

Figure 2.11

5.* Suppose a graph G is such that each vertex has $\leq k$ edges touching it. Show that G can be colored with $k + 1$ colors.

2.6 the relationship between models and problems

There are two important aspects of mathematical modeling that have not been mentioned so far. They are that:

(1) For a given problem there may be a number of mathematical models that could be useful.
(2) Each type of mathematical model will usually be applicable to many different problems.

In this section we want to illustrate these points in the context of graph theoretical models. We begin with point (1) by giving an alternative to the Euler circuit model for the mailman problem.

Example 1

The motivation for this alternative model for the mailman is the notion that it might be wise to take into account the time and effort required to cross intersections. In our earlier model an entire intersection was represented by a single point, which meant that the distance traveled in crossing an inter-section was ignored. In our new model, instead of representing each inter-section by a vertex, we represent each corner of a block by a vertex (Figure 2.12). We join two corners if they are (1) adjacent on the same block or (2) across from one another at an intersection. We assume that if the corners are diagonally across, as at the center intersection, the mailman cannot or will not make a diagonal crossing. For convenience we represent the first type of edge in boldface. In addition to drawing these edges we shall also weight them with numbers that represent the time required to travel them. We formulate the problem this way: Find a circuit in this graph which begins and ends at A, which covers every boldfaced edge at least once but possibly more often, and which takes the minimal amount of time to traverse.

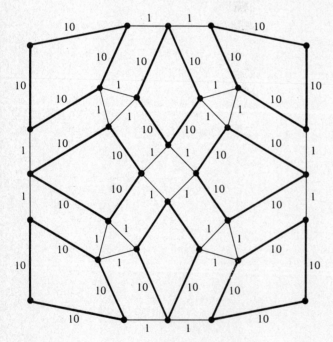

Figure 2.12

Can you find a circuit of this sort that requires only 332 time units to traverse?

This model is more realistic than our earlier one since it takes into account the intersection crossings and since it has time estimates on the edges. However, this doesn't necessarily make it better than the earlier model. In exchange for being more realistic, it is also more difficult to work with. For example, it probably took you longer to find a circuit with time-length 332 in Figure 2.12 than it took you to find the Euler circuit in Figure 2.3. Having found a path of time-length 332, how sure are you that this is the best you can do? Assuming that you are happy with a solution involving 332 time units, how much is really gained over the solution in our earlier model?

To illustrate point (2) about the wide applicability of a given type of

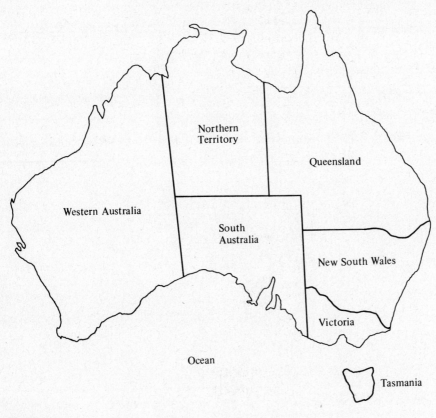

Figure 2.13a

model, we shall describe the problem that historically gave rise to the idea of a graph coloring. This problem seems to have little to do, at least superficially, with our scheduling problem. The two problems are mathematically quite similar, however, since both can be reduced to a graph coloring problem.

Example 2

Map makers traditionally color the various regions and bodies of water in their maps with different colors to aid the reader. In order for this to be helpful, it is imperative that regions that have a common stretch of border receive different colors. This suggests the question of minimizing the number of colors required. Figure 2.13(a) shows a map of the states of Australia, including the island state of Tasmania. By experimenting directly on the map you can probably find a coloring with four colors. Alternatively, we could represent each country and the ocean by a vertex, join two of them if they share a stretch of border, and then attempt to color the vertices of the resulting graph [Figure 2.13(b)].

This problem of coloring the countries and bodies of water of a map gives rise to a very famous mathematical conjecture called the *four-color conjecture*, which seems to be true but whose validity has never been proved.

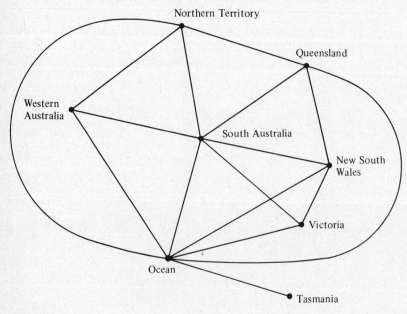

Figure 2.13b

The conjecture is that any conceivable map that could be drawn on the surface of a sphere can be colored with four colors. This problem was invented around 1852 by a student and since then has baffled the world's best mathematicians, an interesting reversal of the usual relationship between students and professors.

Our final example of the versatility of a given modeling approach involves Hamilton circuits.

Example 3

A factory makes eight varieties of guitar pick using just one machine, the Presto Pick Puncher. This machine needs different adjustments in order to make the different styles of guitar pick. The company desires to spend 1 hr of each working day on each style of pick. At the end of each hour the adjustments on the machine need to be converted to those necessary for the next pick style. At the end of the day the machine must be converted back to the style produced at the beginning of the day. The times required for conversion depend on which styles one is converting between and are given below.

	A	B	C	D	E	F	G	H
A		5	2	6	6	5	2	5
B	5		3	4	5	6	1	6
C	2	3		1	6	5	5	6
D	6	4	1		2	4	5	5
E	6	5	6	2		3	6	2
F	5	6	5	4	3		3	6
G	2	1	5	5	6	3		1
H	5	6	6	5	2	6	1	

Since these conversion times are not all the same, it makes a difference in what order the styles are made. Can you arrange an order, beginning and ending with style *A*, so that no conversion ever takes more than 3 min? If not, can you do it so that no task takes more than 4 min?

Solution: We shall draw a graph whose vertices represent the styles and where two styles are joined by an edge, provided the machine can be converted from one to the other in 3 minutes or less. In this graph [Figure 2.14(a)] a path visiting a sequence of vertices indicates that those styles, in the order in which they occur in the path, can be performed in sequence with all conversions requiring only 3 minutes or less. Since we are looking for such a sequence of styles which also has the property that each style occurs exactly once (except that the last style *A* is the same as the first), we clearly need to find a Hamilton circuit in the graph of Figure 2.14(a) which begins and ends at *A*. A little trial and error should convince you that no such Hamilton circuit exists.

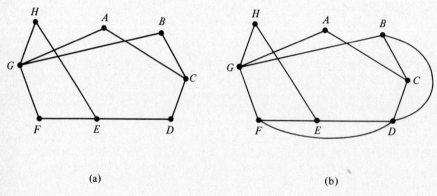

(a)   (b)

Figure 2.14

To determine whether a sequence exists in which no conversion takes more than 4 minutes, we need to work with a different graph. We use the same set of vertices as before, one for each style, but now we join styles where the conversion between them requires no more than 4 minutes. This yields the graph of Figure 2.14(b). Once again the problem is to find a Hamilton circuit beginning and ending at *A*. This time you should be successful in finding one.

EXERCISES 2.6

Each of the exercises that follow can be modeled as a problem involving a graph. Furthermore, each exercise, when interpreted in terms of the graph, involves finding an Euler circuit (as in the mailman problem), a Hamilton circuit (as in the mail distributor problem), or a coloring of the graph (as in the scheduling problem.) For each exercise below, draw the appropriate graph, describing what the vertices represent and on what basis the edges are drawn between vertices. State whether the given problem involves finding an Euler circuit, a Hamilton circuit, or a coloring of the vertices for the graph you have drawn. Then give an estimate of fit between the graph you have drawn and the real-world situation it models. Finally, using the graph you have drawn, solve the problem. If you are unable to see how to carry out any of these steps, consult the hints that follow these exercises.

1. A highway inspector is supposed to check for potholes by driving down each of the streets indicated in Figure 2.15. Assume each street is a two-way street, but he needs to go down it in only one direction to inspect all lanes. Can he arrange a route for himself that covers each street, covers none more than once, and begins and ends at the corner labeled 1?

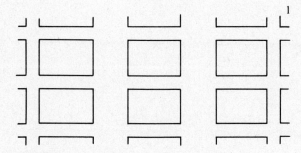

Figure 2.15

2. Figure 2.16 shows a road map of the island of Zanderbecca. Vertices represent cities and edges are roads. In each city one of the following facilities is to be built: a theater, a sports stadium, and a swimming pool. Furthermore, it is decided that when two cities are next to one another on a road, they should not receive the same facility. Is such an assignment of facilities possible?

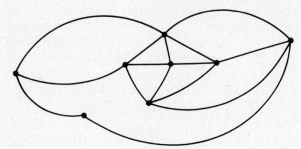

Figure 2.16

3. Figure 2.17 shows a rough sketch of part of Canada. Can you color the provinces so that those that share part of a border are colored differently?

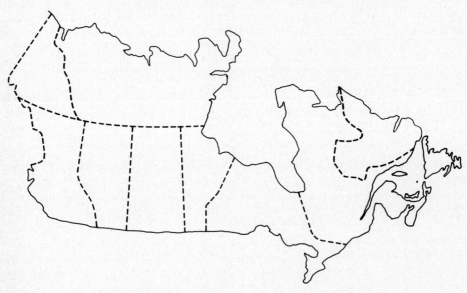

Figure 2.17

4. Mr. Daily's first-grade class is secretly planning an insurrection. The seven key leaders are listed in the chart below. It is decided that during recess the written details of the uprising will be passed among the seven, starting with Abe. The process is complicated by two factors. First, ideological rifts among the seven make even casual contacts between some of them impossible without risk of a major tantrum. Second, in the interest of avoiding interception, it is required to pass the piece of paper, starting and ending with Abe, among all seven without it ever reaching anyone twice (except Abe). The chart indicates which pairs of students have friendly relations (F) and can make contact to pass a message and which pairs are not friends (NF) and cannot have a message passed directly between them. Can the message be passed as required?

	Abe	Tom	Dave	John	Lee	Gerry	Ren
Abe	—	F	NF	NF	F	NF	NF
Tom	F	—	F	F	NF	NF	F
Dave	NF	F	—	NF	F	F	F
John	NF	F	NF	—	F	F	NF
Lee	F	NF	F	F	—	F	F
Gerry	NF	NF	F	F	F	—	NF
Ren	NF	F	F	NF	F	NF	—

5. Mr. Daily puts down the rebellion of Exercise 4 and wishes to punish the seven leaders with an extra graph theory problem for homework. His plan is complicated by the fact that friendly students will collaborate (assume that unfriendly ones will not) if they are given the same problem. Therefore, he would like to make sure that friendly students are given different problems. He would like to minimize the number of different problems he assigns. What is the minimum number?

6. A domino is a rectangle divided into two squares, into each of which is placed one of the numbers between 0 and 6. We shall allow two dominoes to be placed end to end in a row, as in Figure 2.18(a), when the squares at which the dominoes meet are labeled with the same numbers. When the two squares do not have the same labels, as in Figure 2-18(b), we do not allow the configuration to be formed. Similarly, we do not allow L-shaped configurations as in Figure 2.18(c). We do, however, allow a domino to be turned around as in Figure 2.18(d). Suppose you are given dominoes with the following labelings: (1, 2), (1, 4), (1, 6), (2, 3), (2, 6), (3, 5), (3, 4), (4, 5), (5, 6). Can you arrange them all in a row in accordance with the rules above and so that the first number in the row is the same as the last number in the row?

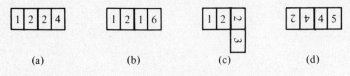

(a)   (b)   (c)   (d)

Figure 2.18

7. An airport radar device designed to assist in the prevention of midair collisions scans eight sectors in the vicinity of the airport. When two planes are detected in close proximity, the number of the sector in which this is occurring is reported, providing a rough "fix" on the possibly dangerous situation. Since computers use numbers written in a binary code, the sectors are labeled 000, 001, 010, 011, 100, 101, 110, 111, as in

Figure 2.19(a). When the scanner is in a particular sector, the computer keeps the number of that sector "in its head," reporting it to the operator only when two planes are detected. As the scanner passes from one sector to another, the number held by the computer changes. Thus, as the scanner goes from sector 0 to sector 1, the computer changes what it holds from 000 to 001. In doing this, only 1 digit (the final 1) needs changing. By contrast, in going from sector 3 to 4 the computer changes its number from 011 to 100, a change involving 3 digits. Such a change involves a higher risk of error due to lack of synchronization than a change involving fewer digits. You may verify that the total number of digit changes in going counterclockwise from 000 back to 000 with the labeling of Figure 2.19(a) is 14. If we relabel the sectors as in Figure 2.19(b), however, we have only 12 digit changes. Can you relabel the sectors so that only 8 digit changes take place; that is, can you relabel the sectors so that in going from any sector to the next only one digit needs changing?

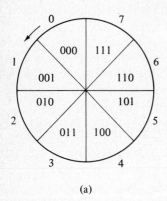

(a)

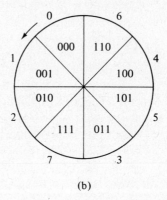

(b)

Figure 2.19

8. In a newly emergent nation, each of the six principal cities is to be given one television station. Four different channels are available. Luckily, it is not necessary to have different channels for different stations. It is necessary, however, that two cities that are within 80 mi or less of one another receive different channels to avoid possible interference problems. Let the cities be named A, B, C, D, E, and F, and let the distances between them be given by the following chart.

	A	B	C	D	E	F
A	0	150	135	96	76	84
B	150	0	96	147	90	102
C	135	96	0	72	76	51
D	96	147	72	0	69	48
E	76	90	76	69	0	24
F	84	102	51	48	24	0

Can you make an assignment of the four channels such that cities within 80 mi or less receive different channels?

9. Modern trends in zoo keeping call for large enclosures of a few acres or more in which a number of different species are allowed to roam freely and through which zoo visitors pass in enclosed or elevated vehicles. Suppose that a zoo wishes to provide such enclosures for 10 species listed in the table below. Things are complicated by the fact that it is thought to be inadvisable to allow species that are natural enemies to live in the same enclosure. The table indicates such incompatibilities by the mark X. No entry in a box indicates compatibility. For reasons of economy it is desired to have as few different enclosures as possible. What is that minimum number?

	1	2	3	4	5	6	7	8	9	10
1			X							
2				X						
3	X			X						
4		X	X							
5						X				X
6					X		X			
7						X		X		X
8							X		X	
9								X		X
10					X		X		X	

10. A scoutmaster is planning to drive the eight members of his troop in cars to a jamboree. To avoid trouble, only boys who are friends (F) are to go in the same car. What is the smallest number of cars that are necessary to transport the boys? If cars could carry only two boys, would you have a different answer?

	1	2	3	4	5	6	7	8
1	—	F	NF	NF	NF	F	NF	F
2	F	—	F	F	NF	NF	NF	F
3	NF	F	—	NF	NF	F	NF	NF
4	NF	F	NF	—	F	F	NF	NF
5	NF	NF	NF	F	—	F	F	NF
6	F	NF	F	F	F	—	F	F
7	NF	NF	NF	NF	F	F	—	NF
8	F	F	NF	NF	NF	F	NF	—

Hints for the Exercises in Section 2.6

1. Let intersections be vertices and join two of them if they are a block apart on the same street. Reread the analysis of the mailman problem.
2. The graph is already drawn. Reread the scheduling problem.
3. Let vertices represent provinces and join two of them if they have a common border. A convenient way to do this is to place the points inside the respective countries and join them across the common border arcs. Now you need to assign colors to vertices subject to a certain condition.
4. Let vertices represent people and join them according to the chart. The chart contains at least two distinct criteria that might be used to determine whether vertices should be joined or not. One is useful; the other is not. Reread the mail distributor problem.
5. Let vertices represent people and use the same graph as in Exercise 4. Reread the scheduling problem.
6. Let vertices represent numbers and join two of them if there exists a domino containing precisely those two numbers. In this graph the edges

can be thought of as representing dominoes. What would using all the dominoes mean in terms of the graph? Conversely, in terms of the dominoes, what is signified by two edges with a common point?

7. Let the binary numbers 000, 001, etc., be represented by vertices and join two of them if only one digit needs to be changed to get from one to the other. The solution involves the same sort of circuit we looked for in the mail distributor problem.

8. Let the cities be vertices and connect two if those cities are within 80 mi. Reread the scheduling problem.

9. Let the species be vertices, joined on the basis of incompatibility.

10. Let the boys be vertices, joined on the basis of friendship or non-friendship. Which?

SUGGESTED READING

BELLMAN, R., K. COOKE, AND J. LOCKETT, *Algorithms, Graphs and Computers,* Academic Press, New York, N.Y., 1970. This is probably the best view of applied graph theory available. The flavor is right and very little is included just because it is cute.

RENYI, A., "A Socratic Dialogue on Mathematics," in *Dialogues in Mathematics,* Holden-Day, Inc., San Francisco, Calif., 1967. This short dialogue examines the relationship of mathematics to the real world. The tone is philosophic but definitely not abstruse.

TUCKER, A., "Perfect Graphs and an Application to Optimizing Municipal Services," *SIAM Review,* **15**, No. 3, July, 1973. Some interesting insights into the relationships between pure and applied mathematics. The article is an expository survey and contains no proofs.

3
graphs and digraphs

3.1   graphs and isomorphism

We have examined some problems that can be usefully modeled by a geometric structure consisting of points and properly chosen connecting lines. Since we shall be discussing such structures at length, we shall give precise definitions of what we are talking about in this section.

DEFINITION 1

A graph G is a finite set of points, called *vertices*, together with a finite set of curved or straight connecting lines called *edges*, each of which joins a pair of vertices. These vertices and edges must satisfy the condition that no edge begins and ends at the same vertex.

If G is a graph, the set of vertices of G is denoted $V(G)$ and is always assumed to be nonempty. The set of edges of G is denoted $E(G)$ and may be empty, as in the graph depicted in Figure 3.1(a). Graphs without edges are called *null graphs*. They are seldom useful and always uninteresting.

At the other extreme, one may have every pair of vertices joined by a single edge. In this case we call the graph a *complete graph* and denote it K_n where n is the number of vertices. Figure 3.1(b) shows the complete graph K_4.

It is important to observe that the definition of a graph imposes restrictions on the types of structures that qualify as graphs. Thus, the structure in Figure 3.2 is not a graph because it violates the condition that no edge may join a vertex to itself.

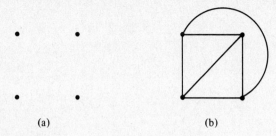

(a)   (b)

Figure 3.1

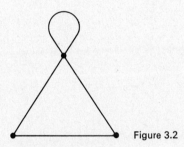

Figure 3.2

When a graph has two or more different edges joining the same pair of vertices, these edges are called *multiple edges*. See Figure 3.3 for an example of such a graph.

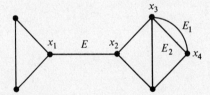

Figure 3.3

We may sometimes refer to an edge of a graph by naming the two vertices that determine it—these vertices are called *end points* of the edge in question. Thus the edge E in Figure 3.3 could be denoted $[x_1, x_2]$ since x_1 and x_2 are the end points of E. In the graph of Figure 3.3, however, the edges E_1 and E_2 both have x_3 and x_4 as their end points. In this case, the notation $[x_3, x_4]$ would be ambiguous and we shall not use it, preferring to refer to the edges as E_1 and E_2.

It is convenient to have some terminology for the number of edges that share the same end point.

DEFINITION 2

Suppose G is a graph and x is a vertex of G. The *valence* of x, denoted val(x), is the number of edges of G that have x as one end point. If val(x) = i, we say x is *i-valent*.

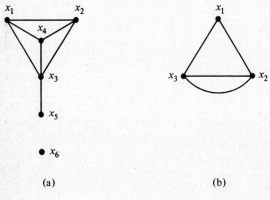

(a)   (b)

Figure 3.4

If G is the graph of Figure 3.4(a), val(x_1) = val(x_2) = val(x_4) = 3, val(x_3) = 4, val(x_5) = 1, while val(x_6) = 0.

If H is the graph with multiple edges in Figure 3.4(b), val(x_1) = 2, val(x_2) = val(x_3) = 3.

A few remarks concerning how to draw graphs are in order. In Figure 3.5 we have reproduced three graphs that three different people used to model the friendship relations among four individuals, Tom, Dick, Harry, and Gideon.

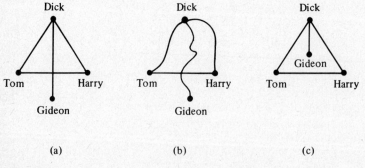

(a)   (b)   (c)

Figure 3.5

All these graphs represent the same situation and, in some sense, the graphs are structurally the same. For example, the graphs in Figure 3.5(a) and (c) can be transformed into one another by moving the vertex representing Gideon up above the edge joining Tom and Harry. It is important to realize that the point where the edge joining Tom and Harry [Figure 3.5(a)] meets the edge joining Dick and Gideon is *not* a vertex. This crossing point is present by accident so to speak and certainly does not represent a person. To avoid possible confusion, we make the following convention:

CONVENTION

All vertices in the graphs drawn in this book will be denoted with dark dots.

As a consequence of this convention, if a pair of edges intersect at a point that is not drawn with a heavy dark dot, we shall regard such a crossing point as "accidental" and shall not take it to be a vertex of the graph.

At this point, we wish to give a more precise idea of what we mean when we say that two graphs are *structurally the same*. The technical term we shall use to indicate this condition is *isomorphic*. The word has Greek origins: *iso* means equal and *morphe* means structure.

DEFINITION 3

Suppose G is a graph with n vertices labeled $x_1, \ldots, x_n$. We shall say that graph G' is *isomorphic* to G (or G' and G are isomorphic) if it is possible to find some labeling of the vertices of G' with $x'_1, x'_2, \ldots, x'_n$ in such a fashion that:

(1)   Each vertex of G' gets one and only one label.
(2)   If the vertices x_i and x_j ($x_i \neq x_j$) of G are joined by k ($k \geq 1$) edges, then the vertices labeled x'_i and x'_j in G' are joined by k edges.
(3)   If the vertices x_i and x_j ($x_i \neq x_j$) of G are not joined by an edge, then the vertices labeled x'_i and x'_j in G' are not joined by an edge.

If it is impossible to label G' as required, then G and G' are not isomorphic.

Example 1

Verify that the labelings of the graphs G and G' in Figure 3.6 show that they are isomorphic graphs.

Solution: We must verify that the number of edges that join any pair of vertices x_i and x_j in G is the same as the number of edges joining vertices x'_i and x'_j in G'. The results are shown in tabular form.

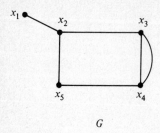

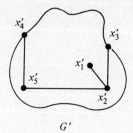

G   G'

Figure 3.6

Vertices in Graph G	Number of Joining Edges in G	Vertices in Graph G'	Number of Joining Edges in G'
x_1 and x_2	1	x_1' and x_2'	1
x_1 and x_3	0	x_1' and x_3'	0
x_1 and x_4	0	x_1' and x_4'	0
x_1 and x_5	0	x_1' and x_5'	0
x_2 and x_3	1	x_2' and x_3'	1
x_2 and x_4	0	x_2' and x_4'	0
x_2 and x_5	1	x_2' and x_5'	1
x_3 and x_4	2	x_3' and x_4'	2
x_3 and x_5	0	x_3' and x_5'	0
x_4 and x_5	1	x_4' and x_5'	1

Since the numbers in columns 2 and 4 are the same, the graphs G and G' are isomorphic.

Notice that in Example 1 the valences of corresponding vertices (e.g. x_1 corresponds to x_1') are the same. This is not an accident. It is an instance of the following theorem, whose proof is an easy consequence of the definition of isomorphism.

THEOREM 1

If G and G' are isomorphic and x_i and x_i' are corresponding vertices in these graphs, then val(x_i) = val(x_i').

The following example shows how this theorem can be used to help determine an isomorphism.

Example 2

Determine whether the graphs G and G' in Figure 3.7 are isomorphic.

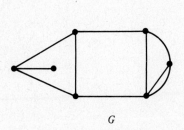 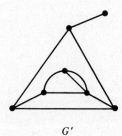

G G'

Figure 3.7

Solution: Since none of the vertices of G in Figure 3.7 are labeled, we begin by labeling G's vertices by $x_1, \ldots, x_7$ in an arbitrary way, for example, as in Figure 3.8.

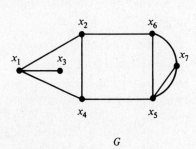 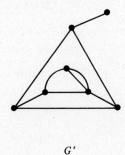

G G'

Figure 3.8

Since vertex x_3 in G is 1-valent and G' has only one 1-valent vertex, by Theorem 1 this vertex must by labeled x'_3 [see Figure 3.9(b)]. Since the vertex x_1 is joined to x_3 in G, we can label the vertex joined to x'_3 in G' by x'_1 [see Figure 3.10(b)].

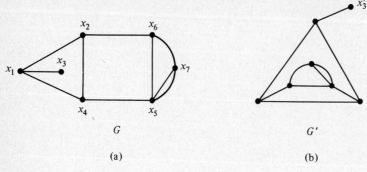

G

(a)

G'

(b)

Figure 3.9

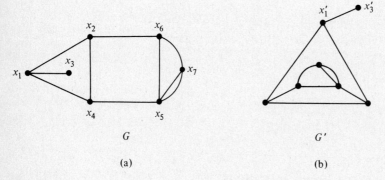

G

(a)

G'

(b)

Figure 3.10

In order to continue the labeling of G', we note that vertex x_1 in G is joined to two 3-valent vertices, namely, x_2 and x_4. This means that we have a choice as to how to continue the labeling of G'. The possibilities are shown in Figure 3.11.

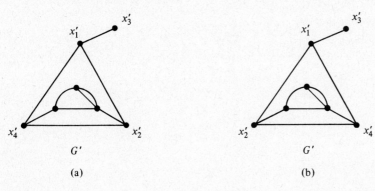

G'

(a)

G'

(b)

Figure 3.11

Suppose we attempt to complete the labeling in Figure 3.11(a). Since x_4 is joined to the 4-valent vertex x_5 in G, x'_4 would have to be joined to a 4-valent vertex in G'. However, there is no 4-valent vertex in Figure 3.11(a) connected to x'_4. Hence, this labeling [Figure 3.11(a)] amounts to having barked up the wrong tree. Abandoning this attempted labeling, let us try to complete the labeling in Figure 3.11(b). In this labeling of graph G', since x'_4 is joined to a 4-valent vertex, we label this 4-valent vertex x'_5 (see Figure 3.12).

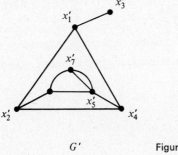

G'

Figure 3.12

Since x_5 is joined by two edges to x_7, it is clear which vertex to label x'_7 in Figure 3.12. The only remaining vertex is now labeled x'_6. One can check, using a table similar to the one in Example 1 (but having 21 lines), that the labeling in Figure 3.13 exhibits an isomorphism.

If neither of the labelings in Figure 3.11 could be completed, then the graphs would not have been isomorphic.

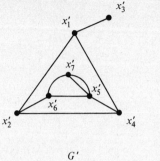

G' Figure 3.13

Example 3

Determine whether the graphs G and G' in Figure 3.14 are isomorphic.

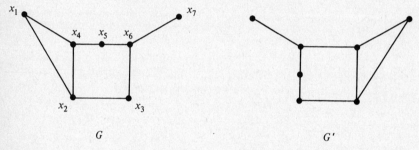

G G'

Figure 3.14

Solution: Since G has precisely one 1-valent vertex x_7 joined to the 3-valent vertex x_6, we begin the labeling of G' as in Figure 3.15. Since x_6 is joined to two 2-valent vertices, the vertex x_6' would have to be joined to two 2-valent vertices to complete the labeling of G'. Since x_6' is not joined to two 2-valent vertices, G and G' are not isomorphic.

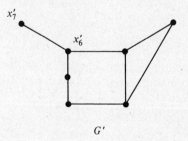

G' Figure 3.15

40

One warning about a common source of error should be made. Consider the labeling of the graphs G and G' in Figure 3.16.

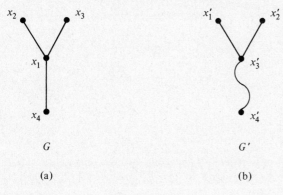

G

(a)

G'

(b)

Figure 3.16

Note that although x_1 and x_2 are joined in G, x_1' and x_2' are not joined in G'. However, G and G' *are* isomorphic. The labeling in Figure 3.16(b) will not demonstrate this, yet the labeling in Figure 3.17 enables one to conclude G and G' are isomorphic.

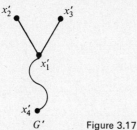

G'   Figure 3.17

With a little practice you will be able to pick out structural properties of a pair of graphs G and G', which, when they are not present in both, will show that G and G' are not isomorphic.

For example,

(1)   If G and G' do not have the same number of vertices and the same number edges, they are not isomorphic.

(2)   If G and G' do not have the same number of k-valent vertices, they are not isomorphic.

EXERCISES 3.1

1.   Explain why each of the structures in Figure 3.18 is not a graph.

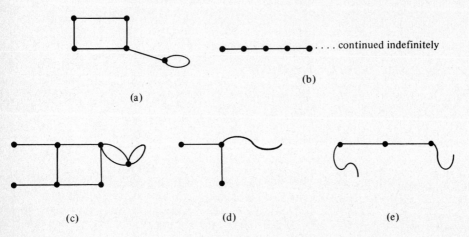

.... continued indefinitely

(b)

(a)

(c)   (d)   (e)

Figure 3.18

2.   (a)   Draw a complete graph with six vertices. How many edges are
           there in the graph you drew?
     (b)   Draw a null graph with eight vertices.
     (c)   If a complete graph has n vertices, how many edges does it have?
     (d)   Can you think of real-world situations where the appropriate
           graph model would be (i) the complete graph? (ii) the null graph?

3.   What are the valences of the vertices in the graphs in Figure 3.19?

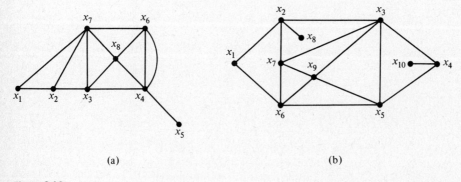

(a)   (b)

Figure 3.19

4.   For each of the pairs of graphs in Figure 3.20, determine whether the
     two graphs are isomorphic. If they are, label the graphs to demonstrate
     the isomorphism. If they are not, can you say why not?

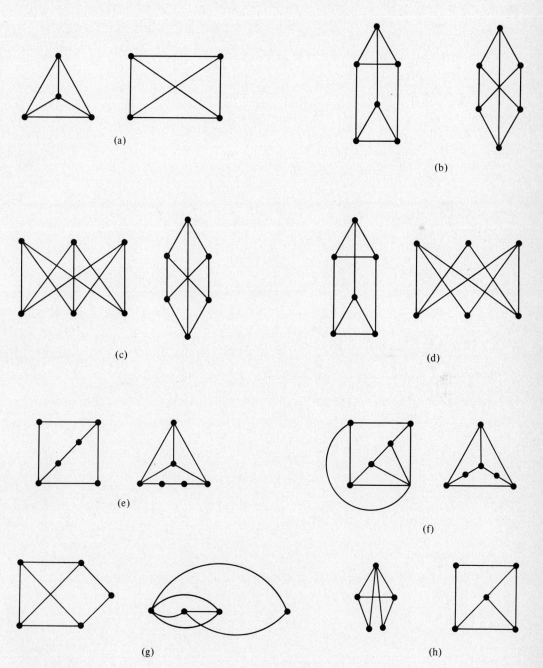

(a)

(b)

(c)

(d)

(e)

(f)

(g)

(h)

Figure 3.20

5.   For each of the structures in Figure 3.21, state whether it is a graph or
      not. For those that are graphs, try to redraw them with as many straight
      edges as possible, changing the number of accidental crossings if
      necessary. The redrawn graph should be isomorphic to the original.

Figure 3.21

6. A graph *G* is called *planar* if it is isomorphic to a graph *G'* that has no accidental crossings. For each of the graphs in Figure 3.22, determine whether or not it is planar. If it is not, draw a graph isomorphic to the given one but with as few accidental crossings as possible. You may introduce curved edges if that is helpful.

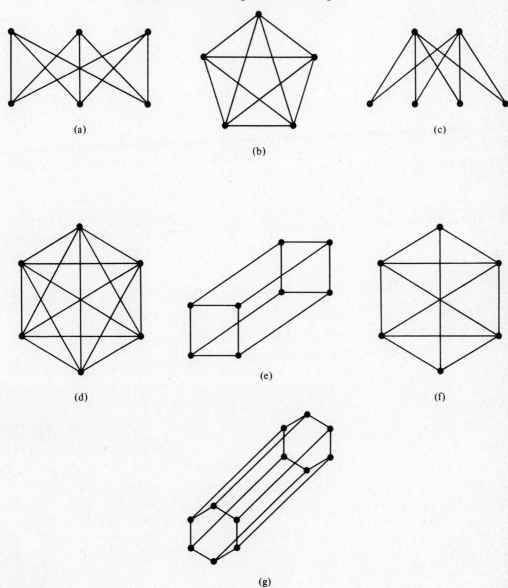

Figure 3.22

7.*   (a)   A graph G is said to have a *biplanar representation* if it can be drawn on a piece of paper with solid lines and dotted lines so that the solid lines meet only at vertices and the dotted lines meet only at vertices. The graph on the left in Figure 3.23 is biplanar, as shown by the drawing on the right.

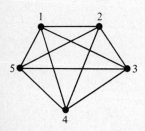 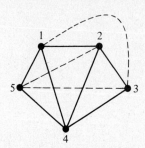

Figure 3.23

      (b)   Do the graphs in Exercises 6(a) and 6(d) have biplanar drawings?

      (c)   Does the complete graph with seven vertices have a biplanar drawing?

      (d)   Does the complete graph with eight vertices have a biplanar drawing?

Remark: The complete graph with nine or more vertices has no biplanar drawing.

      (e)   Explain how the concept of a biplanar drawing has applications in the theory of designing printed circuits for electric components. (*Hint:* Think of the components as vertices, the edges as solder or an electric conductor, and the paper as an insulator board. What happens when two solder strips meet at a point that is not a vertex?)

8.*   Draw as large a collection as you can of graphs with six vertices and three or fewer edges and such that no two graphs in your collection are isomorphic.

9.*   Draw as large a collection as you can of graphs with four vertices and no multiple edges and such that no two graphs in your collection are isomorphic. (*Hint:* There are 11 such graphs.)

10.*   Draw as large a collection as you can of graphs with exactly five vertices and at least five edges but no multiple edges and such that no two graphs in your collection are isomorphic. (*Hint:* Your collection should have exactly 20 graphs in it.)

3.2 counting in graphs and an application to chemistry

We have seen many different examples of graphs—so many that perhaps you may be surprised that there is more underlying structure to a graph than the definition seems to imply. To illustrate this we issue the following challenge: Can you construct a graph with four vertices x_1, x_2, x_3, x_4, with the properties $\text{val}(x_1) = \text{val}(x_2) = \text{val}(x_3) = 2$ and $\text{val}(x_4) = 3$? When you have decided that it is impossible, read the following theorem to see why.

THEOREM 2

If G is a graph with n vertices labeled $x_1, \ldots, x_n$ and G has e edges, then

$$\text{val}(x_1) + \text{val}(x_2) + \cdots + \text{val}(x_n) = 2e \tag{3.1}$$

In particular, in any graph, the sum of the valences, i.e., the left-hand side of Equation (3.1), is an even number.

Proof: Let us take an animated view of the situation. Imagine an observer moving from vertex to vertex, visiting in turn $x_1, \ldots, x_n$ (see Figure 3.24). At each vertex he puts a check mark at every edge that touches that

Figure 3.24

vertex and counts the number of check marks he has made. Thus, at x_1 he counts $\text{val}(x_1)$, at x_2 he counts $\text{val}(x_2)$, etc. If he keeps a running total, the number he arrives at is the left-hand side of the equation in the statement of the theorem; namely, $\text{val}(x_1) + \text{val}(x_2) + \cdots + \text{val}(x_n)$. This number signifies the total number of check marks he has made. Since each edge has received exactly two check marks, however—because our observer occupies each of the two end points for any particular edge—the number of check marks is also $2e$.

It is important to realize that the condition in Theorem 2 is necessary but not sufficient for the existence of a graph with a prescribed number of edges and certain valence conditions. For example, suppose we are asked to find a graph with vertices x_1, x_2 and with valences 5 and 1, respectively. Notice that the sum of the valences is even. Hence, our theorem does not rule out the existence of such a graph. In fact, it even tells us how many edges to expect in the graph if it exists. A little experimentation should convince you that the task is hopeless, however. No graph exists with the required properties.

Some more examples may shed further light on Theorem 2.

Example 1

Does there exist a graph with two vertices, each of which has valence 4?

Solution: First we apply Theorem 2 to see if there is any hope that a graph will exist. If we denote the vertices by x_1 and x_2, then we have $\mathrm{val}(x_1) = 4$ and $\mathrm{val}(x_2) = 4$.

Hence, $\mathrm{val}(x_1) + \mathrm{val}(x_2) = 4 + 4 = 2e$ and $e = 4$. Hence, perhaps such a graph exists. Trial and error leads to the graph in Figure 3.25.

x_1 $\qquad\qquad$ x_2

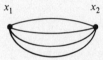

Figure 3.25

Note that if we required a graph with two 4-valent vertices, which had no multiple edges, no such graph would exist.

Example 2

Does there exist a graph with vertices x_1, x_2, x_3, x_4, x_5, having valences 1, 3, 4, 2, 3, respectively?

Solution: Applying the theorem,

$$\mathrm{val}(x_1) + \mathrm{val}(x_2) + \mathrm{val}(x_3) + \mathrm{val}(x_4) + \mathrm{val}(x_5) = 1 + 3 + 4 + 2 + 3 = 2e$$

Hence, $2e = 13$, which is impossible. No such graph can exist.

We can indicate a simple application of Theorem 2 if we take a brief excursion into the world of chemistry. Chemical molecules consist of a number of atoms joined to one another by bonds. Thus, for example, the simple graph in Figure 3.26 is a model for a molecule of ordinary water, written H_2O by the chemist. Chemists call diagrams of this type *structural formulas*. The edges between two atoms represent bonds between them.

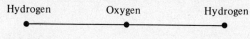

Figure 3.26

The fact that the oxygen atom is involved in two bonds is expressed by saying that the valence of oxygen in this compound is 2. Actually, chemists often refine this concept further by calling some valences negative and others positive but this need not concern us. In general, if an atom is involved in k bonds in a certain compound, then it is said to have a valence k in that compound. For example, in the methane molecule C_1H_4 shown in Figure 3.27(a),

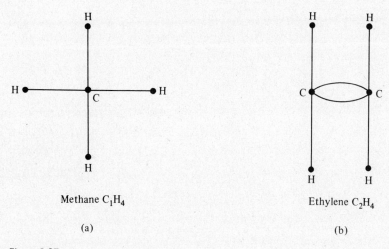

Methane C_1H_4

(a)

Ethylene C_2H_4

(b)

Figure 3.27

the carbon atom has a valence of 4 and each hydrogen atom has a valence of 1. Sometimes we have so-called *double bonds* between atoms, as in the compound ethylene C_2H_4 depicted in Figure 3.27(b). Each carbon atom in ethylene has a valence of 4. The case of ethylene illustrates the fact that in some cases the structural diagram of a molecule has multiple edges. Bonds of even higher multiplicity may also occur.

The parallel with our graph theoretic use of the term *valence* should now be clear: When we make a graphic model of a molecule, the chemical valence of an atom in that molecule is the same as the graph theoretic valence of the vertex representing that atom.

We shall restrict the rest of our discussion to hydrocarbons, which are molecules consisting of only two kinds of atoms, hydrogen and carbon. In hydrocarbon molecules each carbon atom has a valence of 4 and each hydro-

gen atom has a valence of 1. Our counting theorem now helps us to answer
questions like this:

Example 3

Does there exist a hydrocarbon with five carbon atoms and three hydrogen atoms?

Solution: If such a molecule exists, there are three vertices with
valence 1 and five vertices with valence 4. By Theorem 2, $2e = 3(1) + 5(4)$.
Since $3(1) + 5(4) = 23$, an odd number, such a molecule doesn't exist.

We should emphasize that our graphs of molecules are models and that,
as with all models, there has been some simplification. In our case the three
dimensionality of the molecule, as well as the angles of the bonds, have been
ignored. For example, in all water molecules the bonds make an angle of
about 109 degrees (see Figure 3.28), while in a methane molecule the hydrogen

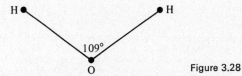

Figure 3.28

atoms, rather than lying in a flat plane as suggested by Figure 3.27(a),
actually form a tetrahedron in three-dimensional space, enclosing the carbon
atom at the center as shown in Figure 3.29. For some purposes in chemistry
it is important to know these things but it was not important for our purposes
so we ignored angles and three dimensionality.

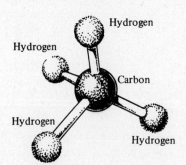

Figure 3.29   A methane molecule

Another point to bear in mind concerning the relation of our graphic models to real molecules is this: Just because one can draw a structural formula (i.e., a graph) that has the right valences does not mean that there exists a real molecule exhibiting that structural formula. There exist many descriptions and drawings of unicorns but no real-life examples of them.

EXERCISES 3.2

1. For each of the following, determine whether the requirements given contradict the condition of Theorem 2. If they do not, try to draw a graph to fit the requirements.
   (a) Three 1-valent vertices, two 2-valent vertices, one 3-valent vertex.
   (b) Four 2-valent vertices, four 3-valent vertices, two 5-valent vertices.
   (c) Three 1-valent vertices, two 2-valent vertices, two 5-valent vertices.
   (d) One 2-valent vertex, five 3-valent vertices, two 4-valent vertices, one 5-valent vertex.
   (e) Two 2-valent vertices, one 6-valent vertex.
   (f) One 2-valent vertex, two 3-valent vertices.
   (g) One 1-valent vertex, one 2-valent vertex, one 3-valent vertex, one 5-valent vertex.

2. In the cases of Exercise 1 in which you were able to draw a graph with multiple edges, can you also draw a graph with the required properties that has no multiple edges?

3. What is the maximum number of edges in a graph with four vertices without multiple edges? What happens if we allow multiple edges? Show that $[n(n - 1)]/2$ is the maximum number of edges in a graph with n vertices and no multiple edges.

4. Given the graph G in Figure 3.30:

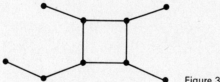

Figure 3.30

   (a) List the valences of the vertices.
   (b) Find a graph H, with valences as in part (a), that is not isomorphic to G.

5.* Let G be a graph without multiple edges and x any vertex of G. Show that val(x) is less than or equal to the number of remaining vertices in G.

6.* Let M be a graph and x any vertex of M. Show by an example that $\text{val}(x)$ may be larger than the total number of remaining vertices in M.

7.* Draw all the graphs without multiple edges that have four vertices and for which the valences of all the vertices are the same.

8.* Can you give a sequence of numbers such that all graphs whose vertices have this collection of numbers as their valences are isomorphic? (*Hint:* One such sequence is 1, 1. The graph   is the only possible one.)

9. Determine if there can exist hydrocarbons with the numbers of hydrogen (H) and carbon (C) atoms listed below:

   (a)   7 carbon atoms, 13 hydrogen atoms.

   (b)   4 carbon atoms, 7 hydrogen atoms.

   (c)   5 carbon atoms, 6 hydrogen atoms.

10. What is the largest number of hydrogen atoms that a hydrocarbon with (a) one, (b) two, (c) three, (d) four carbon atoms can have? Generalize.

11.* If a hydrocarbon has four double bonds and six carbon atoms, what is the largest number of hydrogen atoms it can have?

12. Two hydrocarbons are called *isomers* of one another if they have the same number of hydrogen and carbon atoms but their structural formulas (graphs) are not isomorphic. Figure 3.31 shows two isomers of butane, C_4H_{10}.

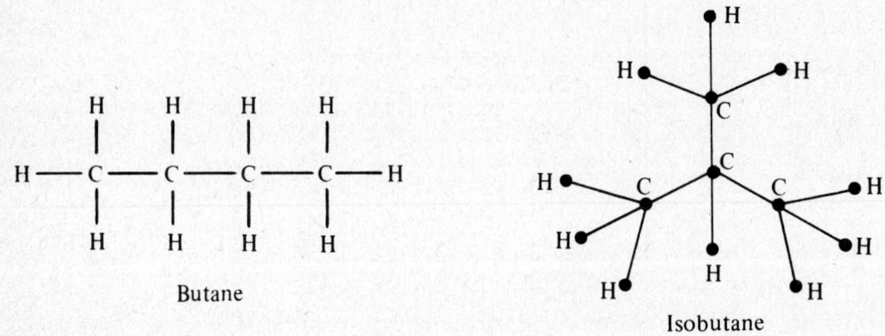

Butane   Isobutane

Figure 3.31

   (a)   Are there any other isomers of butane?

   (b)   How many isomers are there of pentane, C_5H_{12}?

   Remark: In general, a hydrocarbon whose formula is C_nH_{2n+2} is called a *saturated hydrocarbon*. For each value of n it is of interest to know what the isomers of the compound are since different isomers have different properties.

3.3 digraphs

We have used graphs to model a wide variety of real-world situations from chemical molecules to street networks. We can increase the number of situations we can model still further, however, by allowing each edge of a graph to have a direction indicated on it by an arrow.

DEFINITION 4

A *digraph D* is a finite, nonempty set of points, called *vertices*, together with some directed edges joining pairs of these points. These directed edges are subject to one restriction: The initial and terminal vertices of a directed edge may not be the same.

Thus, the structures in Figures 3.32(a) and (b) are digraphs, while the structure in Figure 3.32(c) is not because it violates our definition.

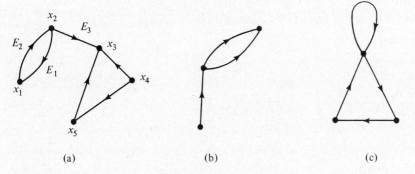

(a)   (b)   (c)

Figure 3.32

The street network of Figure 3.33, with its one-way streets indicated by arrows, is an obvious example where a digraph is called for as a model. If, instead of the digraph of Figure 3.34, we use as our model the graph in Figure 3.35, which results by removing the arrows, we would have a poor model that might mislead us. For example, the graph in Figure 3.35, has a Hamilton circuit. It turns out, however, that the digraph in Figure 3.34 has no Hamilton circuit that "obeys the arrows." (To see why, examine the lower right vertex.)

Most of the concepts concerning digraphs are analogous to concepts we have studied for graphs. We modify the old concepts to take into account the

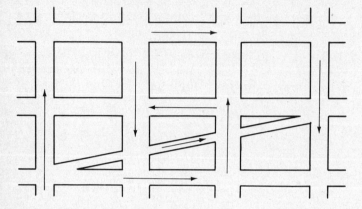

Figure 3.33

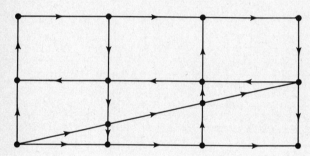

Figure 3.34

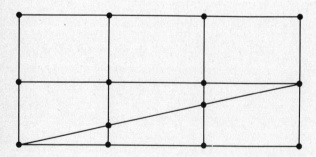

Figure 3.35

54

fact that our edges are now directed. Thus, instead of valence at a vertex, we have two kinds of valence at that vertex, the outvalence and the invalence.

DEFINITION 5

If x is a vertex of the digraph D, the *outvalence* of x, denoted outval(x), is the number of directed edges which touch x and which are directed *away* from x. The *invalence* of x, denoted inval(x), is the number of directed edges which touch x and which are directed toward x.

DEFINITION 6

If we begin at a vertex of D and traverse a sequence of directed edges of D in such a way that each edge is traversed in the direction of the arrow, then we generate a *directed path*. If a directed path returns to its starting vertex, it is called a *directed circuit*.

Example 1

In the digraph of Figure 3.32(a) outval(x_2) = 2, while inval(x_2) = 1. By beginning at x_2 and traversing in turn E_1, E_2, E_3, we obtain a directed path, while if we begin at x_1 and traverse in turn E_1, E_3, we do not obtain a directed path. If we begin at x_2 and traverse the sequence of directed edges E_1, E_2, we obtain a directed circuit. If we begin at x_2 and traverse the sequence of directed edges E_2, E_1, however, we do not obtain a directed circuit.

We shall now give a few more examples to show situations where digraphs are the appropriate model.

Example 2

A group of residents in a commune are asked which of the persons that they are acquainted with in the commune they consider their friends. Since Jack may describe Mary as a friend but Mary might not describe Jack as a friend, the graph model in Figure 3.36(a) would not be appropriate. Digraphs would provide a natural model for these friendship relations. The digraph in Figure 3.36(b) would model the situation that Mary is a friend of Jack but Jack is not a friend of Mary. The digraph in Figure 3.36(c) would model the situation that Jack is a friend of Mary and Mary is Jack's friend.

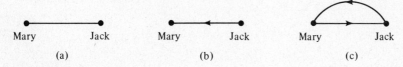

| Mary | Jack | Mary | Jack | Mary | Jack |
| (a) | | (b) | | (c) | |

Figure 3.36

Example 3

In corporations, universities, and other structures, certain people have authority over certain others. This can be illustrated with a digraph in which the people in the organization are represented by vertices and a directed edge goes from person *A* to person *B*, provided *A* has authority over *B*.

Example 4

Suppose we have a collection of statements that are logically dependent; that is, there are implications that exist between them. We can model the logical structure of this set of statements with a digraph whose vertices represent the statements and where a directed edge is drawn from the vertex representing statement *A* to the vertex representing statement *B* provided statement *A* implies statement *B*. Here is a specific example:

(1)   The gross national product (GNP) rises.
(2)   Auto sales rise.
(3)   Employment goes up.
(4)   People take more vacations in the country.
(5)   Air pollution in the cities increases.

(Drawing by Dana Fradon. © 1970 The New Yorker Magazine.)

"Which came first—the wage hike or the price hike?"

Naturally there are numerous cause and effect relations that *might* exist between these. Figure 3.37 shows one possibility.

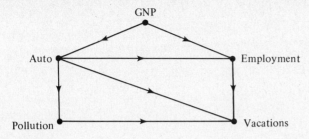

Figure 3.37

Here is an example where logical relations between statements about a graph G are modeled with a digraph (Figure 3.38). The digraph makes apparent the logical relationship among the three statements.

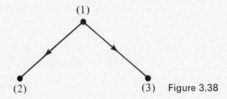

Figure 3.38

(1) G is a complete graph with n (≥ 2) vertices.
(2) G contains no vertices of valence 0.
(3) G has $[n(n - 1)]/2$ edges.

Example 5

Biologists have long been aware that living creatures are dependent on one another for food and that if one species becomes endangered, any species that preys heavily upon it may also be endangered. It is convenient to be able to model this situation geometrically with a digraph. The digraph in Figure 3.39 models the predator relationships among some members of a prairie ecosystem.

A directed edge from species A to another species B indicates that species B preys upon A. Thus, in Figure 3.39, the directed edge from gopher to coyote means that the gopher is one of the coyote's sources of food.

Digraphs such as the one shown in Figure 3.39 are known as *food webs*.

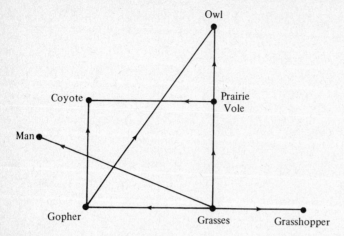

Figure 3.39

EXERCISES 3.3

1. Draw a *reasonable* implication digraph whose points represent the following propositions:
    (a)   Use of atomic energy to generate electricity increases.
    (b)   Middle Eastern tensions rise.
    (c)   Oil companies go out of business.
    (d)   Air pollution declines.

2. For each of the following situations, decide whether a graph or a digraph would be most suitable to model the situation.
    (a)   The won and lost record of the teams in a tournament in which each pair of teams play against one another once and ties do not occur.
    (b)   The influence patterns that exist in the U.S. Senate.
    (c)   The blood relationships among guests at a wedding feast.
    (d)   The best friend relationships in a boy scout troop.
    (e)   The power relations among a group of nations.

    Remark: For some of the above, it is not entirely clear whether a graph or a digraph would be best. It might depend on the situation.

3.* Use a digraph to construct a model for the tournament between four baseball teams:

The Chargers beat the Ringers.
The Chargers beat the Royals.

> *The Royals beat the Ringers.*
> *The Sleuths beat the Chargers.*
> *The Royals beat the Sleuths.*
> *The Ringers beat the Sleuths.*

4. For each of the digraphs in Figure 3.40, determine the invalence and outvalence of each vertex.

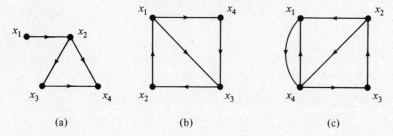

     (a)   (b)   (c)

Figure 3.40

5. (a) A vertex of a digraph D of invalence 0 is called a *transmitter*. Find all the transmitters of the digraph in Figure 3.41:

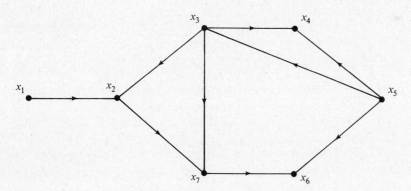

Figure 3.41

    (b) If the digraph D in part (a) were a model for the way a rumor planted with members of a group of seven people $(x_1, \ldots, x_7)$ would be spread, determine if a rumor told to all the transmitters would reach all the people in the group.

6. Use a digraph to model the authority structure of:
    (a) Your family.
    (b) The United States Government.

(c)   The government of the state in which you live.

(d)   The government of the city (town) in which you live.

7.   Here is a suggestion for some experiments in social psychology. For a number of different groups of people, construct the "does listen to" digraphs. A "does listen to" digraph for a group is constructed as follows:

> Represent people by vertices and draw a directed edge from A to B if B generally listens to A's suggestions. Call a pair of points symmetric if neither listens to the other or each listens to the other. Otherwise call them unsymmetric. Measure the extent of symmetry in various groups of people and correlate this with group morale.

3.4   critical path method

We shall now give a more dramatic example of how digraphs can be applied. When faced with a complicated job, it is often wise to resist the impulse to plunge right in to begin the work. A little analysis beforehand may reveal that great savings in time can be achieved by doing things in a certain order. The purpose of this section is to discuss CPM (Critical Path Method), a method of analyzing jobs that are composed of a number of interdependent smaller activities. In what follows, we shall maintain this distinction between the words *job* and *activity*. Job will refer to the total project at hand, while activity will denote one of the constituent tasks.

The method we present will solve the model home problem of Chapter 1. For convenience, however, we shall start with a simpler problem. Suppose we wish to minimize the time an ocean liner must spend in port. We shall simplify matters by assuming that the job of getting under way again involves just the four component activities shown in the table and that the times required for these activities are as indicated.

A_1.	Disembark passengers	3 hr
A_2.	Unload cargo	12 hr
A_3.	Take on new passengers	4 hr
A_4.	Load new cargo	15 hr

How shall we proceed? If we take these activities in serial order, the time required to complete the job is $3 + 12 + 4 + 15 = 34$ hr. Some of these activities can, in theory, be performed simultaneously, however: Cargo operations can proceed independently of passenger operations. The only order that must be maintained is A_1 must precede A_3 and A_2 must precede A_4. A good way to illustrate this is to draw as a model *the activity analysis digraph* as described in the following definition (see also Figure 3.42).

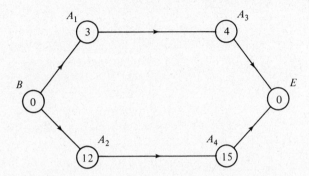

Figure 3.42

DEFINITION 7

Given a series of activities $A_1, A_2, \ldots, A_n$, together with the times required for each of these activities and given complete information about which activities must precede which and which may be simultaneous, we construct the activity analysis digraph in the following way:

(1) Represent each A_i by a vertex and label it with the time required for that activity.

(2) Create two additional vertices, each labeled with the number 0, one representing the job's beginning, denoted B, and the other representing the job's end, denoted E. By an abuse of common sense we shall refer to B and E as activities also, even though they require no time or effort to perform and are really just points in time.

(3) Draw a directed edge from one activity to a second activity only if the first activity precedes the second and if no activity must intervene between the end of the first activity and the beginning of the second.

The digraph in Figure 3.42 results from applying this definition to the information concerning the ocean liner problem.

Can you see that the most efficient execution of the whole project involving the ocean liner would require only 27 hr? We shall shortly discuss a method for finding the most efficient time and the optimal scheduling from the activity analysis digraph. However let's first go back to the model home problem of Chapter 1, whose precedence relationships are shown in the following table, using more suggestive abbreviations than those used in Chapter 1.

Activity	Time (days)	Immediate Predecessors
B (begin)	0	
S (site preparation)	2	Begin
Fo (building foundation)	3	Site preparation
M (building main structure)	14	Building foundation
EW (electric wiring)	5	Building main structure
P (plumbing)	4	Building main structure
IF (interior finishing)	7	Electric wiring and plumbing
EF (exterior finishing)	11	Building main structure
Fu (furnishing)	2	Interior finishing and exterior finishing
L (landscaping)	4	Exterior finishing
E (end)	0	Furnishing and landscaping

The activity analysis digraph for this problem, shown in Figure 3.43, is drawn according to the information in column 3 of the table. Can you determine that the minimum time is only 34 days?

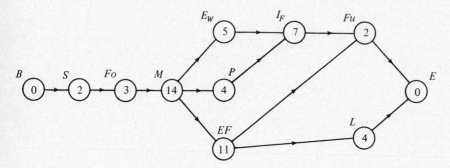

Figure 3.43

Now we need to consider the following question: Given an activity analysis digraph for a job, how can one determine the shortest time for completion of the whole job, and how can one find a scheduling that accomplishes this shortest time? For the sake of simplicity and custom we shall imagine that time, denoted t, will be measured from the starting point B. In other words, B is time 0. In symbols, at B, $t = 0$. Our problem can now be phrased this way: Given an activity analysis digraph for a project, what is the smallest value of t at which E, the end, may occur and at which values of t should the other activities be scheduled to begin in order to achieve this smallest value of t for E? The answer turns upon the significance of a directed path in the activity analysis digraph. Suppose we have a directed path beginning at B and ending at some activity A_i. All the activities along this path must be performed in the order of their occurrence before one can perform A_i. If we add the times for all the activities on this path up to but not including A_i, we arrive at an amount of time that must elapse before A_i can be performed. There may, however, be another directed path from B to A_i (see Figure 3.44). We can

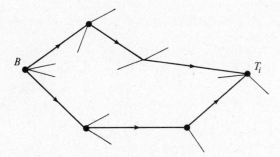

Figure 3.44

apply the same analysis to this path and arrive at another amount of time that must elapse before A_i can be performed. This amount of time may be different from the first. The earliest time at which we can actually begin to perform A_i is equal to the time for the longest path (measured in terms of time). In particular, the earliest time we can arrive at E (the end) is equal to the time for the so-called *critical path* from B to E, the path of longest time from B to E. Thus the answer to our question is as follows: The minimal time for the completion of the whole job is determined from the activity analysis digraph by calculating the time-length of the longest (in terms of time) directed path from B to E, that is, the length of the critical path. The beginning of activity A_i should be scheduled at a time equal to the time-length of the longest (in terms of time) directed path from B to A_i. In calculating the lengths of the paths to A_i, we do not count the time for A_i itself.

Example 1

Let us determine the most efficient scheduling of the ocean liner problem (Figure 3.42). The critical path is obviously B, A_2, A_4, F, with a length of 27. The minimum time for the whole job is thus 27 hr. To determine when the various activities are to be scheduled to begin is quite simple here since, for each activity except for E, there is only one directed path leading to it. Activities A_1 and A_2 can both be scheduled to begin at time 0. Activity A_3 should be scheduled to begin at time 3, right after A_1 is done, while A_4 should be scheduled to begin at time 12. Notice that in this scheduling there is a great deal of slack time between the end of A_3, which occurs at time 7, and the activity E. However we cannot arrive at E until A_2 and A_4 are done, and these are not completed as early as A_3 is completed.

Example 2

Let us determine the optimal scheduling and the minimal time for the slightly more complicated model home problem (see Figure 3.43). There are four directed paths from B to E. Together with their time-lengths they are:

$$
\begin{array}{ll}
B, S, Fo, M, EW, IF, Fu, E & 33 \\
B, S, Fo, M, P, IF, Fu, E & 32 \\
B, S, Fo, M, EF, Fu, E & 32 \\
B, S, Fo, M, EF, L, E & 34
\end{array}
$$

Consequently, the minimal time is 34 days since the longest path (critical path) has time-length 34. An optimal scheduling designed to achieve this minimal time is as follows:

Activity	Start Time	Reason
S	0	There is only one directed path from B to S and it has length 0.
Fo	2	There is only one directed path from B to Fo and it has length 2.
M	5	There is only one directed path from B to M and it has length 5.
EW	19	There is only one directed path from B to EW and it has length 19.
IF	24	Of the two directed paths from B to IF, the longer has length 24.
P	19	There is only one directed path from B to P and it has length 19.
EF	19	There is only one directed path from B to EF and it has length 19.
Fu	31	Of the three directed paths from B to Fu, the longest has length 31.
L	30	There is only one directed path from B to L and it has length 30.
E	34	Of the four directed paths from B to E, the longest has length 34.

Example 3

A large corporation decides on a major sales campaign to promote a product yet to be decided upon. The main activities in this project are:

M. A series of meetings to determine the product to be promoted (1 week).

AC. Creation of advertising copy, jingles, and so on (4 weeks).

ML. Assembling a suitable mailing list for a direct mail campaign (1 week).

SO. Sending out the direct mail appeals (1 week).

SP. Shipping an extra supply of the product chosen to retail outlets (3 weeks).

BR. Holding briefings for regional sales representatives (1 week).

Decide upon suitable precedence relationships, draw an activity analysis digraph, and then find the optimal scheduling.

Solution: There may be some debate concerning what the exact precedence relationships are, but those shown in the digraph in Figure 3.45

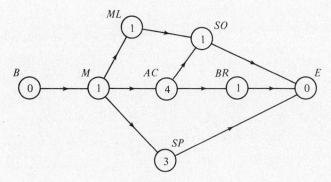

Figure 3.45

seem reasonable. The accompanying chart shows the optimal scheduling for this digraph. Note that there are two critical paths.

Activity	Starting Time	Reason
M	0	
ML	1	There is one path from *B* to *ML* and it has length 1.
AC	1	There is one path from *B* to *AC* and it has length 1.
SO	5	The two paths from *B* to *SO* have lengths 2 and 5.
SP	1	There is one path from *B* to *SP* and it has length 1.
BR	5	There is one path from *B* to *BR* and it has length 5.
E	6	The four paths from *B* to *E* have lengths 3, 6, 6, and 4.

EXERCISES 3.4

1. A group of girls rooming together decide to cook a special dinner for some unsuspecting males. They break down this project into five component activities with time estimates as follows:

K	Clean house	30 minutes
D	Decide on menu	15 minutes
P	Purchase food	60 minutes
C	Cook food	50 minutes
S	Set table	10 minutes
F	Place cooked food on table	4 minutes

Decide upon a reasonable activity analysis for this job. You may decide which activities precede others but be prepared to justify your choices. Determine the optimal scheduling and the minimum time corresponding to the optimum scheduling.

2. The president of a college has just resigned and decides to move to another state. Being a mathematician, he performs an activity analysis of the project of moving, centered around the following activities:

Bo	Procure boxes to pack things	1 day
H	Arrange to purchase new home	14 days
P	Pack	5 days
C	Purchase clothing suitable for new surroundings	1 day
S	Send off boxes to new address	1 day
M	Shop around for a moving company	3 days

An activity analysis digraph is shown in Figure 3.46. Determine an optimum scheduling and the associated minimal time for the move.

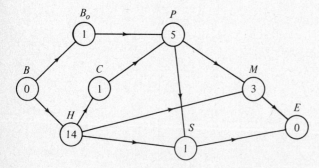

Figure 3.46

3. The activity analysis in Figure 3.47 represents the job of teaching a child to understand and manipulate fractions. The vertices represent various simpler arithmetic topics, such as addition, subtraction, multiplication, equality, and part-whole relationships, that need to be mastered first. The directed edges represent, as usual, the immediate precedences. Find the optimal scheduling and the minimal time.

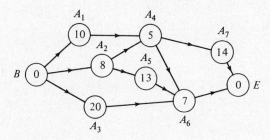

Figure 3.47

4. Under the influence of a new administration in Washington, the Peace Corps is to be reorganized and revitalized. Participating countries have suggested a focus on improving their agricultural technology. The new director's analysis shows that there are nine distinct activities that need to be carried out in order to get such a program under way:

   (a) A new Peace Corps *administration* needs to be appointed.

   (b) A *study* needs to be made of the current state of agriculture in the nations in the participating group.

   (c) Based on the study, goals and methods need to be formulated and *negotiated* with the participating nations' governments.

   (d) The results of the study and the aims of the Peace Corps need to be *disseminated* and advertised in the United States.

   (e) Recruitment and training of *volunteers* (preliminary).

   (f) Arranging for local community participation in the projects in the host countries and making other administrative and technical *arrangements*.

   (g) Proposing and lobbying for *legislation* in the United States Congress that would fund the projects.

   (h) Final selection of volunteers, briefing, and *sending* them off.

   (i) Putting out bids to *purchase* necessary materials and initiating necessary construction.

   Below is a precedence table. Draw the activity analysis digraph from these precedences and then find the optimal scheduling to complete the whole job in the minimal amount of time.

	Activity	Time	Precedences
0.	Begin	0	None
1.	Administration	2	Begin
2.	Study	8	Begin
3.	Negotiation	6	Administration and study
4.	Dissemination	5	Administration and study
5.	Volunteers	8	Dissemination
6.	Arrangements	12	Negotiation
7.	Legislation	10	Negotiation
8.	Send-off	2	Volunteers, arrangements, and legislation
9.	Purchase	4	Legislation
10.	End	0	Purchase and send-off

5.   Here is the precedence table for the job of building a log cabin in the woods.

	Activity	Time	Precedences
B	Begin	0 weeks	None
L	Learn how	8 weeks	B
S	Purchase building site	12 weeks	B
F	Lay foundation	1 week	L, S
T	Cut trees and prepare logs	7 weeks	L, S
C	Construct cabin	10 weeks	F, T
E	End	0 weeks	C

Draw the activity analysis digraph and find the optimal scheduling.

6.* Explain why, in an activity analysis digraph, one never finds three activities connected as in Figure 3.48.

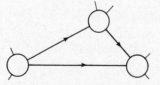

Figure 3.48

SUGGESTED READING

BUSACKER, R. G. AND T. L. SAATY, *Finite Graphs and Networks*, McGraw-Hill Book Co., New York, N.Y., 1965. This is a very straightforward junior-senior level text with one commendable novelty, a chapter devoted entirely to a list and short descriptions of applications of graph theory to a wide range of subjects. Unfortunately, none of the applications is discussed in much detail.

HARARY, F., R. NORMAN, AND D. CARTWRIGHT, *Structural Models*, John Wiley & Sons, Inc., New York, N.Y., 1965. Chapter 11 on tournaments and Chapter 13 on balance are interesting examples of graph theoretic modeling. The book in general is addressed to uses of graph theory in the social sciences.

ORE, O., *Graphs and Their Uses*, Random House, New York, N.Y., 1963. This paperback, written by a world-famous mathematician, is probably the simplest and most engaging introduction to the subject.

WIEST, J. AND M. LEVY, *A Management Guide to PERT/CPM*, Prentice-Hall, Inc., Englewood Cliffs, N.J., 1969. A clear introduction to the Critical Path Method (CPM).

4
paths and connectivity

4.1 definitions and terminology

> Tourist: *"Say, can you tell me how to get to Main Street from here?"*
>
> Local resident: *"Sorry, you can't get there from here."*

The main purpose of this chapter is to discuss Euler circuits and related matters. Before we start, we need to dispose of two preliminaries:

(1) We shall elaborate our intuitive notion of a "circuit."
(2) We shall develop a notion of connectedness.

To be specific, suppose a police patrol car needs to patrol all the streets represented by edges in the graph of Figure 4.1. Notice we use the singular graph instead of graphs. You may think you see two graphs but we can regard Figure 4.1 as containing one graph (admittedly, it comes in two pieces). This graph G models the streets that must be patrolled, even though it consists of two separate mini-routes with no connections between them. It seems reasonable to regard the graph G as a *disconnected* graph, consisting of two connected pieces or *components*. Furthermore, this disconnectedness of G results because it is not possible for the patrol car to travel along a

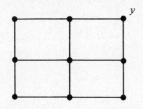

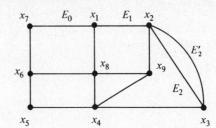

Figure 4.1

route or *path* in the graph that starts in one component (piece) and ends in the other. Doing so would require a *jump* somewhere from one component to another, which would be contrary to the nature of a path. Using the basic concept of a path, we shall later define the concept of a circuit.

DEFINITION 1

If we have a sequence of vertices in a graph (or a digraph) $x_1, x_2, \ldots, x_n$ so that each vertex x_i together with its successor in the sequence x_{i+1} are end points of an edge E_i (directed from x_i to x_{i+1} in the digraph case), then we call the sequence of edges $E_1, E_2, \ldots, E_{n-1}$ a *path* in the graph (digraph) from x_1 to x_n.

Instead of listing the edges in the path, it is often more convenient to list the vertices. Thus the path E_0, E_1 in Figure 4.1 might be listed as x_7, x_1, x_2. Listing the vertices to describe a path can sometimes be ambiguous. For example, x_7, x_1, x_2, x_3 could mean either of these two paths: E_0, E_1, E_2 or E_0, E_1, E_2'. This ambiguity only arises when the path uses one of a set of multiple edges; in such cases we shall stick to the edge listing method. Other examples of paths in Figure 4.1 are

$$x_7, x_1, x_2, x_9, x_4, x_8 \tag{4.1}$$

$$x_6, x_8, x_4, x_9 \tag{4.2}$$

$$x_1, x_2, x_9, x_4, x_8, x_9 \tag{4.3}$$

$$x_9, x_2, x_1, x_2 \tag{4.4}$$

However, x_3, x_6, x_8 is not a path because there is no edge connecting x_3 and x_6.

In (4.3), notice that $x_1, x_2, x_9, x_4, x_8, x_9$ is a path, even though one vertex x_9 is repeated. It is quite legitimate for a path to repeat one or more vertices. Repeating vertices can have advantages since the average police

71

patrol car probably passes through some vertices (intersections) often in one tour. Edges may also be repeated, as shown in (4.4), where the edge joining x_1 and x_2 is repeated.

Examples of paths in the digraph of Figure 4.2 are x_5, x_4, x_2 and x_2, x_3, x_4, x_2, x_3, x_1. However, x_1, x_2, x_3 is not a path because there is no directed edge from x_1 to x_2.

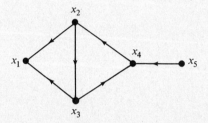

Figure 4.2

Intuitively, a graph or digraph is *connected*, provided that one can walk from any vertex to any other (following the arrows in the digraph case). The formal definition is as follows:

DEFINITION 2

A graph or digraph G is *connected* if, for any two vertices x and y in G, there is a path starting at x and ending at y.

Since connectedness (or the lack of it) is probably the most obvious aspect of a graph's structure, it is not surprising that when graphs are used to model real-world phenomena, the connectedness (or lack of it) reflects important facts about the real-world phenomena being modeled.

Here are some examples:

Example 1

Let G (a graph) model the established acquaintanceship relationships among a group of people. Vertices represent people and edges represent existing acquaintance ties. If G is connected, any two people in G may some day come into contact or become acquainted through knowing other people who form a path between the two. If G is not connected and has more than one component, then these components represent different groups, whose individual members will not come into contact through mutual acquaintance. For instance, suppose the disconnected graph of Figure 4.1 models the friendship relations of people represented by the vertices. Persons x_7 and x_8, while not directly acquainted, may become so through their mutual contact x_1.

Because x_7 may become acquainted with x_8, he may also become acquainted with x_9, as the latter is directly acquainted with x_8. In general, x_7 may become acquainted with anyone with whom he is connected by a path. He will never become acquainted with y because there is no path from x_7 to y.

Conjecture (Kenneth Hoffman): The acquaintanceship graph for the adult United States population is connected.

Example 2

If the graph G is a model of a molecule, it must be connected. If it were not, there would be two or more components, each having no connection with the rest of the molecule. Since no chemical bonds hold these components together, they could not be regarded as part of the same molecule.

As we turn to some examples involving digraphs, the concept of connectedness is a bit more subtle, although the definition is the same as for graphs. Does the digraph of Figure 4.2 look connected? It does give the impression of being in one piece. Unfortunately, while this type of visual impression is useful when dealing with graphs, we must be more careful when we have a digraph. In fact, according to our definition, the digraph in Figure 4.2 is *not* connected because there is no path from x_3 to x_5 (remember, in a digraph a path must "obey the arrows"). If we wish to go in the other direction, that is, from x_5 to x_3, we could use the path x_5, x_4, x_2, x_3; however, we have to be able to go from x_3 to x_5 as well. The definition of connectedness requires that, no matter at which vertex you wish to start, it should be possible to reach any other vertex by some path. Figure 4.3 shows an example of a connected digraph.

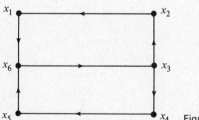

Figure 4.3

We have defined a path in such a way that it may repeat edges or vertices. You may recall that in our mailman and mail distributor problems of Chapters 1 and 2 we wished to avoid repetitions. When we wish to discuss paths in which repetitions of one kind or the other are to be ruled out, we shall refer to a *path without repeated edges* or a *path without repeated vertices*, as the case may be. In Figure 4.1 the path x_1, x_8, x_9, x_4, x_8 has no repeated edges but it has repeated vertices.

DEFINITION 3

A *circuit* is a path whose first and last vertices are the same.

Example 3

In Figure 4.1, the path $x_1, x_8, x_9, x_4, x_8, x_9, x_2, x_1$ is a circuit. The path $x_7, x_1, x_2, x_9, x_4, x_5, x_6, x_7$ is also a circuit. The path x_7, x_1, x_8, x_9 is not a circuit since the first and last vertices are not the same. The sequence of vertices $x_6, x_8, x_3, x_2, x_1, x_2, x_6$ is not a circuit or even a path since x_8 and x_3 are not joined by an edge.

One place where repeated edges or vertices become annoying is in the measurement of the length of a path. How long should we say the path x_8, x_9, x_4, x_8, x_9 (Figure 4.1) is? There are only three distinct edges in this path but one of them occurs twice so maybe we should say the length is 4. This is exactly what we do:

DEFINITION 4

The *length of a path* is the number of edges that appear, each edge being counted once each time it appears as we trace out the path.

The following theorem will be useful later.

THEOREM 1

If G is a connected graph or digraph with n vertices, then between any two distinct vertices of G there is a path of length no greater than $n - 1$.

Proof: Let x_i and x_j be any two vertices of G. Since G is connected, there is some path starting at x_i and ending at x_j. We shall show that if this path has length greater than $n - 1$, then it can be shortened. The reason is as follows: If the length is greater than $n - 1$, then as we trace out the path, we encounter vertices more than n times (counting the starting vertex x_i and the ending vertex x_j). The vertices we encounter cannot all be different since there are only n different vertices in G. Therefore, some vertex is encountered twice, which means that the path has a *loop* in it (see Figure 4.4). We can take a shortcut from x_i to x_j (Figure 4.5) by bypassing the loop. If the shortened path is still longer than $n - 1$, we apply our argument again. We can continue the shortening process until we reach a path of length no more than $n - 1$.

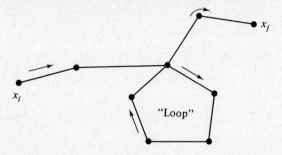

Figure 4.4

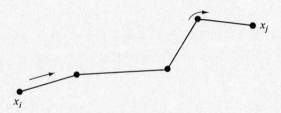

Figure 4.5

EXERCISES 4.1

1. For each of the following sequences of vertices, determine, by referring to Figure 4.6, which ones are paths, paths that repeat no edges, circuits, and circuits that repeat no edges.

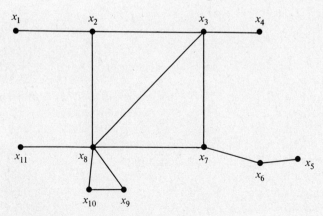

Figure 4.6

(a) x_1, x_2, x_8, x_3, x_7
(b) $x_1, x_2, x_8, x_{11}, x_8$
(c) x_2, x_8, x_7, x_3, x_2
(d) x_4, x_8, x_9, x_{10}
(e) $x_{10}, x_8, x_9, x_8, x_7, x_3, x_8, x_{10}$
(f) $x_{11}, x_8, x_7, x_3, x_2, x_8, x_9$
(g) $x_6, x_7, x_2, x_3, x_7, x_6$
(h) x_8, x_3, x_2, x_1
(i) $x_5, x_6, x_7, x_8, x_2, x_1$

2. Referring to the Figure 4.6, list all the circuits that repeat neither vertices nor edges.

Remark: In a graph it is usual to consider two circuits the same if they arise from one another by reading off the vertices in cyclic order or reverse cyclic order.

3. What is the length of the longest path not repeating vertices or edges in Figure 4.6? How many different paths of this length are there?

4. Consider the graph in Figure 4.7.

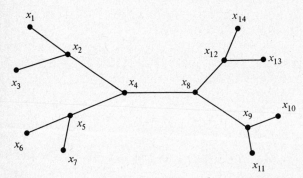

Figure 4.7

(a) How many different paths that do not repeat vertices or edges are there between x_1 and x_3? between x_2 and x_5? or between x_3 and x_5? Can you make a general statement about the number of *paths* without repeated vertices between any given pair of vertices?

(b) How many circuits that repeat no edges are formed when the edge $[x_1, x_3]$ is added to the graph? or the edge $[x_2, x_5]$? or the edge $[x_3, x_8]$? Generalize.

5. In the graph in Figure 4.3, find
(a) a path from x_1 to x_3;

(b)   a circuit that includes x_2, x_3, and x_5;

(c)   a path from x_1 to x_5 that visits x_6.

6.   DEFINITION

The *distance* from vertex x to vertex y in a connected graph G or digraph D is denoted $d(x, y)$ and is equal to the length of the shortest path joining x to y.

(a)   In Figure 4.6, find
   (i) $d(x_{10}, x_5)$
   (ii) $d(x_1, x_{11})$
   (iii) $d(x_{11}, x_6)$

(b)*   For a graph, show that $d(x, y)$ obeys the following rules:
   (i) $d(x, y) \geq 0$
   (ii) $d(x, y) = d(y, x)$
   (iii) $d(x, y) = 0$ if and only if $x = y$
   (iv)   Given any three vertices x, y, z, $d(x, y) + d(y, z) \geq d(x, z)$.

(c)*   Do i, ii, iii, and iv of part (b) hold for digraphs?

7.   For the graph of Figure 4.1, list two paths (without repeating vertices), each of which connects x_6 with x_8. Can you find two such paths that have no vertices in common except for x_6 and x_8?

8.   Explain why a path in a graph or digraph without repeated vertices must also be a path without repeated edges.

9.   In each of the digraphs in Figure 4.8, determine whether or not the digraph is connected.

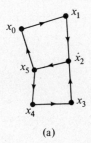

(a)

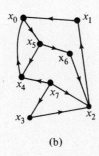

(b)

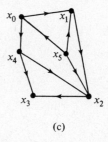

(c)

Figure 4.8

10.   Consider the acquaintance graph of all people in the United States (please do not attempt to draw it). It has been suggested that any two people in the graph can be connected with a path of length at most 4.

Can you find a path of length at most 4 connecting you to the President of the United States? (A possible criterion for acquaintanceship is this: Two people are acquaintances if it would not be very abnormal for either to engage the other in verbal conversation, either personally or on the telephone.) Try this game for other prominent persons besides the President.

11. Explain why, in the context of Exercise 10, it is of no interest to consider paths with repeated vertices.

12. Suppose x is a vertex of a digraph D and outval $(x) = 0$. Can D be connected? If not, say why not. If D can be connected, draw an example.

13. In the context of Exercise 12, suppose inval $(x) = 0$ instead of outval $(x) = 0$. Answer the same question.

14. For each of the graphs in Figure 4.9, determine how many components (maximal connected pieces) the graph has.

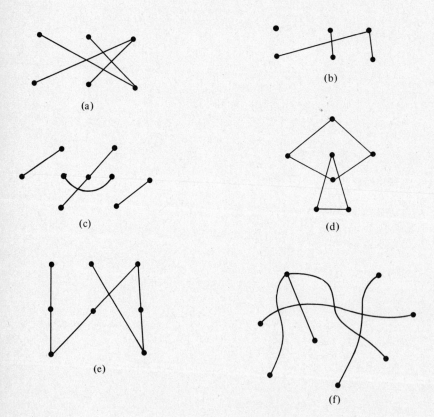

Figure 4.9

15. DEFINITION

Let G be any graph. $G - v$ will denote the graph obtained from G by deleting from G the vertex v together with all the edges at v. Note, however, that the end points of these edges remain.

DEFINITION

Let G be any graph. $G - [x_i, x_j]$ (or $G - E$ where E is the edge $[x_i, x_j]$) will denote the graph obtained from G by the deletion of the edge $[x_i, x_j]$. Note that when an edge is deleted, its end points are not to be removed.

For each of the graphs in Figure 4.10, find the graphs $G - v$ for each vertex v in the graph. In addition, find the graph $G - E$ for each edge E in the graph. State how many components are obtained in each case.

G_1

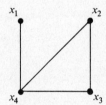

G_2

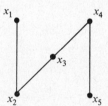

Figure 4.10

16. DEFINITION

Edge E is called a *bridge* if $G - E$ has more components than G. (See Exercise 15).

(a) Construct a graph with at least seven edges, for which every edge is a bridge.
(b) Construct a graph with no bridges.

17. Construct a graph with at least five non 1-valent vertices such that each vertex that is not 1-valent is a cut vertex. [A *cut vertex* is one for which $G - v$ has more components than G (see Exercise 15).]

18. Think of several real-world situations where the graphs that model these situations might have
(a) cut points;
(b) bridges.

19. Suppose G is connected and E is an edge of G. What is the largest number of components $G - E$ can have?

20. Construct a graph G, with a vertex v, such that $G - v$ has
   (a)   three components;
   (b)   four components;
   (c)   five components.

21.* Let M be any graph. Suppose there is a path in M from x_i to x_j. Show there must be a path that repeats no edges or vertices between x_i and x_j.

22.* Let M be a graph. Suppose $[x_i, x_j]$ is an edge of some circuit that repeats no edges in M. Does $[x_i, x_j]$ lie on some circuit that repeats no edges or vertices? What if the circuit repeats only one vertex?

23.* Suppose we define the concept of *mutual reachability* in a graph or digraph like this: Vertices A and B are mutually reachable provided there is a path starting at A and ending at B and a path starting at B and ending at A. Explain why the following is true: If A and B are mutually reachable and B and C are mutually reachable, then A and C are mutually reachable.

24.* Define A and B to be *simply mutually reachable* if they are mutually reachable (Exercise 23) using paths that have no repeated vertices. Explain why the following is true: If A and B are simply mutually reachable and B and C are simply mutually reachable, then A and C are simply mutually reachable.

25.* Show that a graph M, all of whose vertices are 4-valent, cannot have any bridges. Show that M may have a cut vertex even though all its vertices are 4-valent. Generalize both parts of this problem. (See Exercises 16 and 17.)

26.* Show that if a graph with at least two edges has a bridge, then it must also have a cut vertex. Must it have two cut vertices? (See Exercise 17.)

27.* Let G be a graph such that $G - E$ consists of two connected pieces H_1 and H_2. Let x_1 and x_2 belong to H_1 and H_2, respectively. Show that any path in G joining x_1 and x_2 must use edge E.

28.* Suppose that E is a bridge of the graph G. Show that E cannot be on a circuit unless that circuit repeats an edge. (See Exercise 16.)

29.* Suppose M is a graph with every vertex of valence 2. Show that M has neither bridges nor cut vertices. (See Exercises 16 and 17.)

30.* The paraffin compound C_3H_8 has only one isomer. There are several possible graphs, however, that can be drawn having three vertices of valence 4 and eight vertices of valence 1. Explain.

4.2 mailmen, Koenigsberg's bridges, and Leonhard Euler

Now that we have methods for describing routes in graphs, we are prepared to attempt a graph theoretical solution to the mailman problem raised in Chapter 1, for which a model was obtained in Chapter 2. Rephrased in the terminology of this chapter, the mailman can solve his problem provided there exists a circuit passing through each edge of the graph in Figure 2.3 exactly once.

Although we could now attack this graph theoretical problem directly, it is worthwhile to take a short historical excursion. We shall discuss a puzzle which gives rise to the same problem in graph theory as the mailman problem and which was the actual problem that motivated the Swiss mathematician Leonhard Euler to invent the theory of graphs.

Figure 4.11 shows a portion of the map of the city of Koenigsberg as it

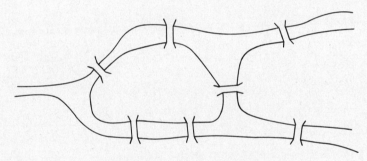

Figure 4.11

appeared in 1735. Koenigsberg was formerly located in Prussia but is now in the Union of Soviet Socialist Republics and has been renamed Kalingrad. Figure 4.11 displays a portion of the Pregel River including Kneiphof, one of the islands in the river. Also depicted are the bridges connecting the island to the city itself.

It was the pastime of some of the citizens of Koenigsberg to attempt the following: One was to start at any point on the shore of the river or on Kneiphof and walk over all the bridges once and only once, returning to one's starting spot.

This problem was posed to Leonhard Euler (1707–1783) who, in a remarkable paper, solved it and all problems in the same genre in one fell swoop. Let us follow in Euler's footsteps and attempt to recover his thinking.

Leonhard Euler (1707–1783), though born in Switzerland, spent a large portion of his life in St. Petersburg, Russia. He was extremely prolific, publishing over 500 works during his lifetime. His collected works are now being published and are expected to exceed 70 volumes. Euler made major contributions in many areas of mathematics, including the theory of functions and algebra. In 1735 he solved the problem of the Koenigsberg bridges and in 1752 expanded Descartes' discussion of the fact that for a three-dimensional solid, $V + F - E = 2$, where V, F, and E denote the number of vertices, faces, and edges of the solid, respectively. A contemporary claimed that Euler could calculate effortlessly "just as men breathe, as eagles sustain themselves in the air."
(Courtesy The Wolff-Leavenworth Collection, George Arents Research Library at Syracuse University.)

We begin by trying to find a model for the Koenigsberg bridge problem using the theory of graphs. Let us represent Kneiphof by a single vertex and each shoreline by a single vertex. This last provision may seem radical but since one can walk from any part of the shoreline to any other part without crossing a bridge, all points of the shoreline are equivalent as far as the problem is concerned. We now join two vertices by an edge, provided that there is a bridge joining the land blocks they represent. When we do this, we obtain the graph shown in Figure 4.12.

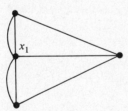

Figure 4.12

To solve the problem of the Koenigsberg bridges, one must construct a circuit which passes exactly once through each edge of the graph. To honor Euler and to avoid constant repetition of terms, we give the following definition.

82

DEFINITION 5

A circuit that traverses every edge of a graph exactly once is called an *Euler circuit* of that graph.

Before we analyze the Koenigsberg bridge problem, one remark about circuits is in order. We have spoken of them as beginning and ending at some vertex but actually we can begin and end at any vertex on the circuit we like.

Imagine that we attempt to construct an Euler circuit starting with vertex x_1 in the graph in Figure 4.12. We first observe that if the graph we are dealing with were not connected, there would be some vertices that could not be joined by a path to x_1. Hence, no Euler circuit could exist. Since the graph in Figure 4.12 is connected, there is some hope that it may have an Euler circuit. Let us imagine that we try to construct such a circuit starting at x_1. We initially leave x_1 using one edge; each time we enter and leave x_1, we use up two edges; finally, at the very end we must use one further edge to return to x_1. This means that if an Euler circuit existed, the valence of x_1 would have to be even. The valence of x_1 is 5, however, implying that the Koenigsberg bridge problem has no solution. What is more important, we see that for any graph whatsoever, unless the valence of every vertex is even, there can be no Euler circuit. On the other hand, a little experimentation with graphs that do have vertices all having even valences indicates that such graphs do seem to have Euler circuits. In fact,

THEOREM 2 (Leonhard Euler, 1735)

A graph G has an Euler circuit if and only if G is connected and the valence of each vertex of G is even.

Proof: There are two facts that are to be established.

(1) If G is connected and all its vertices have even valence, then G contains an Euler circuit.
(2) If G contains an Euler circuit, then G is connected and all its vertices have even valence.

We begin with a demonstration of part (2) because it is simpler. Suppose that G has an Euler circuit. Since this circuit covers all the edges, it must reach all the vertices, so G must be connected. Now let x_i be any vertex of the graph G. As we have remarked earlier, we can imagine that our Euler circuit begins and ends at x_i and that, in between starting and ending there, it revisits x_i a number of times, say, t times. Each of these t times the circuit revisits x_i, it goes in and then out, using two previously unused edges. These intermediate visits use up $2t$ edges at x_i. Starting and ending use up one edge each so a total of $2t + 2$ edges at x_i are used by the circuit. An Euler circuit uses all edges of the graph, however, so these $2t + 2$ edges are all the edges at x_i. Thus, the valence of x_i is even. Since this argument applies to any vertex, we have completed the proof of Part 2.

To prove part (1) we proceed as follows. Pick any vertex of G, say x_i. Leave x_i along any edge and wander along through the graph in a random path, making sure that we never traverse a previously used edge.

Suppose that we reach a vertex x_k by following this procedure and that there is no unused edge at x_k. We claim that x_k is actually the same vertex as x_i. If this were not the case, we would have entered x_k one more time than we were able to depart from x_k; hence the valence of x_k would have to be odd. Since this is contrary to the hypothesis, we conclude that $x_k = x_i$. Two cases now arise.

Case 1

All the edges of G have appeared in the circuit we constructed above.

Case 2

Some edges of G do not appear in the circuit constructed above.

Clearly, in Case 1, we have constructed an Euler circuit. In Case 2, let π_1 denote the circuit we have constructed, which is not an Euler circuit. Delete the edges of the circuit π_1 from G, thereby obtaining a new graph H_1. (*Warning:* The graph H_1 may not be connected.) Since G was connected, some vertex of H_1 has a vertex x_j in common with π_1. Furthermore, each vertex of H_1 has even valence since G had even valence at each vertex and since the circuit π_1 used an even number of edges at each vertex it passed through. Starting at the vertex x_j, we can wander at random, as in the construction of π_1, until we return to x_j, obtaining a circuit π_2. We now piece π_1 and π_2 together as follows: In writing out π_1 when we first reach x_j, we insert the edges of the circuit π_2, returning to x_j, and then continue with the remainder of π_1. Thus, we obtain a circuit $\pi_1 + \pi_2$, which starts at x_i. If this circuit includes all the edges of G, we have the required Euler circuit. If not, we delete $\pi_1 + \pi_2$ from the edges of G to obtain a new graph H_2, and repeat the process above. Since G has only a finite number of edges, after a finite number of repetitions of this process we would obtain the required Euler circuit.

Example 1

By consulting Figure 4.13, one can see how this process proceeds in a typical example. Since the valence of each vertex of the graph G in Figure 4.13

$$\pi_1 = x_1, x_2, x_3, x_4, x_5, x_6, x_1.$$

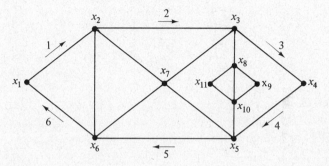

Figure 4.13

is even, we know that there exists an Euler circuit in the graph, by Theorem 2. Starting at x_1 (Figure 4.13), we produce the circuit

$$\pi_1 = x_1, x_2, x_3, x_4, x_5, x_6, x_1$$

If we delete the edges of π_1 from G, we obtain the graph H_1, shown in Figure 4.14. Starting at vertex x_3, we generate the circuit π_2:

$$\pi_2 = x_3, x_8, x_{10}, x_5, x_7, x_2, x_6, x_7, x_3$$

Piecing together π_1 and π_2 gives:

$$\pi_1 + \pi_2 = x_1, x_2, x_3, x_8, x_{10}, x_5, x_7, x_2, x_6, x_7, x_3, x_4, x_5, x_6, x_1$$

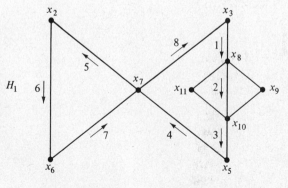

Figure 4.14

If we delete the edges of $\pi_1 + \pi_2$ from G, we obtain H_2, the graph shown in Figure 4.15. In H_2, we construct the circuit π_3 [(4.5)] starting with the vertex x_{10}.

$$\pi_3 = x_{10}, x_{11}, x_8, x_9, x_{10} \tag{4.5}$$

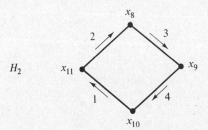

H_2

Figure 4.15

Inserting π_3 into the path $\pi_1 + \pi_2$ at the first occurrence of x_{10} (actually any occurrence of x_{10} will do), we obtain the circuit

$$\pi_1 + \pi_2 + \pi_3 = x_1, x_2, x_3, x_8, x_{10}, x_{11}, x_8, x_9, x_{10}, x_5, x_7, x_2, x_6,$$
$$x_7, x_3, x_4, x_5, x_6, x_1 \tag{4.6}$$

It is important to note that $\pi_1 + \pi_2 + \pi_3$ [see (4.6)] is not the only Euler circuit for the graph in Figure 4.13.

In the proof we have developed a mechanical procedure or *algorithm*, which if followed step-by-step can be used to find an Euler circuit in any connected graph with even-valent vertices. This algorithm can be briefly summarized:

(1)   Start the construction of the Euler circuit at any vertex.
(2)   From this initial vertex, wander at random using only edges not previously traversed. When you cannot proceed further in this manner, go to instruction 3.
(3)   Now either you have traversed all the edges, in which case you are done, or else it will be possible for you to choose an untraversed edge with a vertex in common with edges already traversed.
(4)   Using the vertex referred to in instruction 3, wander at random using only edges not previously traversed. When you cannot proceed further in this manner, take the path just generated and amalgamate it with the last path. Then proceed back to instruction 3.

You may have discovered by now that for the examples we do in this book it is rarely necessary to use the formal procedure just described. Generally, one can generate the Euler circuit all in one fell swoop with just a little effort. Why, then, did we go through the formal procedure involving stitching circuits together? There are at least three reasons:

(1) The procedure formed part of the proof that such circuits could be found under the right conditions.

(2) For very large and confusing-looking graphs it may be necessary to rely on a formal procedure that we know works.

(3) The formal procedure is well-adapted to computer use since computers must be precisely instructed about how to solve problems. One cannot ask a computer to proceed by that mixture of trial and error and judgement that you would use instead of the formal procedure.

THE EXPANDING UNICURSE*

Some citizens of Königsberg
Were walking on the strand
Beside the river Pregel
With its seven bridges spanned.

"O Euler, come and walk with us,"
Those burghers did beseech.
"We'll roam the seven bridges o'er,
And pass but once by each."

"It can't be done," thus Euler cried.
"Here comes the Q.E.D.
Your islands are but vertices
And four have odd degree."

From Königsberg to König's book
So runs the graphic tale
And still it grows more colorful
In Michigan, and Yale.

—*Blanche Descartes*

Having discovered precisely what conditions guarantee the existence of an Euler circuit, it is tempting to leave the problem of edge traversibility in a graph and to go to a different problem. But a mathematician is rarely content when he solves a problem. He always wants to look at his original

*Reprinted with the permission of Academic Press and Professor William Tutte.

problem and its solution from many directions, with the hope that some slightly different perspective will lead him to something new. This is especially true if the new problem requires little additional new work. One way to go about formulating new problems is to examine some of the attributes of the original problem, in this case the Euler circuit problem, and to change some of these attributes.

For example, suppose we can drop the requirement of the mailman problem that forces us to return to the starting vertex. Now we have the following intriguing problem: Let A and B be two vertices of a given graph G. Can we find a path which covers every edge of G exactly once and which starts at vertex A and ends at vertex B? For the graph in Figure 4.16 such a path would be

$$A, x_1, x_4, B, x_5, x_8, A, x_{14}, x_{15}, x_9, x_8, x_7, x_6, x_5, x_{10}, x_{11}, x_{12}, x_{13},$$

$$x_{14}, x_9, x_{10}, x_{15}, x_{13}, x_1, x_2, x_3, x_4, x_{12}, x_{15}, x_{11}, B$$

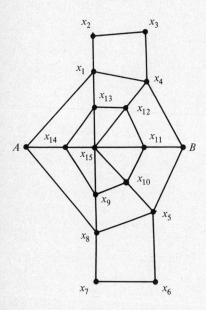

Figure 4.16

We emphasize that for any particular graph we could probably find the answer to such a question by using trial and error. What we really are looking for, however, is some test (preferably a simple one) that when applied to any graph G will tell us whether or not such a path exists.

DEFINITION 6

A graph G has an *Euler path* between vertex x_i and vertex x_j ($x_i \neq x_j$) provided that there is a path from x_i to x_j which repeats no edge and which includes all the edges of G.

It turns out that the problem of finding Euler paths can be attacked by a modification of our earlier work on Euler circuits.

THEOREM 3 (Leonhard Euler)

A graph G has an Euler path between the distinct vertices x_i and x_j if and only if G is connected and x_i and x_j are the only vertices of G having odd valence.

Proof: Suppose that G has an Euler path between the vertices x_i and x_j that are distinct. Since this Euler path exists, G must be connected.

In the graph G, let us add an extra edge E from x_i to x_j and let us call this new graph $G + E$. We can extend our Euler path of G to an Euler circuit of $G + E$ by tacking the edge E onto the end of the Euler path. Since $G + E$ now has an Euler circuit, each vertex has even valence, by Theorem 2. Therefore, each vertex of G is even-valent except for x_i and x_j, which are odd-valent.

Now we shall assume that G is connected and that every vertex of G is even-valent, except for x_i and x_j, which are odd-valent. We wish to prove that G has an Euler path. Let us join x_i and x_j by an edge (an additional edge if they are already joined). We obtain a connected graph, whose vertices all are even-valent, and can therefore conclude that there is an Euler circuit that can be written starting with x_i. The edge that we added, when removed from this Euler circuit, will tell us how to construct an Euler path from x_i to x_j.

New mathematical problems are frequently suggested by old ones. Often the new problems can be shown to model real-world problems of interest. Can you think of a modification of the mailman problem that can be solved by using an Euler path? If we can solve new real problems with little additional work, so much the better. As we shall see in the next section, however, this ideal may not always be obtainable.

EXERCISES 4.2

1.   For each graph in Figure 4.17, determine if it has an Euler circuit, an Euler path, or neither. If it has an Euler circuit or Euler path, draw it.

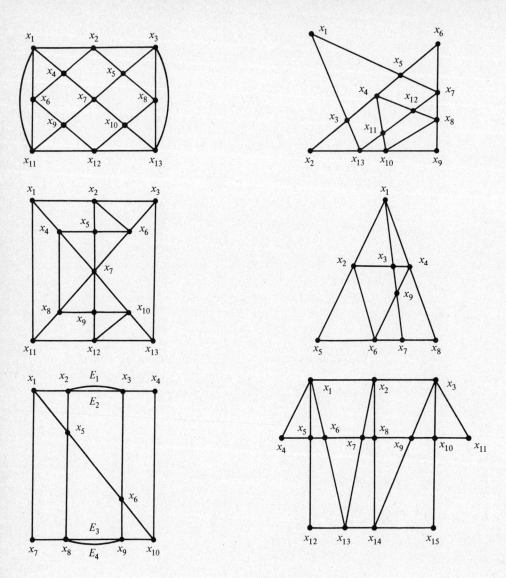

Figure 4.17

2. The graph in Figure 4.18 has been constructed as a model of the hallways of a building that is to be converted into a museum. Assume that the exhibits are to be placed only on one side of the hallway. The entrance and exit exhibits are to be located at A. Indicate how it might be possible to paint numbered arrows on the hallway walls so that if

a person followed the numbered arrows sequentially, he would see all the exhibits once and only once.

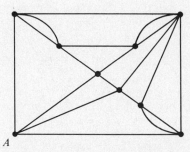

A   Figure 4.18

3.   A highway inspector wishes to save the taxpayers' money. He hopes to find a route starting at x_1 and returning to x_1 that will take him over each section of the highways he must inspect once and only once. Can he find such a route? If so, draw one such route. (See Figure 4.19.)

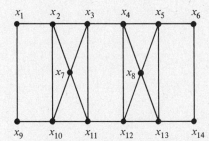

Figure 4.19

4.   Can a person find a route through the floor plan of the house in Figure 4.20 that will take him through each door exactly once?

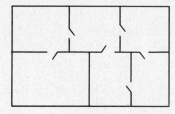

Figure 4.20

5.* Can you find an Euler circuit for the graph in Figure 4.13 with the property that as you traverse the circuit it never cuts across itself?

4.3   variations on Euler circuit problems

A good modeling slogan might be "If at first you don't succeed, try, try again." Typically, one tries the simplest model one can think of to solve a problem and when that proves inadequate (as it often does), one tries again, this time using a more sophisticated model. In this section we shall illustrate this aspect of modeling.

Suppose we need to design an efficient route for a street sprinkler truck that needs to traverse each street (edge) of Figure 4.21. In our first model we

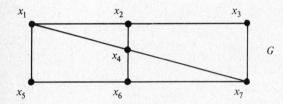

G

Figure 4.21

shall convert this quest for an efficient route into this question: Can we find an Euler circuit in the graph of the street network? Since there are vertices of odd valence, the answer is no; thus, our first modeling attempt produces merely the information that the ideal solution is unattainable.

As our second modeling attempt we shall convert the quest for efficiency into this problem: Find a circuit covering every edge in Figure 4.21 at least once, but having the minimal number of duplications.

For example, by trial and error, one can find a circuit

$$x_1, x_2, x_4, x_1, x_4, x_2, x_3, x_7, x_6, x_4, x_7, x_6, x_5, x_1 \tag{4.7}$$

that repeats three edges, $[x_1, x_4]$, $[x_2, x_4]$, and $[x_6, x_7]$. Is three the minimal number of repetitions? This question cannot be answered as easily as the Euler circuit problem, but in the spirit of applied mathematics we can make enough headway for some practical purposes. First we shall establish a lower bound to the number of duplications. This lower bound will serve as a limit to our optimism in estimating how small a number of duplications we can get away with by designing a circuit cleverly.

THEOREM 4

Let G be a graph with v_0 being the number of vertices having odd valence. Any circuit in this graph that covers every edge at least once must have at least $v_0/2$ duplications.

Proof: Suppose C is a circuit that covers every edge of G. We shall use C to construct a slightly different graph G' and a circuit C' in G'. To form G', follow circuit C in G and each time you come to an edge E previously used, add another edge to G connecting the end points of E. To form circuit C' in G', follow the circuit C except where it duplicates an edge in G. At such an edge, there is a newly added edge in G', which we use before continuing on, as in C. Figure 4.22 shows how G' is constructed for the circuit (4.7) taken as C.

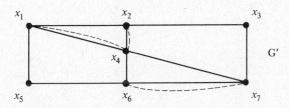

Figure 4.22

The dotted edges are those that have to be added to G in Figure 4.21 to obtain G'.

We add a new edge in G' every time we would be otherwise forced to duplicate an edge in G. Consequently, the number of newly added edges to form G' equals the number of duplications in C. Therefore, we need to determine the number of edges that have been added.

The key fact in determining the number of added edges is that C' is an Euler circuit in G'. Therefore, all the vertices of G' have even valence. At each of the odd-valent vertices of G we need to raise the valence by at least one in order to make the valence even. Consequently the new edges must be added in such a way as to touch every odd-valent vertex. Now each edge added has only two ends and so can touch at most two vertices. The most optimistic possibility is that each edge touches exactly two of the v_0 odd-valent vertices. In this case $v_0/2$ edges have been added.

For any particular graph for which we have found a circuit covering all the edges, we can use this theorem to give us some notion of how good a solution we have found. For example, in the graph of Figure 4.21 there are four odd-valent vertices, so $v_0 = 4$. Consequently, any circuit covering all the edges needs at least two duplications. The solution shown in (4.7) has three duplications. We can conclude that perhaps if we looked harder, we could find a better solution. On the other hand, we know enough not to try to do better than two because that is not possible. Can you find a circuit covering all the edges in Figure 4.21 but with only two duplications?

We can reverse the idea of the proof of the theorem to help us find a circuit covering all edges of a given graph with not too much duplication. In

Figure 4.23(b) we have converted some edges of graph G in Figure 4.23(a) into multiple edges by adding some dotted edges. These edges have been added in a strategic way so as to create a new graph G', all of whose vertices have even valence. The reason this is useful is that G' has an Euler circuit, say C'. When we try to trace out C' on G (not G'), we find a circuit that duplicates some edges. One duplication occurs for each added (dotted) edge. Figure 4.23(b) gives us a solution using eight duplications. In Figure 4.23(c) we have been a bit cleverer and reduced this to four duplications. Can we do still better? According to the theorem we cannot, so we can stop trying.

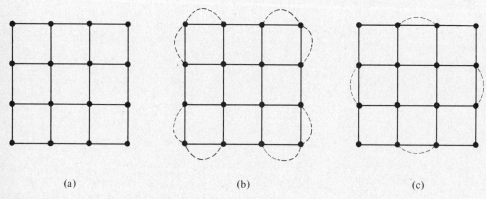

(a)   (b)   (c)

Figure 4.23

For the graph of Figure 4.24(a) the theorem tells us that a path with minimal duplications has at least two duplications. If you look for a solution with two duplications, you will have trouble finding one. You will undoubtedly be able to find solutions with four duplications. Two such solutions are shown in Figures 4.24(b) and (c). Can you find yet another? It turns out that four is the best you can do. This is a case where the lower estimate provided by the theorem is not exact.

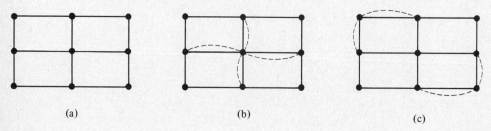

(a)   (b)   (c)

Figure 4.24

In general, it is hard to tell what the minimal duplication is for a given graph. Our theorem, while useful, often doesn't give a good lower estimate. It is not unreasonable, if we are concerned with street sprinklers, to restrict ourselves to rectangular street networks, that is, networks where every block has four corners and every intersection has exactly two streets crossing there. In this case, we can obtain some fairly good and sometimes exact notions of what the minimal duplication is. Suppose our network is h blocks high and m blocks long. (See Figure 4.25.)

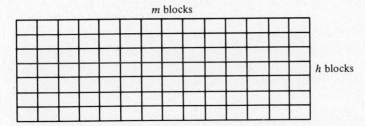

Figure 4.25

There are $2h + 2m - 4$ odd-valent vertices, namely, all those on the outer boundary of the network except for the four corners. According to our theorem, the minimal duplication for a path that covers all edges is at least $h + m - 2$. We shall show that this is a reasonably good estimate of the minimal duplication for rectangular networks, in the sense that we can find paths whose duplications come quite close to this figure, even hitting it on the nose, depending on whether h and m are even or odd numbers.

Case 1

h and m are both odd. Then Figure 4.26 shows how to add $(m - 1)/2$

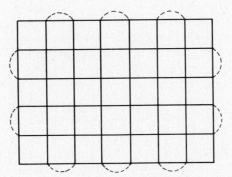

Figure 4.26

edges along the top and along the bottom and $(h - 1)/2$ edges on each of the sides, making an even-valent graph. The number of edges added is $h + m - 2$, and we know that there is a path with this many duplications. Our theorem tells us we can't hope to do better. Therefore, the estimate given by Theorem 4 is exact when h and m are both odd.

Case 2

h and m are both even. Figure 4.27 shows a solution with $m/2$ edges along the top and bottom and with $h/2$ edges added on each side, the resulting graph having $h + m$ extra edges added. These additions produce a graph with all vertices even-valent so we know there is a circuit covering all edges and with $h + m$ duplications. Since the best we could possibly do would be $h + m - 2$, we know that even if our $h + m$ is not the best, it could be improved at most by the removal of two duplications.

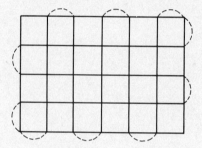

Figure 4.27

Case 3

One of h and m is odd and the other is even. We leave this case for you to analyze.

Until now, our model has not taken into account the different lengths the streets may have. If this is considered too, our problem would be to minimize the total length of duplication. This is not necessarily the same as minimizing the number of duplications because it may happen that duplicating three short streets is better than duplicating two long ones. Since this final level of sophistication in our modeling process poses substantial mathematical problems, we shall not discuss it further.

EXERCISES 4.3

1. For the graph in Figure 4.28, what is the minimal duplication you can find? How does it compare to the lower estimate of our theorem?

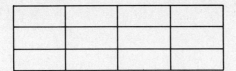

Figure 4.28

2. Can you supply the analysis for Case 3?
3. For the graphs of the radial street networks in Figure 4.29, what are the minimum duplications you can find? How do the numbers compare to the lower estimate of the theorem?

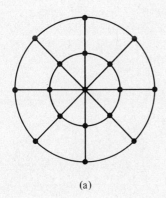

(a)

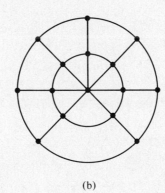

(b)

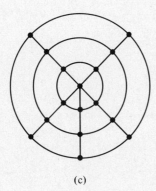

(c)

Figure 4.29

4. For a general radial street network with n streets radiating from the center and r rings around the center [e.g., in Figure 4.29(c), $n = 5$ and $r = 3$], what is the minimal duplication that can be arranged? (*Hint:* The answer depends on whether n is even or odd.)

5.* Show that in any graph v_0 the number of odd-valent vertices is even. Consequently, $v_0/2$ is a whole number.

6.* Can you find a circuit in Figure 4.30 that has eight duplications? Can you prove that this is the minimum number possible for this graph?

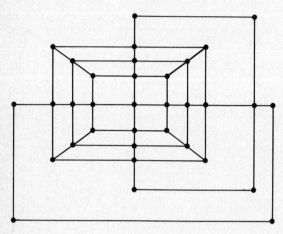

Figure 4.30

4.4   variations on Hamilton circuit problems

In this section we shall continue our study of rectangular street grids but this time from the point of view of finding Hamilton circuits in them. Figure 4.31

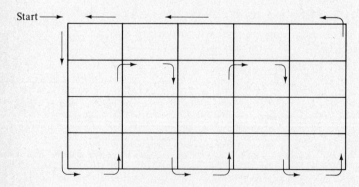

Figure 4.31

shows a simple and fairly general way of trying to find a Hamilton circuit in a rectangular street grid. Unfortunately this method won't work all the time, as illustrated in the grid of Figure 4.32. By comparing these grids you will probably conclude that in an $m \times n$ grid where one of the dimensions is odd, there is always a Hamilton circuit, which can be generated by first orienting the grid so its horizontal dimension is odd and then following the pattern of Figure 4.31. This conclusion does not, however, tell us whether even-by-even grids have Hamilton circuits. It may be that some other method of solution might work for even-by-even grids. We shall shortly see that this is not the case, but we pause for a biographical refreshment followed by a definition.

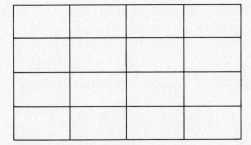

Figure 4.32

Sir William Rowan Hamilton (1805–1865), born in Ireland, showed an early talent for languages. At five, he was able to read Greek, Hebrew, and Latin. He showed talent in mathematics early, too, and while still a student at 22, he was appointed Royal Astronomer of Ireland. Hamilton's major contributions to mathematics were in algebra. He lends his name to the concept of a Hamilton circuit via his invention of the Icosian Game. This game requires a route of edges visiting without repetition the 20 vertices of a dodecahedron (Figure 4.40) which had been labeled with the names of European towns. Hamilton hoped to become rich by marketing this game, but unfortunately the game was not a success.
(Courtesy The Bettmann Archive)

DEFINITION 7

A graph is *bipartite* if it is possible to color its vertices with only two colors, S and C.

Figures 4.33 and 4.34 show examples of bipartite graphs, where the vertices colored C are denoted by circles and the vertices colored S are denoted by squares. Figure 4.35 shows two graphs that are clearly not bipartite; studying them should convince you that no graph having a circuit with an odd number of distinct vertices is bipartite.

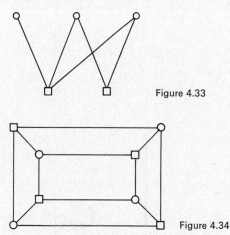

Figure 4.33

Figure 4.34

Figure 4.35

It is not hard to see that rectangular grids are always bipartite: Simply color the vertices across the top row alternately with circles and squares, beginning with a circle at the upper left. Now do the same with the next row down but starting with a square at the left. Evidently we can continue this for all the other rows, thereby coloring all the vertices suitably. Figure 4.36 shows this in an even-by-even grid. In connection with the Hamilton circuit problem,

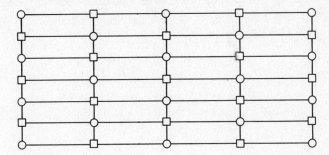

Figure 4.36

for this grid it will be of particular interest to us to see whether or not the number of circle vertices equals the number of square vertices. We note that the top row, which has an even number of edges, has an odd number of vertices. Consequently, there is one more circle vertex than there are square vertices. In the next row there is a reverse imbalance: one more square vertex than circle vertices, thus canceling out the imbalance in the first row. As we go down the rows, keeping a cumulative record of the imbalance, we discover that after counting an even number of rows there is no imbalance, but after counting an odd number of rows there is an imbalance of one circle. Since we have an odd number of rows, there is an imbalance of one circle. Since we have an odd number of rows in an even-by-even grid, we have shown that in an even-by-even rectangular grid where the vertices are divided into circle and square vertices so as to exhibit the bipartite nature of the graph, the number of square vertices is unequal to the number of circle vertices.

With this in mind, the following theorem shows that an even-by-even grid does not have a Hamilton circuit.

THEOREM 5

Suppose G is a graph which is bipartite and which has vertices colored C or S, with no two vertices joined by an edge having the same color. If G has a Hamilton circuit, then the number of vertices colored S is the same as the number of vertices colored C.

Proof: Let us denote the Hamilton circuit in G by

$$x_1, x_2, \ldots, x_i, \ldots, x_1 \tag{4.8}$$

where every vertex of the graph appears in (4.8) exactly once (except for x_1). In the circuit described in (4.8), two vertices that are next to one another cannot both be colored C nor both colored S since in a bipartite graph no edge can join two vertices, both of which have the same color. Hence in (4.8) the vertices must alternate between the two colors. Since no vertex except x_1 is repeated, this alternation means that there must be equal numbers of vertices colored C and S.

Unfortunately, Theorem 5 is one of the few simple theoretical tools useful in studying Hamilton circuits. Actually, however, it is not very useful since it can only be used to demonstrate the *non-existence* of Hamilton circuits. In general, Hamilton circuit theory is much messier than Euler circuit theory, which is more or less contained in Theorem 1. It is partly for this reason that much more attention is currently being paid to Hamilton circuits and allied concepts. The other important reason for the intensive study of these matters is the practical importance of closely related concepts. We end the section with a brief description of one of these concepts.

Suppose that we consider a salesman whose job is to visit each of the cities A through E shown in Figure 4.37. In order to get from one city to

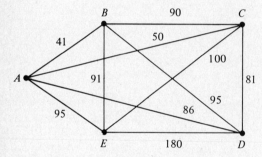

Figure 4.37

another, the salesman must, of course, pay a certain cost (perhaps the price of an airline ticket). The cost of traveling between two cities in Figure 4.37 is indicated by the number on the edge joining them. If you choose, you can interpret the numbers as being the distances between the cities involved; using costs would represent a more general situation. The problem for the salesman is to find a minimal cost tour that visits each of the cities. Restated somewhat differently, the salesman is seeking a minimal cost Hamilton circuit. This problem, called the *traveling salesman problem*, has attracted a tremendous amount of attention because of its many applications in this and various other formulations. Unfortunately, no simple procedure for solving it exists. This may seem surprising since there is a simple method of attempting a solution. If the salesman is at a given city, he should travel to a city that has not already been visited, for which the cost is cheapest. This seemingly reasonable rule gives the route (Figure 4.37) $ABCDEA$ for a salesman whose home is *A*, yet the route *ADCBEA* is cheaper. This example illustrates the idea that doing the best thing available at a given time may not give an optimal solution to a whole problem. Although many fine mathematicians have worked on the traveling salesman problem, no solution to the general problem has been found. However, there are some good computer-oriented methods for finding approximate solutions.

EXERCISES 4.4

1. Using Theorem 5, determine whether the graph used to model the mail distributor problem from Chapter 1 has a Hamilton circuit.

2. Suppose *G* is a connected graph with the property that its vertices can be divided into two sets, *A* and *B*, with the following properties:
   (a) If *x* and *y* are vertices that are both in *A* or both in *B*, then there is no edge that joins *x* and *y*.
   (b) The number of vertices in set *A* equals the number of vertices in set *B*.

   Must *G* have a Hamilton circuit? Try some experiments.

3. For each of the graphs in Figure 4.38, determine if the graph has a Hamilton circuit. Can this be proved by Theorem 5?

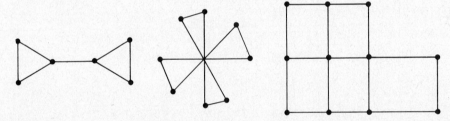

Figure 4.38

4. DEFINITION

   A *Hamilton path* in a graph *G* is a path whose beginning and end vertices are distinct, and which contains each vertex of *G* exactly once.

(a)   Which graphs in Figure 4.39 have Hamilton paths?

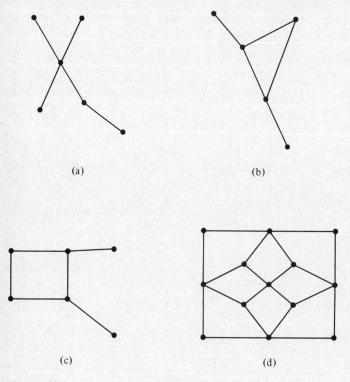

Figure 4.39

(b)   Show that if *G* has three 1-valent vertices, it cannot have a Hamilton path.

5.   Obtain a solution to Hamilton's "great icosian" game by determining if there is a Hamilton circuit for the graph consisting of the vertices and the edges of the solid shown in Figure 4.40.

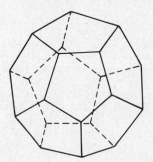

Figure 4.40

6. Find the minimal cost Hamilton circuit for the graph in Figure 4.37.

7. Find the solution of the traveling salesman problem, using air distances, for the four largest cities in your home state.

8. Find the solution to the traveling salesman problem, using air distances, for the cities of Chicago, San Francisco, Los Angeles, and New York.

9. Would the answers to Exercises 7 and 8 be considerably changed if air distance were replaced by rail or auto distance?

10. Solve the traveling salesman problem for the graphs in Figure 4.41.

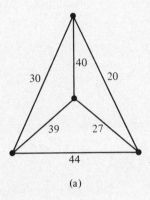

(a)

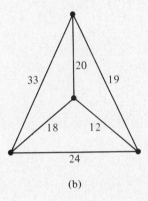

(b)

Figure 4.41

11. Imagine you are planning to take an automobile trip in which you visit Chicago, Houston, and San Francisco, starting from your hometown. Explain how your knowledge of the traveling salesman problem might result in a cheaper vacation.

12. For the model shown in Figure 4.37, find the minimum cost of a salesman who is required to start in A, finish in D, and visit B, C, and E.

13.* Figure 4.42 shows a portion of a small town where Mrs. Fabricant lives. Her home is located at P. Nearly every day she must make stops at the bakery (B), the supermarket (S), the laundry (L), the newstand (N), and the bank (BK).

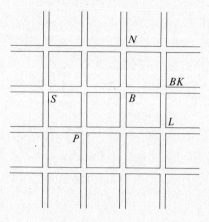

Figure 4.42

(a) What is the length of the shortest route that starts at her home, visits each of the required stops, and returns her home?

(b) Suppose Mrs. Fabricant insists that her first stop be at L? Does this affect the length of her shortest route?

Arthur Cayley (1821–1895) was an English mathematician whose most important work was in algebra. Cayley worked on the four-color conjecture (1878), but his major contribution to graph theory lay in his investigations about trees. This work, dating to papers written in 1856 and 1874, dealt with enumerating various isomers by counting the number of non-isomorphic trees with various properties. In particular, he calculated the number of isomers of the saturated hydrocarbons $C_nH_{2n^-2}$ for values of n up to 13. Cayley also spent 14 years as a lawyer, but this hardly seemed to interfere with his prodigious mathematical output, as he published several hundred papers during those years.
(Courtesy The Bettmann Archive.)

SUGGESTED READING

BELLMAN, R., K. COOKE, AND J. LOCKETT, *Algorithms, Graphs and Computers*, Academic Press, New York, N.Y., 1970. This is probably the best view of applied graph theory available. The flavor is right and very little is included just because it is cute.

EULER, L., "The Koenigsberg Bridges," *Mathematics in the Modern World*, W. H. Freeman and Company, San Francisco, Calif., 1968. Available in paperback. A translation of Euler's original article on Euler circuits.

Computer Used as Guide to Expert Seeking Way Out of Labyrinth of Urban Problems

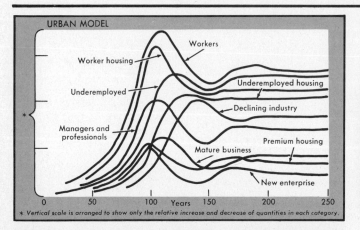

URBAN MODEL

Workers

Worker housing

Underemployed housing

Underemployed

Declining industry

Managers and professionals

Premium housing

Mature business

New enterprise

0   50   100   Years   150   200   250

* Vertical scale is arranged to show only the relative increase and decrease of quantities in each category.

Jay W. Forrester

FORRESTER CHART

New enterprise

Mature business

Declining industry

Premium housing

Worker housing

Under-employed housing

Managers and professionals

Workers

Under-employed

UJ — Jobs for underemployed
TC — Tax collections
UR — Ratio of jobs to underemployed
UJM — Attractiveness of job market
UM — Movement of underemployed into labor class
UAMM — Attractiveness of opportunity to move into labor class
UHM — Attractiveness of housing to underemployed
UHR — Ratio of housing to underemployed
AMM — City's real attractiveness to underemployed migrants
AMMP — City's attractiveness as perceived by underemployed outsiders
UA — Arrival of underemployed migrants
TPCR — Tax per capita
PEM — Public expenditures
UHPR — Rate of underemployed housing construction
UHPM — Attractiveness of housing available

By WILLIAM K. STEVENS

To Jay W. Forrester, whose specialty is grappling with complexity, the modern city is a diabolically complicated system that not only hides its real nature but even sets up decoys for those who want to learn how it functions.

So he has tried to cut through the urban tangle and evade the decoys by using a computer to keep track of and simulate the interaction of what he believes to be the major factors that cause a city to rise, flourish and then decay.

This model of a functioning metropolis, expressed in mathematical equations, is a general, theoretical one that is not designed to apply to any particular city. But, Mr. Forrester, a professor at the Massachusetts Institute of Technology, has used it to test the efficacy of such traditional urban "cure" policies as job training and housing construction for the poor.

5

computers and related topics

5.1 introduction

In this chapter we examine some aspects of computers and the mathematics relevant to them. This subject is so vast that we shall be in the position of the three blind Indians who examined an elephant by touching him. The Indian examining the trunk thought the beast must be long and tubular like a snake; the one touching the side figured the elephant was broad and massive like a wall; the one investigating the legs concluded he was dealing with a round, tall animal in the shape of a tree. Even when their information was pooled, it would have been hard to comprehend the nature of the elephant.

We shall discuss the computer briefly from different vantage points—electrical circuits, binary arithmetic, programming, encoding of information—and hope that the total picture, while inevitably disjointed, will nevertheless be helpful.

It is sad but true that many practical problems are too long for hand computation. Think back to the mailman problem of Chapters 1 and 2 where we were led to look for an Euler circuit in a graph with 13 vertices and 32 edges. This was a small graph, so trial and error did the trick for us. But imagine pushing your pencil through a graph with 13,000 vertices and 32,000 edges in search of a trial and error solution. Here it is perhaps some comfort that we found a step-by-step procedure guaranteed to find an Euler circuit in a graph that has one. If we use this procedure, it will still take us a long time but at least we know we won't waste time on trials that turn out to be errors. Our procedure is even more of a convenience if we attempt to instruct a computer to solve this large problem for us. This then is the first thing to know about computers: One of their great virtues is that we use them to avoid human drudgery in cases where we can give the computer a specific procedure to follow.

Although everyone is familiar with many uses of computers, a display of the tremendous scope of the computer "revolution" may, nevertheless, come as a surprise. Considering that this revolution began near the end of World War II, a rather short time ago, the computer has already had a vast impact on our lives.

Some of this impact is illustrated by the following list of applications.

(1) Preparing payroll checks; processing personal checking accounts; and being useful in many other situations in the areas of banking, finance, and insurance.

(2) Billing by large retail firms. Credit card accounts.

(3) Processing subscriptions and large mailings of national organizations and magazines.

(4) Sorting of mail.

(5) Analyzing and tabulating census data.

(6) Constructing inventories.

(7) Airline, train, and hotel reservation systems.

(8) Traffic and air traffic control.

(9) Translating materials from foreign languages.

(10) Diagnostic medicine. Cardiograms and encephalograms (brain wave records) are being analyzed by computer. Patient data are computer processed.

(11) Computing orbits for satellites and the launching and guidance of rockets and missiles.

(12) Teaching machines.

(13) Long- and short-term weather forecasting.

The list given above could be extended almost indefinitely. Its purpose is to remind you of the wide variety of ways in which computers enter our daily lives.

Computer Finds Isaiah Was Written by 2 Men

Jerusalem, March 30 (Reuters)—Electronic computers have proved virtually beyond doubt that two Isaiahs, probably living 100 years apart, wrote the Biblical book of Isaiah, according to a Hebrew University doctoral thesis presented here.

Tests analyzing stylistic and linguistic details showed that Chapters 40 to 60 of the prophet's work were written by a second Isaiah, believed to have lived in the sixth century B.C.

Hebrew University professors say the probability of the first Isaiah's also having written the chapters attributed to the second Isaiah is 1 in 100,000.

If generally accepted, the thesis, presented by a 57-year-old Biblical scholar, Yehuda Radday, would end a 150-year-old dispute among Biblical scholars.

Mr. Radday, who is in charge of Biblical teaching in the department of general studies at the Haifa Technion, set out on his research certain there was only one Isaiah.

EXERCISES 5.1

1. Add to the list of applications of computers by commenting on the uses to which computers have been put in the following situations and institutions:
   (a) the stock exchange
   (b) the internal revenue service
   (c) the social security administration
   (d) registration at your college
   (e) dating service companies
   (f) space program
   (g) election returns

5.2 the major components of a computer

In Figure 5.1 we show a simplified schematic diagram for the most important components of a modern scientific computer. A sketch of the work handled by each of the components of such a computer follows.

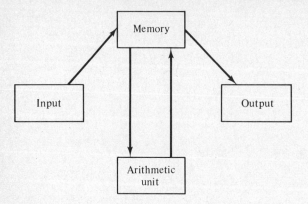

Memory

Input

Output

Arithmetic
unit

Figure 5.1

INPUT AND OUTPUT DEVICES: The input device is used for transferring information from a human user of the machine into the computer so that it can act on or manipulate that information. The output device is the means of transferring information contained within the computer back to the human operator. At the most basic level the computer carries on its work with electrical currents flowing through its complicated circuitry. We cannot control this electrical activity directly by having our brain or any other organ send electrical signals so we need some middlemen between our thoughts and the computer's circuitry. Typically we would first put our instructions on paper. This list of instructions would be converted to a deck of IBM cards punched with holes whose patterns reflect what we wrote on the paper. This deck would then be fed into a piece of equipment called a *card reader*, where a light shines upon the cards and the holes on each card determine pulses of electricity in a photoelectric cell stationed on the other side of the card from the light. These pulses may be fed directly into the main part of the computer. Other input methods that are often used instead of or in conjunction with this basic process are special typewriters, magnetic tapes, and cathode-ray tubes. The same or similar devices are used for output devices to transform the computer's electrical "information" into something understandable to a human being.

Much of the equipment which we just discussed as well as other equipment which you may have seen in computer installations or in pictures of them is so-called *peripheral equipment*. Peripheral equipment is used in preparing punched cards, sorting such cards, and transferring data either from cards to tape or from tape to printed form. To try to appreciate the nature of the computer from the peripheral equipment would be like trying to understand the human brain by examining books, phonograph records, television sets, and other media by which the brain receives its sensory inputs. If you

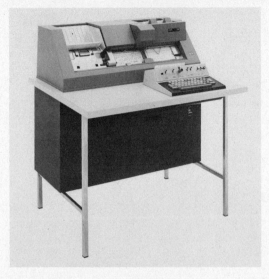

Some peripheral equipment. (*Courtesy of IBM.*)

ever find yourself working with computers, you will need to know more about the peripheral equipment but since our goal is a general understanding of computers, we shall not discuss this equipment further.

MEMORY UNIT: The memory unit of a computer is a device where information is stored. The computer memory is divided into cells, each of which contains one unit of information. Each cell has an *address* (or number)

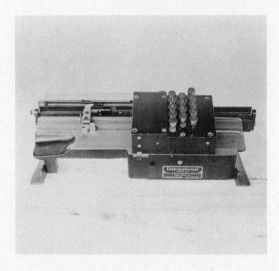

A Keypunch. *(Courtesy of IBM.)*

attached to it. When a particular unit of information is needed, it can be located easily by giving the address (or number) of the cell where it is being stored at that moment. Some memory cells will contain input data, others will contain the answers to a problem that are waiting to be sent to an output device, and still other cells will contain (in numerical form) the instructions telling the computer how to go about solving the problem. Often memory cells will be used for storing the results of intermediate calculations that come up in the course of obtaining the final answer to a problem. The most common types of memory units are magnetic drums, magnetic cores, and magnetic tapes. The different types of memory vary according to their accuracy and their *access time*. The access time refers to the amount of time required to get a piece of information to or from the memory unit.

ARITHMETIC UNIT: The arithmetic unit of the computer is that part where the computations occur. The additions, subtractions, multiplications, divisions, and logical operations that must be done to solve a complicated problem are performed in the arithmetic unit.

As suggested in Figure 5.1, when doing a calculation, there is a transfer of information between memory and the arithmetic unit. The situation would be similar if your brain were doing a calculation. For example, if you want to compare the price per ounce of two different brands of coffee, what you would do is:

(1)   Divide the price of the first can by the contents of the can in ounces.
(2)   Shift the answer to your memory, thereby permitting your conscious attention to shift to the calculation for the next can.
(3)   Perform the calculation for the second can.
(4)   Recall the first answer from memory and compare it to the second.

You may be curious about how the computer compares with the human brain in such aspects as size, speed, and memory capacity. The results of the comparison, using a powerful computer available in 1970, are shown in the table below:

	Computer	Brain
Size	2000 cubic feet (10 tons)	$\frac{1}{27}$ cubic foot
Speed	10 to 100 million operations per second	1 to 10 seconds for an addition
Memory	1 million items	10 billion items

The brain characteristics apply to almost any brain, say your own. Note that the brain wins in two areas, but when the computer wins, it wins big.

Now that we have outlined the various components of the computer, you may be wondering how it solves a problem. Does the computer solve problems by thinking out its own solution or must some human being instruct the computer how to go about solving the problem? This question, in turn, raises the question of what it means for a machine to think.

Design model for a computer installation. *(Courtesy of IBM.)*

A replica of the difference engine of Charles Babbage.
(Courtesy of IBM.)

Alan Turing, one of the pioneers of logical design for computers, once suggested a test along the following lines to determine whether a computer could think. Imagine that there were two identical rooms, one containing a computer and the other containing a human being. Imagine that you didn't know which room contained the computer and which contained the human being and that you communicated with the occupants of the rooms with a teletype. Your task is to send questions to both rooms and to try to determine from the answers given by the computer and the human being which room contains the computer. If you couldn't tell, no matter how hard your questions, Turing would conclude that for all practical purposes the computer was thinking. Turing's test is ingenious but not everyone would accept it. For example, suppose you had a computer whose memory was so large it could be fed prepared answers to almost any question it could be asked. Do you think this is conceivable? Would you say such a computer was thinking? Another problem that could be raised concerning Turing's test is that it doesn't seem to involve that peculiar capability that humans have of making spontaneous connections in their heads without any apparent prodding by external stimuli. It seems clear that before we can ask "Do computers think?" we should have to spend a good deal of time discussing what we would accept as "thinking." This is a deep question that we won't discuss any further, but it may cheer you to know that at present there are no computers that could pass Turing's test.

The inability of the computer to match the human brain is basically due to the fact that the computer cannot do anything spontaneously but must be specifically instructed by a human programmer in order to solve the simplest problem. As a matter of fact it is a well-known observation in computer work that "computers never do what you *want* them to do but only what you *tell* them to do." This means that computers are ideal for carrying out instructions that have been given to them. If the computer is obtaining nonsensical answers, more than likely some human being fed in nonsensical instructions. As computer programmers sometimes say, GIGO! (garbage in, garbage out!).

5.3 *algorithms with decision points and feedback loops*

Perhaps the greatest difference between the human brain and a computer is that while the brain can solve many problems in an intuitive way, sometimes without appearing to think about what it is doing, a computer can do nothing without a painfully logical, step-by-step algorithm. Such an algorithm, when written in a form usable by a computer, is called a *program*. Before looking at a program designed for a computer solution to a problem, we shall study some examples from outside the world of the computer. These examples illustrate in more familiar settings the way in which simple tasks (washing clothes, fixing radios) need to be broken down into constituent operations and decisions that must be properly sequenced so that the task can be performed without any further exercise of intelligence or judgment.

In our first example, consider the situation of Hilda, a housewife with two bundles of clothes to be laundered. Hilda's washing machine has controls for water temperature and number of rinses and she sets these controls differently for the two bundles, as shown in Figure 5.2. When Hilda sets the

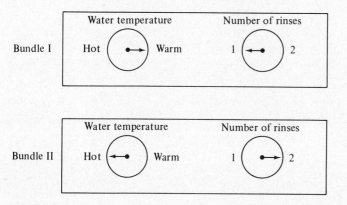

Figure 5.2

dials to achieve a certain combination of water temperature and number of rinses, she is programming her machine. What this means is that she is giving the machine advance instructions that will be exactly and precisely followed.

In Figure 5.3 we show a flow chart for this washing machine that illustrates graphically the sequence of operations the machine will perform. The square boxes describe operations that the machine performs, while the diamond-shaped boxes are so-called decision boxes because they indicate places where the machine must consult the dial settings in order to find out which of two possible operations to perform next.

The chart in Figure 5.3 illustrates one of the two important characteristics of computer programs; namely, it has decision points. The other characteristic, one which is missing in our flow chart, is the feedback loop. We can, however, revise our flow chart to contain a feedback loop. As motivation for this, notice that the flow chart in Figure 5.3 has three boxes representing the same rinse operation. If our machine could rinse up to 50 times, there would be many more rinse boxes—too many to fit in a neat diagram. In Figure 5.4 we revise the last part of our flow chart by "feeding back" the rinse operation into the decision point just above it. The decision box concerning the number

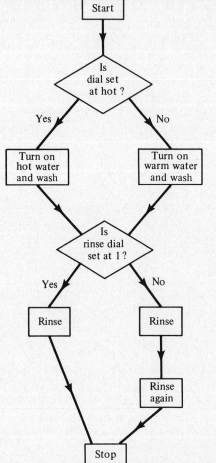

Figure 5.3

of rinses and the rinse operation box, together with the arrows linking them, form what is called a *feedback loop*. Notice that with this loop we can now deal with any number of rinses (if our machine is constructed for them).

Next we shall examine another process from everyday experience. This example involves fixing a radio; decision points and feedback loops are an essential part of the process. The radio repair flow chart (Figure 5.5) should be self-explanatory except for the role of K, which has nothing to do with the radio directly but is merely a counter that keeps track of what tube we are working on at any moment.

To illustrate the operation of this program, we take a particular broken radio, symbolized in Figure 5.6, and show what happens to it at various stages of the program. The diagrams are to be interpreted like this. Each circle represents a tube, a bad one if there is an "x" in the circle and a good one otherwise. The numbers below are labels for the tubes. The x outside the tubes represents the fact that in this particular radio there is an additional defect present in some component other than a tube. The little box at the right of each radio contains the current value of the counter *K*.

    The value of this kind of testing of hypothetical radios is that it gives us a "feel" for how the program works. To some extent, an understanding of how the program works can be achieved just by looking at it and reading it. Often it is hard to grasp how it will work in all contingencies, however; there may also be subtleties about its behavior. For example, did you notice when you first examined the flow chart that this program does a silly thing if you try it on a radio that has nothing wrong with it? To see what happens,

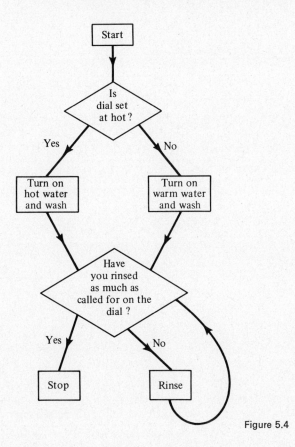

Figure 5.4

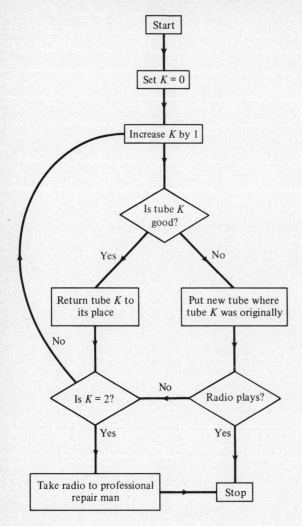

Figure 5.5

try running through the program with such a radio (i.e., no x's anywhere).

We can also present the same information given by the flow chart in the nonpictorial form of a list. To make this conversion, one goes through the flow chart, starting at the beginning and using "go to" statements where there are decision points (for example, statement 3 below). The reverse process of converting from a list to a flow chart follows similar commonsense rules.

120

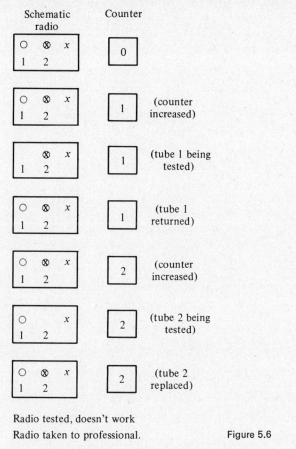

Schematic radio   Counter

0

1   (counter increased)

1   (tube 1 being tested)

1   (tube 1 returned)

2   (counter increased)

2   (tube 2 being tested)

2   (tube 2 replaced)

Radio tested, doesn't work
Radio taken to professional.   Figure 5.6

INSTRUCTIONS FOR RETURNING A BROKEN RADIO TO WORKING ORDER

(The instructions are carried out one after another, unless some instruction specifies what instruction to go to next.)

0.   Start.

1.   Set $K = 0$.

2.   Increase K by 1.

3.   Remove tube K and test it. Is it good?
        Yes—go to 4.
        No—go to 6.

4.   Return tube K to the radio.

5. Is $K = 2$?
    Yes—go to 8.
    No—go to 2.

6. Put a new (good) tube in the K place.

7. Turn on radio. Does it play?
    Yes—go to 9.
    No—go to 5.

8. Take radio to a professional repairman.

9. Stop.

One final thing needs to be said by way of a caution about feedback loops. If one is not careful about constructing a flow chart or list presentation for an algorithm, one may inadvertently construct an infinite feedback loop. For example, you would be ill-advised to hold your breath waiting to reach instruction 4 in the program below:

0. Start.

1. Go to 3.

2. Go to 1.

3. Go to 2.

4. Smile.

5. Stop.

In our last example we shall finally consider a problem for which computers come in handy, namely, the problem of sorting a list of numbers. Suppose that each day 1,349 weather stations across the country report their maximum temperatures to a central meteorological agency that publishes the highest of all these temperatures. Since the reports from the weather stations arrive in no particular numerical order, a computer is used to find the largest entry in the list. Here is the flow chart (Figure 5.7) that shows how the computer might go about it.

You should test this program to see if it really works. When you do so, you will doubtless find it inconvenient to work with a list of 1,349 numbers. If there is an error in the program, this should become apparent when a shorter list of, say, 5 numbers is used. If the flow chart in Figure 5.7 is modified so that "Is $K = 1,348$?" is changed to "Is $K = 4$?" we can easily check the program using lists of 5 numbers. If the new program works for 5 numbers, then the original one probably will work for 1,349 numbers.

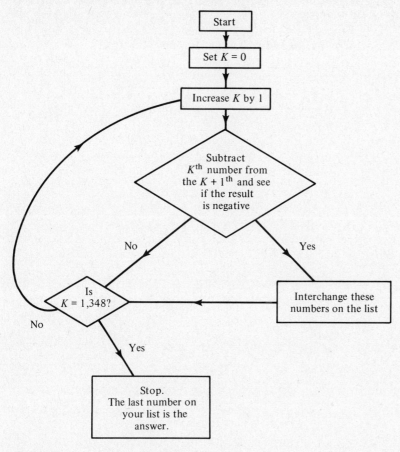

Figure 5.7

EXERCISES 5.3

1. Put the flow chart for Hilda's washer into list presentation form.
2. Suppose Hilda's washer has an agitator speed dial (with a fast and a slow position) and suppose the agitator goes into action after the water is turned on but before the rinse cycle. Draw a flow chart for the operations of this improved washer. Put your flow chart into list presentation form.
3. Below are listed three separate algorithms for fixing a radio. In each case, make a flow chart to describe the algorithm. For each algorithm, decide whether you think it works (use hypothetical radios if necessary). If an algorithm does not work, show a radio for which it does not do the correct thing. Show what actually happens to such a radio as one applies the steps of the algorithm.

(a) 0. Start.
1. Set $K = 1$.
2. Test tube K. Is it good?
   Yes—go to 5.
   No—go to 3.
3. Replace tube K with a new one.
4. Go to 1.
5. Increase K by 1.
6. Go to 2.
7. Take radio to professional repairman.
8. Stop.

(b) 0. Start.
1. Set $K = 1$.
2. Test tube K. Is it good?
   Yes—go to 5.
   No—go to 3.
3. Replace tube K with a new one.
4. Radio plays now?
   Yes—go to 7.
   No—go to 5.
5. Increase K by 1.
6. Go to 2.
7. Take radio to a professional repairman.
8. Stop.

(c) 0. Start.
1. Set $K = 1$.
2. Test tube K. Is it good?
   Yes—go to 5.
   No—go to 3.
3. Replace tube K with a new one.
4. Radio plays now?
   Yes—go to 8.
   No—go to 5.
5. Increase K by 1.
6. Go to 2.
7. Take radio to a professional repairman.
8. Stop.

4. Can you modify the radio repair algorithm so that it deals with three-tube radios instead of two-tube radios? Use either the flow chart or list presentation method.

5. Can you modify the radio repair algorithm to do the sensible thing when presented with a radio that is working? Use the method of hypothetical radios to check that your modified program works for good or bad radios.

6. The radio repair algorithm works by checking one tube at a time. Can you give an algorithm in which all the tubes are checked at once?

7. Suppose you found it desirable, in describing the radio repairing procedure, to give a more detailed algorithm in which "remove the back of the radio" and "replace the back of the radio" must appear where appropriate. Assume that tubes cannot be removed without the back being off and that the radio cannot be turned on with the back off. Construct the appropriate algorithm in both flow chart and list presentation form.

8. Below is an algorithm that is designed to pick out the largest number from a list of four numbers in a vertical column. Does the algorithm do what it is supposed to? Test it for each of the following lists.

(a)	(b)	(c)
1	4	1
3	1	4
2	2	3
4	3	2

1. Start.
2. Set $K = 0$.
3. Increase K by 1.
4. Is the number on line K greater than that on line $K + 1$?
   Yes—go to 5.
   No—go to 6.
5. Interchange the numbers on lines K and $K + 1$.
6. Is $K = 3$?
   Yes—go to 7.
   No—go to 3.
7. Number on line 4 (bottom line) is the answer.
8. Stop.

9. Write a flow chart for the algorithm of Exercise 8.

10. Suppose we change instructions 4 and 7 of Exercise 8 so they now read
   4. Is the number on line K greater than that on line $K + 1$?
      Yes—go to 6.
      No—go to 5.
   7. Number on line 1 (top line) is the answer.
   Does this modified program work?

11. Make a flow chart for the modified algorithm in Exercise 10.

12. The flow chart in Figure 5.8 is designed to do the following: Given any list of four numbers in a vertical column, following these instructions is supposed to rearrange them into numerical order, largest first. However, there is something wrong. Find a column of four numbers for which it does not succeed in doing the required rearrangement. Can you fix the program?

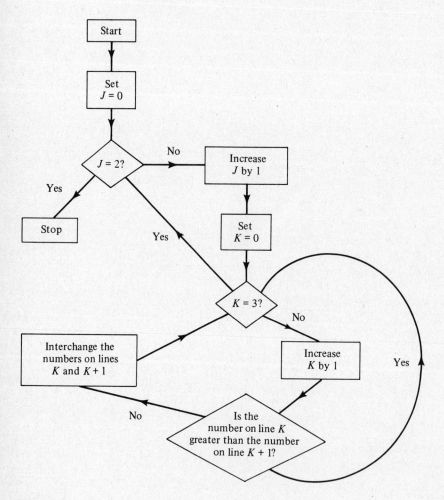

Figure 5.8

13. Does the flow chart described in Figure 5.7 succeed in putting the whole list of numbers in numerical order or does it merely move the largest one to the last position?

14. In our sorting program, suppose we also wanted to find the smallest number on the list. Can you see how to modify the program in the text in order to give both the largest and smallest numbers?

15. Draw a flow chart for the following list presentation of a program designed to alphabetize a list of initials (each entry in the list is a pair of initials, e.g., C. K. for Clark Kent). They should be alphabetized first by last initial and, in case of a tie, by first initial. Does the program do its job? Explain.

    0. Start.

    1. Set $K = 0$.

    2. Increase K by 1.

    3. Does last initial of Kth name on the list come before last initial of name in position $K + 1$?

        Yes—go to 6.

        No—go to 4.

    4. Interchange Kth name with name in $K + 1$ position.

    5. Go to 7.

    6. Does first initial of Kth name come before first initial of name in position $K + 1$?

        Yes—go to 2.

        No—go to 4.

    7. Stop.

16.* What are some of the items that you must program to take a picture with an elaborate camera? (*Hint:* color film versus black and white, shutter speed, etc.)

17.* What are some of the items that you must program to use a stereo system?

18.* Copy a recipe from a cookbook. What are the similarities between the recipe and a list presentation of a program? When would a recipe have a loop?

19.* Obtain a copy of the federal income tax form. Is it a program in list presentation form?

5.4 machine language

Up to now we have discussed algorithms at a level where they describe the general plan of the calculations or procedures to be carried out. There is a more detailed level at which the description of the algorithm can be given using *machine language*, so called because it describes more closely how the machine actually operates. For example, in a previous program we had occasion to instruct the computer to subtract one number from another. In machine language we would go into more detail about how this is to be done. Machine language is thus an additional step toward converting our thoughts into a form the computer can use to carry out our wishes.

Shortly we shall give an example of a payroll program at both levels of detail: the programming language level and the machine language level. Before doing this it may be instructive to keep an analogy in mind. Suppose you are a businessman and you discover that your whole office staff goes to lunch at the same time, leaving the office empty and lowering efficiency. As a first step toward relieving this problem you flip on your dictaphone and dictate a message for your secretary: "Evelyn, figure out a lunch schedule so the office is never empty." This message describes the general plan but it is not specific enough to be acted upon by the office staff. Evelyn will now translate this message into a detailed memo containing a schedule. In our analogy, her memo corresponds to machine language.

Here is a program in programming language that calculates the payroll for a firm. The basic rule is simple enough: The hourly rate is multiplied by hours worked except when there is overtime, in which case an alternate formula involving a time and a half rate for overtime is used.

PROGRAM TO CALCULATE PAYROLL

1. Start.

2. Read data from top card (name, hours, rate) and dispose of card.

3. Is number of hours greater than 40?
   Yes—go to 6.
   No—go to 4.

4. Compute PAY by PAY=HOURS*RATE   (* = multiplication).

5. Go to 7.

6. Compute PAY by PAY=(1.5)*(HOURS-40)*RATE+40*RATE.

7. Print: NAME, PAY.

8. Is there another card left?
   Yes—go to 2.
   No—go to 9.

9. Stop.

Instead of expanding this whole program into machine language, we shall focus on the two formulas that occur in steps 4 and 6. First we work with the formula PAY=HOURS∗RATE that occurs in the payroll program and show how one person's pay is computed from this formula, using his number of hours and basic rate of pay.

A basic idea that must be understood is that conceptually the computer consists in part of a series of "boxes" (these are actually sections of electronic circuitry), some of which are memory boxes (also thought of as storage boxes) and one of which is an arithmetic box or unit where arithmetic is performed (see Table 5.1).

For example, to calculate PAY from our formula we would first have to have the number of HOURS and the RATE stored in some memory boxes (one piece of data to a box). For convenience we shall call these boxes by the words HOURS and RATE, respectively. Table 5.1 shows the setup schematically. For illustration we have filled in the boxes with an HOURS value of 40 and a RATE value of $2.50. The way the calculation will take place is this:

(1) One of the boxes will have its contents copied into the arithmetic unit.

(2) The second box will have its number sent to the arithmetic unit also, but with the instruction to multiply it by what is already there. After this multiplication the arithmetic unit will contain only the product just calculated.

(3) This number will be sent to the storage location labeled PAY.

These steps are described on the left in Table 5.1 in machine language. The right-side of the table shows the contents of the various boxes at the various stages of the machine language description. This is not itself part of the machine language description. The abbreviations used in the machine language are explained right after the program.

Table 5.1   Machine Language Program to Compute HOURS∗RATE (The Result to be Called PAY).

	HOURS	RATE	PAY
Memory	40	2.50	
Arithmetic Unit			

Program

1. CLA   HOURS

40	2.50	
40		

Table 5.1   Continued

	HOURS	RATE	PAY
2. MPY RATE	40	2.50	
	100		

	HOURS	RATE	PAY
3. STO PAY	40	2.50	100
	100		

CLA means *clear and add*, which in turn means clear the previous contents of the arithmetic unit and then put into the arithmetic unit the contents of the box listed after the symbol CLA. *Clear and copy* might be a better description of this process but *add* is used for technical reasons. (Notice that when the 40 is copied from the HOURS box to the arithmetic unit it still remains in the HOURS box.)

MPY means *multiply the contents of the arithmetic unit by the contents of the box listed after the symbols* MPY. The result of this multiplication replaces the previous occupant of the arithmetic unit.

STO means *store the number in the arithmetic unit in the box whose name follows the symbol* STO. In our example we stored in an occupied box labeled PAY. We could equally well store in an occupied box, say, HOURS, but then the previous contents of the box would be wiped out to make room for the number being stored.

There are a few other operations of a similar nature which are used in programs of this type but which are not used in Table 5.1: They are ADD, SUB, and DIV, which stand for addition, subtraction and division, respectively, and they are used just like MPY. The next example uses ADD and SUB. DIV, which is not illustrated, works like this: The number in the arithmetic unit is the one being divided *into*, while the number coming from the memory box is the divisor.

Our next example (Table 5.2) involves a machine language version of the other PAY formula in the payroll program (see p. 128). In addition to using the instructions ADD and SUB, there are two other aspects that are novel. We need to use the numbers 40 and 1.5 in our computation as well as the numbers for HOURS and RATE. These numbers, 40 and 1.5, must therefore be stored previously in memory boxes. We'll call these boxes the 40—BOX and the 1.5—BOX, respectively. Another complication, which you will see as you read the program, is that we need to make use of a storage box, TEMP, for an intermediate result of our calculations. Our illustration, to the right of the program, uses the values 60 and $2.00 for HOURS and RATE, respectively.

Table 5.2 Machine Language Program for Calculating
(1.5)*(HOURS−40)−RATE+40*RATE (to be Called PAY).

	HOURS	RATE	40-BOX	1.5-BOX	TEMP	PAY
Memory	60	2	40	1.5		
Arithmetic Unit						

Program

1. CLA HOURS

HOURS	RATE	40-BOX	1.5-BOX	TEMP	PAY
60	2	40	1.5		
60					

2. SUB 40-BOX

HOURS	RATE	40-BOX	1.5-BOX	TEMP	PAY
60	2	40	1.5		
20					

3. MPY 1.5-BOX

HOURS	RATE	40-BOX	1.5-BOX	TEMP	PAY
60	2	40	1.5		
30					

4. MPY RATE

HOURS	RATE	40-BOX	1.5-BOX	TEMP	PAY
60	2	40	1.5		
60					

5. STO TEMP

HOURS	RATE	40-BOX	1.5-BOX	TEMP	PAY
60	2	40	1.5	60	
60					

6. CLA 40-BOX

HOURS	RATE	40-BOX	1.5-BOX	TEMP	PAY
60	2	40	1.5	60	
40					

7. MPY RATE

HOURS	RATE	40-BOX	1.5-BOX	TEMP	PAY
60	2	40	1.5	60	
80					

8. ADD TEMP

HOURS	RATE	40-BOX	1.5-BOX	TEMP	PAY
60	2	40	1.5	60	
140					

9. STO PAY

HOURS	RATE	40-BOX	1.5-BOX	TEMP	PAY
60	2	40	1.5	60	140
140					

In this example we used six memory boxes in all, more than in the previous example. For the exercises in this book, we shall assume there are as many different locations available as we need so you may draw as many as you want when doing problems. We shall restrict ourselves to one arithmetic box however.

There is a more basic form of machine language that is also worth knowing about. It is exactly like the form we have just presented except that the various instructions and storage boxes are referred to by a numerical code. For example, storage boxes might be given numbers like 0000, 0001, 0010, and so on, while the instructions could be coded according to the following table.

Instruction	Numerical Code
CLA	000
ADD	001
MPY	010
SUB	011
DIV	100
STO	101

Suppose now that the boxes labeled HOURS, RATE, and PAY in Table 5.1 were given numbers 0000, 0001, and 0010, respectively. Then the statement "CLA HOURS" would be written in the numerical code as "000—0000". The sequence of instructions in Table 5.1 would be

$$000—0000$$
$$010—0001$$
$$101—0010$$

This numerical code is a little closer to the basic operating level of the computer because with it both the input data (hours and rates in our example) and the instructions are in the same form, namely, in the form of numbers. As we shall see in the next section, numbers are readily represented electronically.

A final question that may deserve an answer is: Why bother with general descriptions using flowcharts or list presentations (as in Section 5.3) if we ultimately have to translate them into machine language? The reason is that it is not "we" who have to translate our program into machine language. This task taken over by the computer itself, leaving us free to think at the more comfortable level of the programming language. This is the same division of labor that exists between the businessman and his secretary.

EXERCISES 5.4

1. In each of the following cases, complete the entries in the boxes on the right of Table 5.1 assuming that the initial contents of the HOURS and RATE boxes are
   (a) HOURS=20   RATE=4
   (b) HOURS=30   RATE=2
   (c) HOURS=40   RATE=1.50

2. In each of the following cases, complete the entries in the boxes on the right of Table 5.2 assuming that the initial contents of the HOURS and RATE boxes are
   (a) HOURS=70   RATE=30
   (b) HOURS=40   RATE= 2
   (c) HOURS=20   RATE= 4 (Is the program applicable in this
                                              case?)

3. The formula PAY=(1.5)*(HOURS−40)*RATE+40*RATE can be rewritten (by simple algebra) PAY=(1.5*HOURS−20)*RATE. Write a machine language program for this calculation. Is it more or less efficient than the original formula in terms of how many machine language instructions are required?

4. Write a machine language program for calculating PAY according to the formula PAY+(1.5*HOURS−20)*RATE+BONUS−WITHHOLD.

5. Write a machine language program for calculating PAY according to the formula PAY=SALARY−WITHHOLD+BONUS.

6. Write a machine language program for converting centigrade degrees to Fahrenheit degrees using the formula FAHR=($\frac{9}{5}$)*CENT+32.
   Do the problem two ways: In the first solution, suppose that memory boxes have been loaded with the numbers 32, 9, and 5 and a value for CENT; in the second solution, assume that memory boxes have been loaded with the numbers 32 and $\frac{9}{5}$ and a value for CENT.

7. For each of the following formulas, write a machine language program to calculate the formula. In each case you may change the algebraic form of the expression on the right to an equivalent form if you feel it is more convenient.
   (a) $M=(A+B)*C*D$
   (b) $M=A*A−B*B+A+B$
   (c) $M=A*A*A−2*A*A+3*A−7$

8. For each of the following exercises for this section, do the exercise using the numerical code form of machine language. Use the code in the text for the various instructions and choose your location codes from the numbers 0000, 0001, 0010, 0011, 0100, 0101, 0110, 0111, and so on.

(a)   Exercise 4
(b)   Exercise 5
(c)   Exercise 6
(d)   Exercise 7.

5.5   *binary notation*

When a human being does arithmetic, he customarily does it using the ten symbols 0, 1, 2, 3, 4, 5, 6, 7, 8, 9, which make up the *decimal system*. But the electric and magnetic devices that are used to construct a computer do not deal easily with ten symbols. Rather, for an electric or magnetic device, there are usually only two configurations possible, on or off, plus voltage or negative voltage, magnetized or not magnetized. Is it possible that one can do all ordinary arithmetic using only two states or symbols, say, 0 and 1, which would thus constitute a *binary system*?

A good place to start might be to examine the decimal numbers 15 and 51, which are both formed from the digits 1 and 5. The reader who can appreciate that these two numbers are different despite their being formed from the same symbols has survived his first lesson in *placevalue notation*. The second lesson, only slightly more difficult, is to discover exactly how the difference comes about. The key to the question is the fact that the significance of a symbol derives not only from the meaning of the symbol itself but also the place that the symbol occupies. For example, in 15, the 1 refers to a block of ten units, while the 5 denotes five individual units, making fifteen units altogether. In 51, the 5 denotes five blocks of ten units, comprising fifty in all, while the 1 designates a single unit.

15 means $1 \times 10 + 5$

Pictorially:

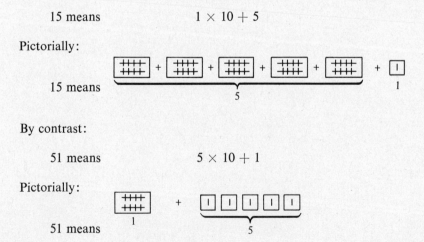

By contrast:

51 means $5 \times 10 + 1$

Pictorially:

For larger numbers we may need a third or fourth place, or perhaps even more. The place in which a symbol appears determines whether it refers to units, blocks of tens, blocks of hundreds, blocks of thousands, etc. In general the nth place refers to the blocks that contain a number of units equal to ten raised to the $n - 1$ power. Thus

$$a_n a_{n-1} \cdots a_1 = (a_n \times 10^{n-1}) + (a_{n-1} \times 10^{n-2}) + \cdots + a_1 \qquad (5.1)$$

Thus, for example,

$$3642 = (3 \times 10^3) + (6 \times 10^2) + (4 \times 10^1) + 2$$

while

$$2463 = (2 \times 10^3) + (4 \times 10^2) + (6 \times 10^1) + 3$$

The importance of the number ten in the decimal system derives from the fact that we have ten distinct numerical symbols available, 0, 1, 2, 3, 4, 5, 6, 7, 8, 9. Ten is the first number to require the placevalue system for its designation. There is no reason, in principle, why one could not invent another separate symbol, say, Δ, for the number ten and thus postpone the use of placevalue until we reach the number eleven. Were we to do this, eleven would be denoted 10 (1 block of eleven + no units) and the first twenty-two numbers would be denoted 1, 2, 3, 4, 5, 6, 7, 8, 9, Δ, 10, 11, 12, 13, 14, 15, 16, 17, 18, 19, 1Δ, 20. Although all this is perfectly possible, it is of no practical value so we shall not consider it further.

Instead of inventing extra symbols to start our placevalue system later, we shall do away with most of the ten customary numerical symbols, leaving only 0 and 1, and begin our placevalue system much earlier. The number system using only the digits 0 and 1 is called the *binary system*. Since we have only two digits available, we must denote the number two, using the placevalue notation, as 10. Of course, now this symbol means 1 block of two units plus 0 single units. Similarly, three is denoted 11, meaning 1 block of two units plus 1 a single unit. Now what about four? We need another place that will denote the number of blocks with four units. Thus four will be written 100, meaning 1 block of four units, no blocks of two units, and no single units. In general, the nth place (counting from right) refers to the blocks that contain a number of units equal to two raised to the $n - 1$ power. Thus:

$$a_n a_{n-1} \cdots a_1 = (a_n \times 2^{n-1}) + (a_{n-1} \times 2^{n-2}) + \cdots + a_1 \qquad (5.2)$$

Here are the numbers from zero up to ten written in binary notation.

zero	0
one	1
two	10
three	11
four	100
five	101
six	110
seven	111
eight	1000
nine	1001
ten	1010

Before going on, it is well to point out what we have and have not done. In the binary notation, we have not banished numbers like four, five, three hundred forty-two, etc. We have, however, found new ways of *designating* them. What has been banished are the symbols 2, 3, 4, . . . , 9.

On the basis of the discussion above, you may already be able to convert a decimal expression of a number to the binary expression of that same number and vice versa. It is useful, however, to have quick mechanical ways to do this. Algorithms for these processes are indicated by the following examples.

CONVERTING BINARY NOTATION TO DECIMAL NOTATION: The method is a straightforward adaptation of our discussion of placevalue. Each digit in the given binary expression denotes a certain power of two, as indicated by the scheme in (5.2) and illustrated below:

$$ \ldots \quad 1 \qquad 0 \qquad 1 \qquad 1 \qquad 0 \qquad 1 $$
$$ \ldots \quad 2^5(=32) \quad 2^4(=16) \quad 2^3(=8) \quad 2^2(=4) \quad 2^1(=2) \quad 2^0(=1) $$

The dots on the left indicate that there may be more digits of the number farther to the left. As we move to the left, the powers of two on the second line increase. Certain of these powers of two are to be added together to give the correct decimal expression. The expressions that are added are those for which the corresponding place in the binary expression is occupied by the symbol 1. The expression corresponding to places occupied by the symbol 0 are not added into the sum. For the example above, we find

$$ 2^5 + 2^3 + 2^2 + 2^0 $$

or

$$ 32 + 8 + 4 + 1 = 45 $$

Thus the binary notation 101101 denotes the same number as the decimal notation 45.

You probably realize that what we have done here simply amounts to saying that

$$101101 \text{ means } (1 \times 2^5) + (0 \times 2^4) + (1 \times 2^3) + (1 \times 2^2) + (0 \times 2^1)$$
$$+ (1 \times 2^0) \quad (2^0 = 1)$$

CONVERTING DECIMAL NOTATION TO BINARY NOTATION: The essence of the method is to try to construct the given number by putting together blocks of various standard sizes. For example, 83 can be achieved with one 64 block, one 16 block, one 2 block, and one 1 block. We use no blocks of the other sizes, namely, sizes 4, 8, 32, or anything larger than 64. Thus 83 is written 1010011. To do this most efficiently:

(1) Write the block sizes up to the largest size that does not exceed the given number. In the case of the number 83, we have

$$64 \quad 32 \quad 16 \quad 8 \quad 4 \quad 2 \quad 1$$

(2) Place a 1 over the largest block size that does not exceed the given number.

$$\begin{array}{ccccccc} 1 & & & & & & \\ 64 & 32 & 16 & 8 & 4 & 2 & 1 \end{array}$$

(3) Subtract this block size from the given number to see how much still needs to be built up using the remaining blocks.

$$83 - 64 = 19$$

(4) Return to step 2, using the number calculated in step 3 in place of "the given number."

$$\begin{array}{ccccccc} 1 & & 1 & & & & \\ 64 & 32 & 16 & 8 & 4 & 2 & 1 \end{array}$$

(5) Repeat steps 3 and 4 until the calculation in step 3 produces a zero.

$$19 - 16 = 3$$

$$\begin{array}{ccccccc} 1 & & 1 & & & 1 & \\ 64 & 32 & 16 & 8 & 4 & 2 & 1 \end{array}$$

$$3 - 2 = 1$$

$$\begin{array}{ccccccc} 1 & & 1 & & & 1 & 1 \\ 64 & 32 & 16 & 8 & 4 & 2 & 1 \end{array}$$

$$1 - 1 = 0$$

(6)   Fill in the remaining places with zeros.

$$
\begin{array}{ccccccc}
1 & 0 & 1 & 0 & 0 & 1 & 1 \\
64 & 32 & 16 & 8 & 4 & 2 & 1
\end{array}
$$

The answer is 1010011.

EXERCISES 5.5

1. Write the numbers from eleven to twenty in binary form.
2. Convert the numbers below from binary to decimal notation:
    (a)   11101
    (b)   11011101
    (c)   10110111
    (d)   1011011011
3. Write in English words and in decimal notation what numbers are represented by the binary notations 1000000, 1010101, 111100, and 11011101.
4. Convert the numbers below from decimal to binary notation:
    (a)   1034
    (b)   1962
    (c)   847
    (d)   1324
5. What is the largest decimal number that can be represented in the binary system using twenty digits?
6.* Continue the list of numbers in the base eleven system, using the Δ symbol, to the number 30.
7.* Ternary notation uses three digits, 0, 1, and 2, and placevalue notation to designate numbers. What do the ternary notations 211, 102, 222 mean in decimal notation? Write the numbers from one through ten in ternary notation.
8.* Write the ternary representation of the numbers from eleven to fifteen.
9.* Suppose we create a tertiary number system with only four digits, 0, 1, 2, 3. Using only these digits and placevalue notation, write the representations for the numbers from one through twenty.
10. Develop a method of converting a number from decimal to a tertiary (base four) expression. Use your method on:
    (a)   349
    (b)   273
    (c)   842
11.* When fractions must be introduced in the decimal system, a decimal point is introduced and digits to the right of the decimal point are

interpreted with place notation. Thus:

$$1.14 = 1 + [1 \cdot (\tfrac{1}{10})] + [4 \cdot (\tfrac{1}{100})]$$

(a) Discuss how the binary point should be used.
(b) Change the binary number 11.101 to decimal.
(c) Change the binary number 101.1011 to decimal.
(d) Change the decimal number 14.75 to binary.
(e) Change the decimal number 175.125 to binary.

12.* Here is an alternate way of converting a decimal number to binary notation. Divide the given number by two and record the remainder (which will be either 0 or 1) as the units digit (rightmost digit) of the binary notation. Take the quotient of the last division and divide by two, recording the remainder as the next binary digit. Repeat this process until a zero occurs as a quotient. When this happens, the remainder that is left at that stage is the last remainder to be recorded.
Can you show why this process really works?

5.6 *binary arithmetic*

It is said that our decimal system evolved because we have ten fingers. If so, we would perhaps be better off with two fingers—at least as far as arithmetic is concerned, for arithmetic operations are a good deal simpler in the binary system than in the decimal system, as we shall now discover.

In the decimal system, in order to do addition we needed to learn one hundred addition rules, beginning with $0 + 0 = 0$ and going all the way up to $9 + 9 = 18$. For sums where one or both of the numbers is greater than 9, we use the method of "carrying" to reduce everything to our basic addition rules. In the binary system there are four addition rules that, together with the method of carrying, suffice to do any addition. These addition rules are listed in the table below with their meanings next to them in words. We supply the linguistic translation to reassure you that it is the same old arithmetic we are doing, even if it sometimes seems disguised.

0	zero	0	zero	1	one	1	one
+ 0	+ zero	+ 1	+ one	+ 0	+ zero	+ 1	+ one
= 0	= zero	= 1	= one	= 1	= one	= 10	= two

The method of carrying is exactly the same as in the decimal system and is illustrated in the example below, in which we add the numbers five and seven to obtain twelve.

(1) Add the rightmost digits to get 10. Enter the 0 and carry the 1.

$$\begin{array}{r} 1\ 0\ 1 \\ +\ 1\ 1_1 1 \\ \hline 0 \end{array}$$

(2) Add the next digits (including what is carried) to get 10. Enter the 0 and carry the 1. Note that here we had to add three numbers, a 0 and two 1's.

$$\begin{array}{r} 1\ 0\ 1 \\ +\ 1_1 1_1 1 \\ \hline 0\ 0 \end{array}$$

(3) In the final step we have three 1's to add. Taking two of them, according to the table, we get 10. Adding the remaining 1 gives 11, which we enter.

$$\begin{array}{r} 1\ 0\ 1 \\ +\ 1_1 1_1 1 \\ \hline 1\ 1\ 0\ 0 \end{array}$$

For another example:

$$\begin{array}{r} 1\ 0\ 1\ 1\ 0\ 1 \\ +\ 1_1 1_1 0_1 1\ 0_1 1 \\ \hline 1\ 1\ 0\ 0\ 0\ 1\ 0 \end{array}$$

Doing multiplication is just as simple. The basic multiplication facts, only four in number, are listed in the table below. With these facts and the usual carrying process, we can do multiplication following the same procedure as with decimal numbers: Do a series of multiplications of the top number by the individual digits of the bottom number, indent the results properly, and then add the results.

0	zero	0	zero	1	one	1	one
$\times 0$	$\times$ zero	$\times 1$	$\times$ one	$\times 0$	$\times$ zero	$\times 1$	$\times$ one
$= 0$	$=$ zero	$= 0$	$=$ zero	$= 0$	$=$ zero	$= 1$	$=$ one

Example 5.1

$$\begin{array}{r} 1\ 0\ 1 \\ \times\ 1\ 0 \\ \hline 0\ 0\ 0 \\ +\ 1\ 0\ 1 \\ \hline 1\ 0\ 1\ 0 \end{array} \qquad \begin{array}{r} 1\ 1\ 1 \\ \times\ 1\ 1 \\ \hline 1\ 1\ 1 \\ +\ 1\ 1\ 1 \\ \hline 1\ 0\ 1\ 0\ 1 \end{array}$$

At this point we can hint at the connection between two facts that we have stressed but not explained completely:

(1) The computer is an electronic device that does all its calculating electronically.

(2) The binary number system is especially well adapted for doing arithmetic electrically.

The fact that in the binary system there are two symbols, 0 and 1, corresponds very neatly to the fact that in an electrical network a given circuit may be in one of two states: It may be carrying current or not. We shall represent binary numbers in an electrical network by having no current in a circuit stand for 0 and having current in a circuit stand for 1. For example, in Figure 5.9 if we leave the switch at S open as shown, there will be no current flowing from the battery at B. This situation would represent 0. If we close the switch, current flows, representing 1. Of course, most binary numbers need many places to represent them. Thus, for example, a binary number such as 10011 needs a bank of five circuits for its representation, one circuit for each of the five places in the number.

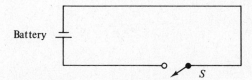

Battery

S

Figure 5.9

We shall now set ourselves the following task. Suppose we have two "input circuits" such as in Figure 5.10(a), each circuit having the possibility of being in state 0 (no current) or state 1 (current), depending on the position of the switches. We want to arrange some electrical hookup (represented for the moment by the shaded box in Figure 5.10(a)) involving these circuits so that, in another pair of circuits called the *output* circuits, we get current or no current in a pattern that represents the sum of the numbers in the input circuits. To be specific, if the input circuits are 0 and 0, we want the output circuits to represent the sum 0. To do this, each output circuit needs to be 0 (no current). This will therefore represent the addition rule that $0 + 0 = 0$ or, more precisely $0 + 0 = 00$, since we have two output circuits. Figure 5.10(a) shows the setup we want. Figures 5.10(b), (c), and (d) show how the output circuits are supposed to look for the other three input possibilities. Be sure to note the pattern of which switches are open and closed.

What we are asking for is some way of having the positions of the input switches determine the positions of the output switches using purely electrical and mechanical means and requiring no human intervention. The solution of this problem is entirely a matter of electrical engineering rather than mathematics but we offer it here to illustrate that it is possible to do arithmetic electrically. The actual solution we shall give, using relays and batteries, resembles the way parts of modern computers work but the resemblance is in the flavor rather than the details.

Before explaining the solution, which is depicted in Figure 5.12, we must explain the relay symbol (Figure 5.11), which appears a number of times,

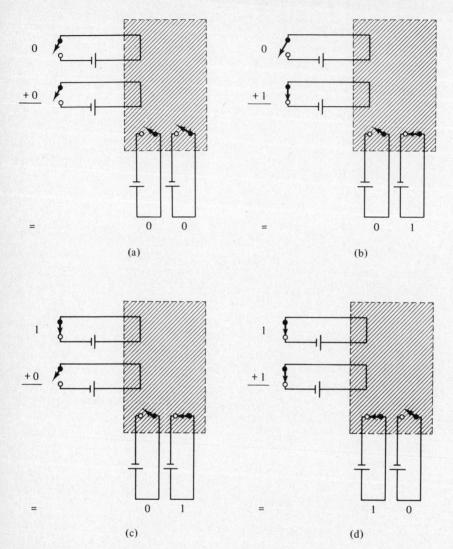

Figure 5.10

the bar with a wire coiled around it. The bar is an electromagnet and it will move when current passes through the coil. We shall suppose that the relays we have drawn are set up to give a rightward motion when energized by a current in the coil. If the electromagnet is attached to an open switch, this movement of the bar will close the switch. The same principle can be used to open a switch. Figure 5.11 shows how a relay can be used to have a switch in

142

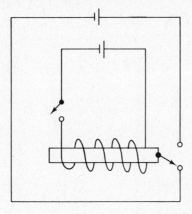

Figure 5.11

one circuit control the position of a switch in another circuit. If we close the switch in the inner circuit, the relay is energized and the switch in the outer circuit closes due to the motion of the bar to the right and current flows in the outer circuit.

One more comment about Figure 5.12 before we see how it works. Most of the circuits have been drawn so they don't seem to be circuits, merely to simplify the picture. For example, the line between the two battery symbols marked A (near the top of the figure) is really to be thought of as a circuit. Symbols marked A represent the same battery. Elsewhere in the diagram there are battery symbols with the same label; in such cases, also, battery symbols with the same label represent the same battery.

Now to see that this arrangement really does allow the X and Y switches to control the V and W switches in the proper way, it is just a matter of checking the various combinations. As Figure 5.12 is set up, the X and Y switches are both 0 (no current) and V and W are both 0, which is the way we want it ($0 + 0 = 00$). Suppose we close X and Y so that the inputs are now 1 and 1. We should find W open and V closed since $1 + 1 = 10$. To see this, we have to examine what effect closing X and Y has on the various relays and circuits. Relays 1 and 3 will close their switches. Thus, current will flow from C through relays 8 and 5. Switch 8 closes making the left-hand output circuit V have the value 1. When 5 closes, current flows from D through relay 7, opening switch 7. Now there will be no current flowing from E through 9. Hence, output circuit W will be zero. Thus we have verified that our hookup adds $1 + 1$ correctly, giving 10. We leave it to you to show that the other addition rules come out correctly; namely, $1 + 0 = 01$ (X closed, Y open causes W to close but leaves V open) and $0 + 1 = 01$ (X open, Y closed causes W to close but leaves V open).

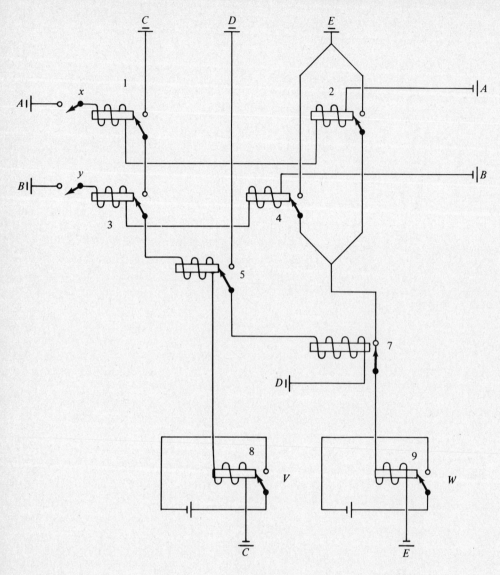

Figure 5.12

144

EXERCISES 5.6

1. Perform the following additions and multiplications in binary:

Add:

(a)
$$\begin{array}{r} 1\ 1\ 0\ 1 \\ 1\ 0\ 0\ 1\ 1 \\ \hline \end{array}$$

(c)
$$\begin{array}{r} 1\ 1\ 0\ 1\ 1 \\ 1\ 0\ 1\ 0\ 1\ 0 \\ \hline \end{array}$$

(b)
$$\begin{array}{r} 1\ 1\ 0\ 1 \\ 1\ 1\ 1\ 0\ 1 \\ 1\ 0\ 1\ 1\ 0 \\ \hline \end{array}$$

(d)
$$\begin{array}{r} 1\ 1\ 0\ 1 \\ 1\ 0\ 1\ 1 \\ \hline \end{array}$$

Multiply:

(e)
$$\begin{array}{r} 1\ 1\ 0\ 1 \\ 1\ 0\ 1 \\ \hline \end{array}$$

(g)
$$\begin{array}{r} 1\ 1\ 0\ 1 \\ 1\ 1\ 0 \\ \hline \end{array}$$

(f)
$$\begin{array}{r} 1\ 0\ 1 \\ 1\ 1 \\ \hline \end{array}$$

(h)
$$\begin{array}{r} 1\ 1\ 1\ 0\ 1 \\ 1\ 0\ 1\ 1 \\ \hline \end{array}$$

2. Add the following column of binary numbers:

$$\begin{array}{r} 1\ 1\ 1\ 0\ 1 \\ 1\ 0\ 1\ 1\ 1 \\ 1\ 0\ 1\ 1\ 0 \\ 1\ 1\ 0\ 0\ 0 \\ 1\ 1\ 0\ 1\ 1 \\ \hline \end{array}$$

3.* Multiply 1.1101 by 1.01. Express your answer in decimal.

4.* Develop an addition table and a multiplication table for a base three number system (i.e., the only symbols are 0, 1, 2). Use these tables to solve

(a)
$$\begin{array}{r} 1\ 2\ 1 \\ +\ 1\ 0\ 2 \\ \hline \end{array}$$

(b)
$$\begin{array}{r} 1\ 1\ 2 \\ +\ 2\ 1\ 0 \\ \hline \end{array}$$

(c)
$$\begin{array}{r} 1\ 1\ 2 \\ \times\ \ 1\ 1 \\ \hline \end{array}$$

(d)
$$\begin{array}{r} 1\ 2\ 1 \\ \times\ \ 1\ 2 \\ \hline \end{array}$$

5.* Develop an addition table and a multiplication table for a base four number system (i.e., the only symbols are 0, 1, 2, 3). Use these tables to perform the calculations below:

(a)   1 3 2   (b)   1 3 1
           + 2 1 3   + 2 1 3

(c)   1 0 3   (d)   1 3 0
           + 2 1 0   × 3 1 2

(e)   1 0 2   (f)   1 0 3
           × 1 1 1   × 2 0 1

6.* (a) Subtract 101101 (binary) from 10111011 (binary). (Check your answer by converting the given numbers to decimals, performing the subtraction and changing your decimal answer to binary.)

(b) Subtract 11011101 (binary) from 1101101111 (binary).

7.* Construct an electrical circuit that multiplies two one-digit binary numbers.

8.* Devise an arrangement of circuits, relays, batteries, and switches capable of adding a two-digit binary number to a one-digit binary number. (*Hint:* The arrangement of Figure 5.12 can be used to do the addition in the units place; then another copy of Figure 5.12, properly hooked up to the first, takes care of the "carry" and the remaining digit in the two-digit number.)

5.7 graphs and matrices

A great deal of practical mathematics is handled by computers today, and graph theory is no exception. This raises the question of how one can make a computer "see" a graph or digraph and draw conclusions about it without the sense of sight upon which we humans rely. For example, let us start with the graph in Figure 5.13. We have five vertices and we need to state which pairs of vertices are connected and by how many edges. The following table will do the trick.

	x_1	x_2	x_3	x_4	x_5
x_1	0	2	1	1	0
x_2	2	0	1	0	0
x_3	1	1	0	0	0
x_4	1	0	0	0	0
x_5	0	0	0	0	0

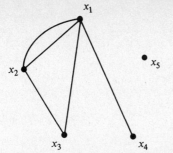

x_1

x_2

x_5

x_3 x_4 Figure 5.13

In this table, 0 indicates no connecting edge, 1 indicates a connecting edge, 2 indicates two connecting edges, and so on. For example, the 0 in row x_3 and column x_4 indicates that x_3 and x_4 are not connected by an edge. The entry in the ith row and the jth column is called the (i, j) entry. A rectangular array of numbers, when written without the row and column labels and enclosed in brackets, is called a *matrix*. If G is a graph, its matrix is denoted $M(G)$. The matrix for the table above is shown below.

$$\begin{bmatrix} 0 & 2 & 1 & 1 & 0 \\ 2 & 0 & 1 & 0 & 0 \\ 1 & 1 & 0 & 0 & 0 \\ 1 & 0 & 0 & 0 & 0 \\ 0 & 0 & 0 & 0 & 0 \end{bmatrix}$$

What if we have a digraph such as that of Figure 5.14? What we do in

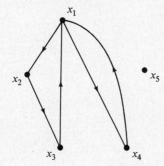

x_1

x_2

x_5

x_3 x_4 Figure 5.14

$$\begin{bmatrix} 0 & 1 & 0 & 1 & 0 \\ 0 & 0 & 1 & 0 & 0 \\ 1 & 0 & 0 & 0 & 0 \\ 1 & 0 & 0 & 0 & 0 \\ 0 & 0 & 0 & 0 & 0 \end{bmatrix}$$

147

the case of a digraph is to make the (i, j) entry equal to the number of directed edges *from* x_i *to* x_j. Thus, the $(1, 2)$ entry of our matrix will be 1 when we wish to describe Figure 5.14 since there is an edge from x_1 to x_2. But the $(2, 1)$ entry will be 0 since there is no edge from x_2 to x_1. The full matrix is shown below Figure 5.14.

Although we have used graphs and computers to introduce the concept of matrices, matrices are useful throughout both pure and applied mathematics. We shall use matrices later in Chapters 6, 8, and 9. In some of the other applications of matrices, the matrices that come in handy are not square in appearance; that is, they may not have as many rows as columns. Here are some more examples of matrices

$$\begin{bmatrix} 3 & -1 & 0 & 1 \\ 1 & -4 & 2 & 3 \\ 2 & 4 & -1 & 0 \\ 1 & 7 & 9 & -1 \end{bmatrix} \quad \begin{bmatrix} 1 & 2 & -3 & -5 \end{bmatrix} \quad \begin{bmatrix} 3 \\ -1 \\ -2 \\ 0 \end{bmatrix}$$

The first of these is called a *square matrix* because it has as many rows as columns. The second, consisting of only a single row, is called a *row matrix* (or sometimes a *row vector*), while the last is a *column matrix* (or *column vector*).

If A and B are two matrices of the same size (e.g., both $m \times n$), we can define the sum of these two matrices to be a new matrix, denoted $A + B$, whose (i, j) entry is obtained by adding the (i, j) entries of A and B. Here are some examples:

$$\begin{bmatrix} 1 \\ 2 \end{bmatrix} + \begin{bmatrix} 2 \\ -1 \end{bmatrix} = \begin{bmatrix} 3 \\ 1 \end{bmatrix} \quad \begin{bmatrix} 1 & 2 \\ 4 & -7 \end{bmatrix} + \begin{bmatrix} 3 & -1 \\ -2 & 6 \end{bmatrix} = \begin{bmatrix} 4 & 1 \\ 2 & -1 \end{bmatrix}$$

Note that we add only matrices of the same dimension.

It should be clear from the definition that some of the usual rules for arithmetic hold also for matrices:

(1) $A + B = B + A$ (Commutative law)
(2) $A + (B + C) = (A + B) + C$ (Associative law)

Example 1

As an illustration of the commutative law (Rule 1),

$$\begin{bmatrix} 2 & 3 \\ 4 & 2 \end{bmatrix} + \begin{bmatrix} 1 & -4 \\ 1 & -2 \end{bmatrix} = \begin{bmatrix} 1 & -4 \\ 1 & -2 \end{bmatrix} + \begin{bmatrix} 2 & 3 \\ 4 & 2 \end{bmatrix}$$

We can calculate the sums on both sides of the equals sign and see that they are equal.

Example 2

As an illustration of the associative law (Rule 2),

$$\begin{bmatrix} 1 & 3 \\ -1 & 0 \end{bmatrix} + \left(\begin{bmatrix} 1 & 2 \\ -1 & 4 \end{bmatrix} + \begin{bmatrix} 0 & 1 \\ 2 & 1 \end{bmatrix} \right) = \left(\begin{bmatrix} 1 & 3 \\ -1 & 0 \end{bmatrix} + \begin{bmatrix} 1 & 2 \\ -1 & 4 \end{bmatrix} \right) + \begin{bmatrix} 0 & 1 \\ 2 & 1 \end{bmatrix}$$

Again, we can calculate the sums on both sides of the equals sign and see that they are equal.

In connection with the addition of matrices, certain matrices have the special property of not changing a matrix when added to it. For example, if we have an $m \times n$ matrix A and we add to it an $m \times n$ matrix Z, all of whose entries are zero, then $A + Z = Z + A = A$. For obvious reasons, Z is called a *zero matrix*. For example,

$$\begin{bmatrix} 1 & -1 & 3 \\ 2 & 1 & 0 \end{bmatrix} + \begin{bmatrix} 0 & 0 & 0 \\ 0 & 0 & 0 \end{bmatrix} = \begin{bmatrix} 1 & -1 & 3 \\ 2 & 1 & 0 \end{bmatrix}$$

and

$$\begin{bmatrix} 1 & 2 \\ -1 & 3 \end{bmatrix} + \begin{bmatrix} 0 & 0 \\ 0 & 0 \end{bmatrix} = \begin{bmatrix} 1 & 2 \\ -1 & 3 \end{bmatrix}$$

By analogy with the addition of matrices, one might imagine that a sensible way to multiply two matrices might be to multiply corresponding entries. As it happens, a more peculiar form of "multiplication" proves more useful. The mechanics of this multiplication are simple enough but rather a mouthful to describe so let us start with a very simple case, the multiplication

$$\begin{bmatrix} 1 & 2 & 3 \end{bmatrix} \begin{bmatrix} 1 \\ 3 \\ 4 \end{bmatrix} = ?$$

Moving from left to right in the first matrix, the elements we encounter will be multiplied by the elements encountered as we move from top to bottom on the second matrix. Thus, we form the products 1×1, 2×3, 3×4. Having found these products, we add them to find the end result:

$$\begin{bmatrix} 1 & 2 & 3 \end{bmatrix} \begin{bmatrix} 1 \\ 3 \\ 4 \end{bmatrix} = (1 \times 1) + (2 \times 3) + (3 \times 4) = 19$$

The general process is illustrated below, the arrows indicating which elements are to be multiplied together. Notice that in order for this process to make sense the first matrix must be as long as the second is tall. That is, they must have the same number of entries.

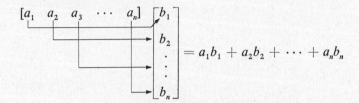

Now let us look at matrices that are neither row matrices nor column matrices. For example, let us multiply the following:

$$\begin{bmatrix} 1 & 2 \\ -1 & 4 \end{bmatrix} \begin{bmatrix} 3 & 1 \\ 2 & -3 \end{bmatrix} = ? \tag{5.3}$$

To accomplish this, we use the method we have just discussed as a building block, performing a number of these simplified multiplications and recording the resulting numbers properly into a new matrix. Each row of the first matrix will be multiplied, in the manner previously described, by the various columns of the second matrix. The results of these multiplications are recorded as follows: The product of the ith row of the first matrix by the jth column of the second becomes the (i, j) entry of the product matrix. The various simplified multiplications involved in (5.3) are indicated below, as is the recording of these into the product.

$$[1 \quad 2] \begin{bmatrix} 3 \\ 2 \end{bmatrix} = 7 \qquad [1 \quad 2] \begin{bmatrix} 1 \\ -3 \end{bmatrix} = -5$$

$$[-1 \quad 4] \begin{bmatrix} 3 \\ 2 \end{bmatrix} = 5 \qquad [-1 \quad 4] \begin{bmatrix} 1 \\ -3 \end{bmatrix} = -13$$

$$\begin{bmatrix} 1 & 2 \\ -1 & 4 \end{bmatrix} \begin{bmatrix} 3 & 1 \\ 2 & -3 \end{bmatrix} = \begin{bmatrix} 7 & -5 \\ 5 & -13 \end{bmatrix}$$

Notice that, just as before (and for reasons that should be clear), the first matrix needs to have the same length (number of columns) as the second one has height (number of rows). To put it another way, we can only perform the multiplication $A \times B$ provided A is an i by j matrix while B is a j by k for some values of i, j, and k. Here are more examples of matrix multiplications:

$$\begin{bmatrix} 1 & 2 & -1 \\ 2 & 3 & 4 \end{bmatrix} \begin{bmatrix} 2 & 1 \\ 3 & -2 \\ -1 & -3 \end{bmatrix} = \begin{bmatrix} 9 & 0 \\ 9 & -16 \end{bmatrix}$$

Here are examples of multiplications that cannot be performed:

$$\begin{bmatrix} 8 & 3 \\ 4 & 7 \end{bmatrix} \begin{bmatrix} 2 & 4 & 19 \end{bmatrix} \qquad \begin{bmatrix} 3 & 7 \end{bmatrix} \begin{bmatrix} 2 & 4 \end{bmatrix}$$

Can you see why?

One of the most peculiar aspects of matrix multiplication is the fact that the order in which one multiplies them usually makes a great difference. For example, let us take the two matrices discussed on p. 149 and multiply them in reverse order. [As you check this out, you will notice that all the "building block" multiplications are just ordinary numerical multiplications since the length (number of columns) of the first matrix and the height of the second (number of rows) are equal to 1.] The result is clearly different from the earlier one.

$$\begin{bmatrix} 1 \\ 3 \\ 4 \end{bmatrix} \begin{bmatrix} 1 & 2 & 3 \end{bmatrix} = \begin{bmatrix} 1 & 2 & 3 \\ 3 & 6 & 9 \\ 4 & 8 & 12 \end{bmatrix}$$

You may recall that numbers obey the commutative laws of both addition and multiplication. Multiplication of matrices shows that we cannot always take commutative laws for granted. Matrices do, however, satisfy numerous familiar algebraical laws, some of which we shall mention here. For example, if one has three matrices A, B, and C where the multiplications AB and BC are both possible, it is possible to multiply all three together. We can form the products $A(BC)$ or $(AB)C$; furthermore, both of these products give the same result, which is simply denoted ABC. This is the so-called *associative law* for multiplication. We shall not prove it but the example below illustrates it.

Example 3

Compute ABC where

$$A = \begin{bmatrix} 1 & 2 \\ -1 & 3 \end{bmatrix} \quad B = \begin{bmatrix} 0 & 0 \\ 2 & 1 \end{bmatrix} \quad C = \begin{bmatrix} -1 & 3 \\ 4 & 1 \end{bmatrix}$$

Solution:

$$AB = \begin{bmatrix} 4 & 2 \\ 6 & 3 \end{bmatrix} \quad BC = \begin{bmatrix} 0 & 0 \\ 2 & 7 \end{bmatrix} \quad A(BC) = (AB)C = \begin{bmatrix} 4 & 14 \\ 6 & 21 \end{bmatrix}$$

When we discussed matrix addition, we mentioned the zero matrices, which could be added to given matrices without changing the given matrices. It makes sense to ask whether there are such matrices with respect to the multiplication operation. Here is a specific example:

$$\begin{bmatrix} 1 & 2 \\ -1 & 3 \end{bmatrix} \begin{bmatrix} & ? & \\ & & \end{bmatrix} = \begin{bmatrix} 1 & 2 \\ -1 & 3 \end{bmatrix}$$

The matrix that works is

$$\begin{bmatrix} 1 & 0 \\ 0 & 1 \end{bmatrix}$$

since

$$\begin{bmatrix} 1 & 2 \\ -1 & 3 \end{bmatrix} \begin{bmatrix} 1 & 0 \\ 0 & 1 \end{bmatrix} = \begin{bmatrix} 1 & 2 \\ -1 & 3 \end{bmatrix}$$

This matrix is called an *identity matrix*. Any $n \times n$ square matrix with 1's on the main diagonal and 0's elsewhere is an identity matrix because it behaves similarly:

$$\begin{bmatrix} a_{11} & \cdots & a_{1n} \\ \cdot & & \cdot \\ \cdot & & \cdot \\ \cdot & & \cdot \\ a_{n1} & \cdots & a_{nn} \end{bmatrix} \begin{bmatrix} 1 & 0 & \cdots & 0 \\ 0 & 1 & & \cdot \\ \cdot & \cdot & & \cdot \\ \cdot & \cdot & \cdot & \cdot \\ 0 & 0 & \cdot & 1 \end{bmatrix} = \begin{bmatrix} a_{11} & \cdots & a_{1n} \\ \cdot & & \cdot \\ \cdot & & \cdot \\ \cdot & & \cdot \\ a_{n1} & \cdots & a_{nn} \end{bmatrix}$$

EXERCISES 5.7

1.   For each of the graphs in Figure 5.15, find $M(G)$.

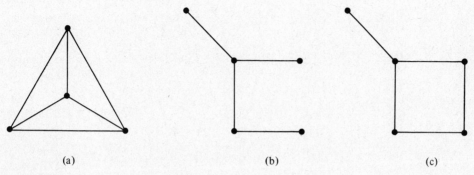

(a)   (b)   (c)

Figure 5.15

2.   Find $M(D)$ for the digraphs in Figure 5.16.

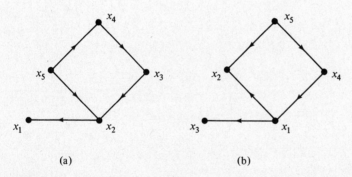

(a)   (b)

Figure 5.16

3.   The matrix below is $M(G)$ for a graph G. Can you draw G?

$$\begin{bmatrix} 0 & 1 & 0 & 1 \\ 1 & 0 & 1 & 0 \\ 0 & 1 & 0 & 1 \\ 1 & 0 & 1 & 0 \end{bmatrix}$$

4. Draw the digraph D whose matrix is

$$\begin{bmatrix} 0 & 1 & 0 & 1 \\ 1 & 0 & 1 & 0 \\ 1 & 0 & 0 & 1 \\ 0 & 1 & 1 & 0 \end{bmatrix}$$

5. In the matrix below, certain entries have been lost. Assuming that the matrix is $M(G)$ for a certain graph G, can you supply the missing entries?

$$\begin{bmatrix} 0 & ? & ? \\ 1 & 0 & 1 \\ 0 & ? & ? \end{bmatrix}$$

6. How can you tell the valences of G from $M(G)$?

7. Can you explain why the main diagonal of $M(G)$, where G is a graph, has only zeros? Is the same true for $M(D)$ where D is a digraph?

8. Find $A + B$, $B + A$, AB, and BA for the matrices below:

$$A = \begin{bmatrix} 1 & 3 \\ 1 & 4 \end{bmatrix} \qquad B = \begin{bmatrix} 2 & -1 \\ 3 & 4 \end{bmatrix}$$

9. Calculate $A + B$, AB, BA, and $A^2 B$ if

$$A = \begin{bmatrix} 1 & 2 \\ -1 & 3 \end{bmatrix} \quad \text{and} \quad B = \begin{bmatrix} 3 & 1 \\ 4 & -4 \end{bmatrix}$$

10. Which of the following multiplications are possible? When a multiplication is possible, compute the product.

$$\begin{bmatrix} 0 & 2 & -4 & 7 \\ 8 & 2 & 9 & -3 \end{bmatrix} \begin{bmatrix} 1 \\ 2 \\ 3 \end{bmatrix} \qquad \begin{bmatrix} 2 & 4 \\ -7 & 8 \end{bmatrix} \begin{bmatrix} 2 & 0 \\ 3 & 1 \\ -1 & 0 \end{bmatrix}$$

$$\begin{bmatrix} 2 & 3 \\ 1 & 4 \end{bmatrix} \begin{bmatrix} 1 & 3 & -1 \\ 0 & 1 & 1 \\ 1 & 0 & 1 \end{bmatrix} \qquad \begin{bmatrix} 1 & 2 & 1 \end{bmatrix} \begin{bmatrix} 1 \\ 4 \\ 1 \end{bmatrix}$$

$$\begin{bmatrix} 1 \\ 3 \\ 2 \end{bmatrix} \begin{bmatrix} 1 & 1 & 0 \end{bmatrix}$$

11.* Suppose G is a graph whose picture is not given but you are told G is connected. If you are given $M(G)$, how can you tell, without drawing the graph G, whether G has an Euler circuit? Apply your test to the matrix in Exercise 3.

12.* If

$$M(G) = \begin{bmatrix} 0 & 3 & 1 & 0 & 0 \\ 3 & 0 & 6 & 0 & 0 \\ 1 & 6 & 0 & 0 & 0 \\ 0 & 0 & 0 & 0 & 1 \\ 0 & 0 & 0 & 1 & 0 \end{bmatrix}$$

why is G not connected? You should answer without drawing G. The dotted lines are a hint.

13.* Find a pair of 2×2 matrices A and B that have no zero entries, yet $AB = 0$. Compute BA for this pair of matrices—is it zero?

14.* If

$$I = \begin{bmatrix} 1 & 0 \\ 0 & 1 \end{bmatrix}$$

show that if A is any 2×2 matrix, then $AI = IA = A$.

15.* Prove $A(B + C) = AB + AC$ in the case where A, B, and C are 2×2 matrices.

5.8 an application of matrix arithmetic

One thread of this chapter still needs to be put in place. We initially introduced matrices so as to be able to describe a graph or digraph to a computer. Then we studied matrix multiplication and addition but with no apparent connection to graphs or digraphs. We shall now supply such a connection as our first (but not only) application of matrix arithmetic.

Let us consider the problem of contamination of living species by pollutants such as mercury or DDT. Typically, one species will become contaminated and then others that prey on the contaminated species also become contaminated. The process of passing on the pollution continues. We shall construct a mathematical model for the diffusion of contamination.

We shall begin our modeling process by considering a food web digraph in which vertices represent species and a directed edge from x to y means that species y preys upon species x. In our food web of Figure 5.17, x_3 preys upon x_1. This means that x_1 will pass its contamination on to x_3. In fact, the con-

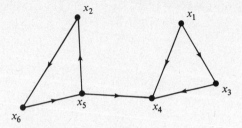

Figure 5.17

tamination will be passed on to any species connected to x_1 with a directed path *from* x_1.

In our example, contamination will spread to x_3 and x_4. Note that there are two ways it can reach x_4, directly from x_1 and indirectly via x_3. The other species will not become contaminated from x_1. This is easy enough to determine visually but if we are keeping track of all this with a computer, we need another method. One idea involves using the matrix of the digraph and the powers of this matrix and is based upon the following theorem.

THEOREM 1

If D is a digraph and M is its matrix and if n is a positive integer, then the (i, j) entry of M^n, which we shall denote by $c_{ij}^{(n)}$, has this significance: $c_{ij}^{(n)}$ is the number of directed paths in D that go from x_i to x_j in exactly n steps.

We shall give the proof only in the case where $n = 2$ and where each entry of M is 0 or 1. The general case is similar but it is a little more complicated.

Proof: Let us recall how the (i, j) entry $c_{ij}^{(2)}$ arises. It is the matrix product of the ith row of M with the jth column of M.

$$[a_{i1} \quad a_{i2} \quad \cdots \quad a_{ip}] \begin{bmatrix} a_{1j} \\ a_{2j} \\ \cdot \\ \cdot \\ \cdot \\ a_{pj} \end{bmatrix} = c_{ij}^{(2)}$$

In multiplying this row and column, we have p separate products to add, and the sum is the (i, j) entry of M^2, $c_{ij}^{(n)}$:

$$c_{ij}^{(2)} = a_{i1}a_{1j} + \cdots + a_{ip}a_{pj}$$

Each of these products is either 0 or 1. The zeros make no contribution to the sum so we are only interested in the products $a_{ik}a_{kj}$, which are equal to 1.

What we need to show is that the number of these is equal to the number of sequences of length two from v_i to v_j. In order for one of these products, say, $a_{ik}a_{kj}$, to be 1, the two numbers making up the product $a_{ik}a_{kj}$ must each be 1. When can this happen? If $a_{ik} = a_{kj} = 1$ for some value of k. This means that there is an edge from v_i to v_k (since $a_{ik} = 1$) and an edge from v_k to v_j (since $a_{kj} = 1$). Putting these directed edges together gives a sequence of length two from v_i to v_j. Thus every 1 in the sum corresponds to a sequence of length two from v_i to v_j. Conversely, given a sequence of length two from v_i to v_j, say, v_i, v_k, v_j, there must then be an edge v_iv_k and another directed edge v_kv_j. Consequently, the entries a_{ik} and a_{kj} in M are both 1 and the product $a_{ik}a_{kj}$, which is one of the products to be summed in forming $c_{ij}^{(n)}$, is 1. Thus, there are as many 1's in the sum as there are sequences of length two in D from v_i to v_j. This concludes the proof of the theorem.

Theorem 1 can be interpreted in terms of the food web digraph of Figure 5.17 and its matrix. The matrix for this digraph and the various powers of this matrix are shown below:

$$
M^1 = \begin{bmatrix} 0 & 0 & 1 & 1 & 0 & 0 \\ 0 & 0 & 0 & 0 & 0 & 1 \\ 0 & 0 & 0 & 1 & 0 & 0 \\ 0 & 0 & 0 & 0 & 0 & 0 \\ 0 & 1 & 0 & 1 & 0 & 0 \\ 0 & 0 & 0 & 0 & 1 & 0 \end{bmatrix}
\qquad
M^2 = \begin{bmatrix} 0 & 0 & 0 & 1 & 0 & 0 \\ 0 & 0 & 0 & 0 & 1 & 0 \\ 0 & 0 & 0 & 0 & 0 & 0 \\ 0 & 0 & 0 & 0 & 0 & 0 \\ 0 & 0 & 0 & 0 & 0 & 1 \\ 0 & 1 & 0 & 1 & 0 & 0 \end{bmatrix}
$$

$$
M^3 = \begin{bmatrix} 0 & 0 & 0 & 0 & 0 & 0 \\ 0 & 1 & 0 & 1 & 0 & 0 \\ 0 & 0 & 0 & 0 & 0 & 0 \\ 0 & 0 & 0 & 0 & 0 & 0 \\ 0 & 0 & 0 & 0 & 1 & 0 \\ 0 & 0 & 0 & 0 & 0 & 1 \end{bmatrix}
\qquad
M^4 = \begin{bmatrix} 0 & 0 & 0 & 0 & 0 & 0 \\ 0 & 0 & 0 & 0 & 0 & 1 \\ 0 & 0 & 0 & 0 & 0 & 0 \\ 0 & 0 & 0 & 0 & 0 & 0 \\ 0 & 1 & 0 & 1 & 0 & 0 \\ 0 & 0 & 0 & 0 & 1 & 0 \end{bmatrix}
$$

$$
M^5 = \begin{bmatrix} 0 & 0 & 0 & 0 & 0 & 0 \\ 0 & 0 & 0 & 0 & 1 & 0 \\ 0 & 0 & 0 & 0 & 0 & 0 \\ 0 & 0 & 0 & 0 & 0 & 0 \\ 0 & 0 & 0 & 0 & 0 & 1 \\ 0 & 1 & 0 & 1 & 0 & 0 \end{bmatrix}
$$

The (2, 4) entry of M is 0; this means that there is no 1-step path (i.e., an edge) from species 2 to species 4. Similarly, if we examine the (2, 4) entry of M^2, we note that it is 0, indicating that there is no 2-step path from species 2 to species 4. Examining species 2 and 4, we see from M and M^2 that there are no possibilities of contamination reaching species 4 from species 2 in one or two steps. However, M^3 has 1 as its (2, 4) entry. Thus, there is a 3-step path (via x_5 and x_6) from species 2 to species 4. Hence, species 4 can be contaminated by species 2, but only after 3 steps. Will there be any contamination reaching x_2 from x_1? Examination of our matrices (above) shows that it cannot happen in 1, 2, 3, 4, or 5 steps. But how do we know it won't happen, for example, in 29 steps? Do we need to calculate every possible power of M (an impossible task)? If not, how many powers will be sufficient? The answer is provided by Theorem 1 of Chapter 4, which can be restated:

THEOREM 2

In a digraph (or graph) having n vertices, if two distinct vertices can be joined by a path, then they can be joined by a path that is no longer than $n - 1$ edges, i.e., $n - 1$ steps.

Applying this theorem to our case, with $n = 6$, we can conclude that any contaminations that occur will occur in 5 or fewer steps. Hence, we need not compute any power of M greater than the fifth.

A convenient way to keep track of the information contained in these matrices is to keep a cumulative sum of the powers as they are calculated. Thus, along with M^2, M^3, M^4, etc., we calculate the sums $M + M^2$, $M + M^2 + M^3$, $M + M^2 + M^3 + M^4$, etc. The reason for this is that the (i, j) entry of $M + M^2 + \cdots + M^r$ is the total number of paths of length no greater than r from x_i to x_j. In particular, when $r = n - 1$, the matrix $M + M^2 + \cdots + M^{n-1}$ tells us, for any given pair of vertices, whether one can be reached from the other. If the (i, j) entry is 0, then j cannot be reached from i by any path, but if the (i, j) entry is different from 0, then j can be reached from i by a path. In our example,

$$M + M^2 + \cdots + M^5 = \begin{bmatrix} 0 & 0 & 1 & 2 & 0 & 0 \\ 0 & 1 & 0 & 1 & 2 & 2 \\ 0 & 0 & 0 & 1 & 0 & 0 \\ 0 & 0 & 0 & 0 & 0 & 0 \\ 0 & 2 & 0 & 2 & 1 & 2 \\ 0 & 2 & 0 & 2 & 2 & 1 \end{bmatrix}$$

Examining this matrix we can conclude that contamination at x_1 will spread only to x_3 and x_4. Contamination at x_2, however, will spread to x_4, x_5, and x_6 (actually, the x_2 row also has a nonzero entry in the (2, 2) place indicating that contamination would spread from x_2 back to x_2—but this is, to some extent, superfluous information). There is no species that will spread contamination to all others because every row has at least two zeros in it. Note that a zero on the diagonal in row i means that contamination starting at x_i does not cycle back to x_i in five steps or less. Furthurmore, if we are interested in learning if contamination at x_i returns to x_i, we would have to consider M^6 in addition to the lower powers of M, since Theorem 2 applies only to paths between distinct vertices. In a six-vertex digraph, it may be there is a path of length 6, but no shorter path between a vertex and itself.

The main theorem we have been working with applies to graphs in addition to digraphs. Its applications, therefore, are a good deal wider than we shall indicate here. Even with digraphs, there are many other situations where we could usefully apply this theorem. As another example, suppose Figure 5.17 modeled a rumor network. Each vertex represents a person; a directed edge from x to y would indicate that x tells any rumors he hears to y (but y does not necessarily reciprocate—if he does, there will have to be another directed edge running the other way). It may be of interest to know how far a rumor will spread if it is planted with a certain person. In addition, one might like to know what is the minimum number of persons who have to be given the rumor in order to be sure it will spread to everyone. Examining the matrix $M + M^2 + M^3 + M^4 + M^5$, we can determine that planting the rumor with x_1 and x_2 will suffice to spread it to everyone.

The model we have created here and the matrix method for dealing with it verge on being useful but have the drawback of not being quantitative. For our food web, for example, it would be more useful if we knew not only whether a certain species would become contaminated but also how much contamination would be passed on to it. As another example, suppose one were interested in studying the dynamics of public opinion, particularly with a view toward determining how various aspects of that public opinion influenced one another. One might isolate the following as the important ingredients in determining public opinion:

(1) Statements of government leaders.
(2) Editorial policies and possible biases of the news media.
(3) Voice of the people, as expressed through election returns.
(4) Opinions of artists and intellectuals as expressed in art, literature, and scholarly works.

Suppose we construct a digraph model by representing each of these opinion sources by a vertex and connecting one vertex to another if the first influences the second. The flow of influence in this digraph is analogous to the flow of contaminants in our food web digraph. Our model is rather shallow, however, since almost certainly each of the opinion sources above influences the other directly and so the digraph would be the one in Figure 5.18. Qualitatively there is nothing to study here, yet it would be useful to know how strong the various influences are. If we knew this, we could assign weights to the various directed edges and perhaps do some useful modeling.

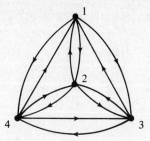

Figure 5.18

EXERCISES 5.8

1. For each of the digraphs D in Figure 5.19,
   (a) Write the matrix M for D.
   (b) Compute M^2 and M^3.
   (c) Compute $M + M^2 + M^3$.
   (d) State by looking at your answer to part c whether or not D is connected.

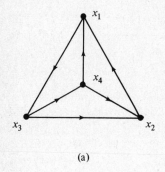

(a)

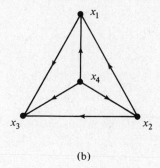

(b)

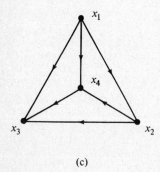

(c)

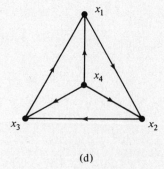

(d)

Figure 5.19

160

2. Given the food web in Figure 5.20, determine entirely by matrix methods how many steps it would take to contaminate the various species if contaminant is introduced at x_2.

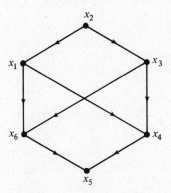

Figure 5.20

3. For the rumor network in Figure 5.21, determine by matrix methods how many persons must be told the rumor if it is to spread to everyone.

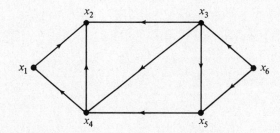

Figure 5.21

4. Suppose D is a digraph whose matrix is

$$\begin{bmatrix} 0 & 1 & 0 & 0 \\ 0 & 0 & 1 & 0 \\ 0 & 0 & 0 & 1 \\ 1 & 1 & 0 & 0 \end{bmatrix}$$

Is it true that from any vertex it is possible to reach any other with a path?

5.* A digraph is called *acyclic* if it contains no directed circuits. How could you tell from the powers of its matrix whether a digraph was acyclic?

SUGGESTED READING

BOLT, A. B., ED., *We Built Our Own Computers,* Cambridge University Press, New York, New York, 1966. This paperback describes the construction of some simple computers by English schoolboys.

HULL, T. E., DAY, D. D. F., *Computers and Problem Solving,* Addison-Wesley, Reading, Mass., 1971. This book contains many interesting problems for computer solution.

KLINE, M., ED., *Mathematics in the Modern World,* W. H. Freeman and Company, San Francisco, Calif., 1968. Available in paperback. This book contains many provocative expository articles about computers.

MESSICK, D., ED., *Mathematical Thinking in Behavioral Sciences,* W. H. Freeman and Company, San Francisco, Calif., 1968. Available in paperback. This book also contains many provocative expository articles about computers.

VON NEUMANN, J., *The Computer and The Brain,* Yale University Press, New Haven, Conn., 1958. One of the great pioneers in the computer sciences discusses the analogies and differences between the computer and the human brain.

6
functional models

6.1 functions

All of us are familiar with statements in books, magazines, and newspapers that describe the change in one quantity due to the change in another quantity. We read of the increase of world population with the passage of time, the decrease in temperature that a rocket's instruments measure as it gains altitude above the earth, the day-to-day variations of the Dow-Jones stock averages, the variation in profits of a major corporation from year to year, and the relationship of the volume of a sphere to the size of its radius.

Sometimes the way one quantity varies with another is known exactly, for example, the variation of the area of a circle as the size of its radius changes. For other situations it would be of great interest to know the exact way two quantities vary with one another. For instance it would be extremely valuable to have a formula that would tell us what the world's population will be for each year in the future. If it seems strange that a formula might exist that predicts the future behavior of population, recall that it is possible, using formulas derived from Newton's laws, to predict the future locations of planets, thereby making it possible to send rockets to Mars or Venus.

Motivated by the need to examine the relationship between how one quantity varies as another quantity (or quantities) varies, mathematicians have invented a special type of model known as a *function*. Before we give a formal definition of this type of model, let us consider a few examples.

Example 1

The owner of a parking lot wishes to charge different rates for parking different sized cars in his lot. Three categories of size have been set up:

Small (S)   Medium (M)   Large (L)

The rates charged for a car will vary with the size of the car. The lowest rates will be charged for small cars since they use less space, cause less congestion, and are easier to accommodate. Table 6.1 shows one possible set of rates. Each category of car size has a rate assigned to it.

Table 6.1

Size	Charge Assigned per Day
Small	$2
Medium	$3
Large	$4

Example 2

Every 10 years, as mandated by the United States Constitution and implemented by an act of Congress, a census of the United States population is taken. The size of the population varies with time. Table 6.2 shows this variation for the censuses taken since 1900.

Table 6.2

Year	Population
1900	75,994,575
1910	91,972,266
1920	105,710,620
1930	122,775,049
1940	131,669,275
1950	150,697,361
1960	179,323,175
1970	200,251,326

Example 3

A manufacturer of corn plasters wants to decide how many boxes of corn plasters to produce next year and what price to set for each box. On the basis of past experience, the manufacturer sets up Table 6.3, indicating the variation of sales with price per box.

Table 6.3

Price per Box	Sales per Month in Boxes
44¢	20,000
46¢	18,600
48¢	18,100
50¢	18,000
52¢	17,400

Note that Table 6.3 is not merely of descriptive interest because some practical questions can be answered on the basis of the information contained there. For example, if a box of corn plasters costs 39¢ to produce, then using Table 6.3 one can compute the price that the manufacturer should set to obtain his greatest profit.

With these examples in mind we make the following definition:

DEFINITION 1

Let X and Y be sets, and suppose that there has been assigned (corresponded) to each element of X a single element of Y. The assignment that is used is called a *function*. The sets X and Y are known as the *domain* and *range* of the function, respectively.

If we give the function a name, say, the symbol f, we can write symbolically $f: X \longrightarrow Y$ to suggest that there is an assignment of the elements of Y to those in X. The element y in Y that is assigned to the element x in X is called the *image* of x under f; we write this $y = f(x)$. The symbol $f(x)$ (read f of x) does not mean f times x; rather, the symbolism denotes as a whole the element y of Y corresponding to x under the assignment f. To repeat, $f(x)$ denotes that element y of the range that we correspond to the element x of the domain. Intuitively, we think of the elements of the range as depending on or varying with the elements of the domain.

To illustrate the ideas above, we shall rewrite the first three examples using *function notation*.

Example 1 (Revisited)

The domain is the set $X = \{$S, M, L$\}$.
The range is the set $Y = \{2, 3, 4\}$.
$f: X \longrightarrow Y$ is given by

$$f(S) = 2$$

$$f(M) = 3$$

$$f(L) = 4$$

Example 2 (Revisited)

The domain is the set $X = \{1900, 1910, 1920, 1930, 1940, 1950, 1960, 1970\}$.

The range is the set $Y = \{75,994,575,\ 91,972,266,\ 105,710,620,\ 122,775,049,\ 131,669,275,\ 150,697,361,\ 179,323,175,\ 200,251,326\}$.

$G: X \rightarrow Y$ is given by

$$G(1900) =\ \ 75,994,575$$

$$G(1910) =\ \ 91,972,266$$

$$G(1920) = 105,710,620$$

$$G(1930) = 122,775,049$$

$$G(1940) = 131,669,275$$

$$G(1950) = 150,697,361$$

$$G(1960) = 179,323,175$$

$$G(1970) = 200,251,326$$

Example 3 (Revisited)

The domain is the set $X = \{44, 46, 48, 50, 52\}$.
The range is the set $Y = \{20,000, 18,600, 18,100, 18,000, 17,400\}$.
$H: X \rightarrow Y$ is given by

$$H(44) = 20,000$$

$$H(46) = 18,600$$

$$H(48) = 18,100$$

$$H(50) = 18,000$$

$$H(52) = 17,400$$

Two things should be emphasized about the definition of a function. First, we do not allow a single element of the domain to have two elements of the range associated with it. This is motivated by the way quantities vary with one another in the real world. For example, at a given time, say, 1940, the United States cannot have a population of 131,669,275 and also a population of 200 million. Similarly, at a certain age, a child has only one weight. Examples such as these make it reasonable to require that to each element of the domain there be corresponded exactly one element in the range set.

Second, we do allow different domain elements to have the same range element assigned to them. Again, the reason for permitting several different domain elements to be assigned to the same element of the range is motivated by the way that quantities vary with one another in the real world. For example, if we are relating the altitude of a firework with time, the firework may be at the same altitude at two different times (domain elements), once on its ascent and once on its descent. Table 6.4 illustrates the function which might describe how the height of a firework depends on the number of seconds which have elapsed since its launching.

Table 6.4

Time (sec)	Height (ft)
0	0
1	32
2	40
3	50
4	40
5	32
6	0

Although the elements of the domain and range of a function can belong to any sets whatsoever, in all our subsequent discussion we shall be concerned with functions whose domains and ranges are sets of numbers.

Often in cases where the domain is very large, it is not convenient to describe a function by explicitly stating which element in the range is assigned to a given element in the domain. In such cases it is often convenient to specify the function under consideration by giving a formula that tells which element of the range is assigned to a given domain element.

Example 4

The domain is the set $X =$ the set of real numbers. The range is the set $Y =$ the set of real numbers. $F: X \longrightarrow Y$ is given by

$$F(x) = 2x$$

The function given by the formula above assigns to each number in the domain the number that is twice as large as that number. Thus, the number assigned to 2, $F(2)$, is equal to 4. The number assigned to -10, $F(-10)$, is equal to -20.

Example 5

The domain is the set $X =$ the set of real numbers. The range is the set $Y =$ the set of positive real numbers together with zero. $G: X \rightarrow Y$ is given by

$$G(x) = x^2$$

The function given by the formula above assigns to each number in the domain the number that is the square of that number. Thus, the number assigned to 4, $G(4)$, is equal to 16. The number assigned to -4, $G(-4)$, is equal to 16. Note that for this function two different elements of the domain often have assigned to them the same number in the range.

Example 6

The domain is the set $X =$ the set of real numbers. The range is the set $Y =$ the set of positive real numbers. $H: X \rightarrow Y$ is given by

$$H(x) = 2^x$$

Example 7

The domain is the set $X =$ the set of real numbers. The range is the set $Y =$ the set of real numbers greater than or equal to -4. $L: X \rightarrow Y$ is given by

$$L(x) = x^2 + 4x - 4$$

Since it is often inconvenient to specify the domain and range of a function defined by a formula, the following convention is usually adopted:

CONVENTION 1

If f is a function defined by a formula, and if the domain and range are not specified, then the domain of the function is the largest set of real numbers for which the formula yields a real number, and the range is the set of all those real numbers obtained by substituting values from the domain into the formula.

Example 8

$$f(x) = \frac{1}{(x-1)}$$

This formula will yield a number for every value of x except 1. Hence, the domain is all numbers except 1. The range consists of all numbers except 0.

Example 9

$$g(x) = \sqrt{x}$$

This formula yields a real number for every number greater than or equal to 0. The range consists of all the real numbers greater than or equal to 0.

EXERCISES 6.1

1. For each of the following tables, describe in function notation the function f that is represented by that table.

(a) Domain	Range	(b) Domain	Range	(c) Domain	Range
A	1.5	0	3	0.1	10
B	1.7	1	9	0.2	20
C	1.9	2	11	0.3	20
D	2.0			0.4	40

2. If $f(x) = (5x + 3)/2$, what are $f(1)$, $f(2)$, and $f(100)$?
3. Convert each of the following verbal descriptions of functions to formula notation.
   (a) The function f assigns to any given number the number 10.
   (b) The function g assigns to any given number another number that is one and a half times that given number.
   (c) The function h assigns to a given number another number that is 10 more than the square of the given number.
4. Construct a function f whose domain and range are $\{1, 2, 3, 4, 5, 6\}$.
5. For each bale of cotton delivered to the textile mill, 10 bolts of cloth can be produced. If $f(x)$ is the number of bolts that can be produced from x bales, describe the function f in formula notation.
6. A manufacturing company hires three types of employees: machinists grade 1, machinists grade 2, and supervisors. Each employee works a 40-hr week. Machinists grade 1 are paid \$7/hr, machinists grade 2

are paid $5/hr, and supervisors are paid $10/hr. The company hires x, y, and z, respectively, of these types of workers.

    (a)   Express the total weekly payroll (p) as a function of x, y, and z.

    (b)   Compute the total hourly wage when $x = 10$, $y = 15$, and $z = 4$.

7.   The Public Relations Division of The Anin Manufacturing Company has prepared the following chart showing the relation of advertising to sales:

Advertising costs (c) in 1000's of dollars	5	10	15	20
Sales (s) in 10,000's of dollars	12	22	32	42

    (a)   Express the relationship between s and c by a formula.

    (b)   Use your answers in part (a) to compute s when $c = 12$, when $c = 22$, and when $c = 40$. Which, if any, of the resulting values of s do you believe are meaningful in the sense that the investment of the given number of advertising dollars will result in approximately the amount of sales "predicted" by the formula?

8.   A large retailer of men's pants discovers that sales volume S is related to sales price P as in the chart below:

P	10	20	30	40	50
S	400	300	250	200	171

    (a)   Verify that S and P are related by the formula

$$S = \frac{12{,}000}{20 + P}$$

    (b)   If a single pair of pants costs $20 to produce, find a formula for the profit in terms of sales.

9.   Answer the question on p. 165 dealing with the selling price that yields maximum profit for the corn plaster manufacturer.

10.* Construct a function whose domain and range are as in Exercise 4 but with the additional property that $f(f(x)) = x$ for all x in the domain.

11.* Beginning in a certain year all new cars are required by law to be outfitted with Cleano, an antipollution device. Since many people keep their cars for some years before buying a new one, it will take awhile for all cars on the road to have Cleano. Let us take a simple model in which the number of cars on the road is 50 million each year and each year 10% of all old cars (i.e., those cars purchased during the previous year or earlier) are removed from circulation (junked) and replaced by new cars. Describe in functional notation the function with domain {1, 2, 3, 4} where $f(t)$ is the number of cars with Cleano on the road at the end of t years after Cleano first becomes available. (*Hint*: During the first year all discarded cars are without Cleano, but in succeeding years some cars with Cleano are discarded along with cars without Cleano, according to our model.)

12.* Can you describe the following functions in formula notation?

(a)
$$f(0) = 0$$
$$f(1) = 1$$
$$f(2) = 4$$
$$f(3) = 9$$

(c)
$$h(0) = -4$$
$$h(1) = -1$$
$$h(2) = 2$$
$$h(3) = 5$$

(b)
$$g(0) = 1$$
$$g(1) = 3$$
$$g(2) = 5$$
$$g(3) = 7$$

(d)
$$k(0) = 0$$
$$k(1) = 2$$
$$k(2) = 6$$
$$k(3) = 12$$

[*Hint*: The answer to part (a) is $f(x) = x^2$.]

13.* By consulting a world almanac, construct a descriptive functional model for each of the following situations:

  (a) The population of the state you live in for the years 1900–1970. Use intervals of 10 years.

  (b) The population of the city (town) you live in for the years 1940–1970. Use intervals of 10 years.

6.2 graphing functions

In the previous section we discussed how functions can be used as models for the way that one quantity varies with another. The tables and formulas that are used to define a function in principle contain all the information about the functional relationship. It is often easier, however, to comprehend information in geometric or pictorial form rather than in algebraic form. Work of the French philosopher and mathematician René Descartes helped make it possible to convert the algebraic formulas and tables defining functions into a pictorial diagram.

As previously stated, for the rest of this chapter we restrict ourselves to functions whose domain and range are sets of numbers. For such a function, the pictorial representation or model that we shall create will be drawn in a two-dimensional plane. We consequently need a more precise way of

René Descartes (1596–1650) had a varied career as philosopher, soldier, writer, physicist, and mathematician. His best known work in mathematics, *La géometrie,* developed algebraic methods for solving geometric problems. Although this book represented a major step towards the development of analytic geometry as we know it, it did not include the use of orthogonal axes and points plotted in four quadrants. Rather, Descartes used oblique axes and, since negative numbers were not completely accepted in his day, he restricted his graphs to what we call the first quadrant. *(Courtesy The Wolff-Leavenworth Collection, George Arents Research Library at Syracuse University.)*

thinking about a plane than as a large flat expanse. To arrive at this precision we first introduce a pair of perpendicular lines into the plane, a horizontal line called the X axis and a vertical line called the Y axis (see Figure 6.1). These two axes will provide a frame of reference with respect to which any point P in the plane can be described by an ordered pair of numbers [for example, $(2, 3)$ or $(-1, 4)$]. The first number of such a pair is obtained by measuring the perpendicular distance from P to the vertical Y axis and attaching a minus sign in the event that P is to the left of the vertical Y axis. The second number is obtained by measuring the perpendicular distance

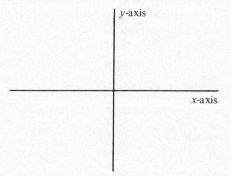

Figure 6.1

from P to the horizontal X axis and attaching a minus sign in the event that P is below the X axis. The first number is called the x coordinate of P and the second is called the y coordinate of P (see Figure 6.2). In this way every point of the plane can be represented by an ordered pair of numbers (x, y).

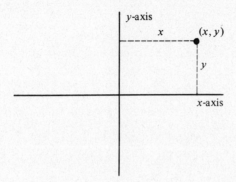

Figure 6.2

Not only can every point be assigned an ordered pair (x, y), but also any ordered pair of numbers (x, y) will have a point of the plane corresponding to it. To find this point we simply measure x units from the Y axis, going to the right if x is positive and to the left if x is negative, and then measure y units from the X axis, going up if y is positive and down if y is negative. Figure 6.3 shows how to plot the points $(-2, 3)$, $(-2, -3)$, $(0, 0)$, $(1, 2)$, $(2, 1)$, $(3, -2)$, $(\frac{1}{2}, -1)$, $(\frac{3}{2}, 2)$, and $(-\frac{5}{2}, 2)$.

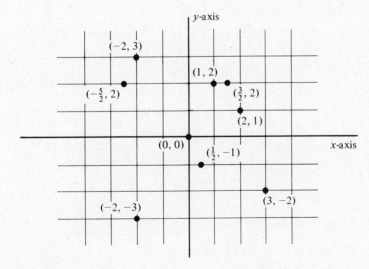

Figure 6.3

The following example describes how we make a pictorial representation of a function whose domain and range consist of numbers. We shall refer to this diagram as a *graph of the function*, even though the terminology may be confusing because the present usage of *graph* is different from our earlier meaning for the word.

Example 1

Suppose f is the function described in function notation by

$$f(-2) = 4 \qquad f(-1) = 1 \qquad f(0) = 0 \qquad f(1) = 1 \qquad f(2) = 4$$

We make the graph of this function by plotting (marking) all the points in the plane that correspond to the ordered pairs in this set:

$$\{(x, y) \mid y = f(x)\}$$

For the case of our example, this set of ordered pairs consists of the ordered pairs listed in Figure 6.4. Note how these ordered pairs correspond to the five parts of the function notation description of f.

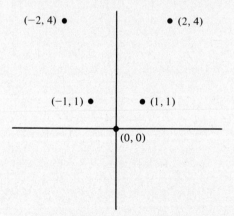

Figure 6.4

In referring to the graph in Figure 6.4, we may describe it, as we have thus far, as the graph of the function $f(x)$ or we may refer to it as the graph of the *equation* $y = f(x)$. The reason for the latter terminology is that if (x, y) is one of the ordered pairs on the graph, it must be true that $y = f(x)$. Given an equation $y = f(x)$ that describes a function, we define the *solution set* of the equation to be the set of ordered pairs (x_1, y_1) such that $y_1 = f(x_1)$.

The method used in Example 1 also applies in the case where the function is defined by a formula, except that we need to add one small modification: Begin by finding the functional notation form for the function; then proceed as before.

Example 2

Consider a function g defined by the formula $g(x) = x^2 + 1$ and defined on the domain $\{-2, -1, 0, 1, 2\}$. The description of g in functional notation is shown in the left-hand column in Figure 6.5. The next column lists the ordered pairs that we need to plot; the graph follows at the right.

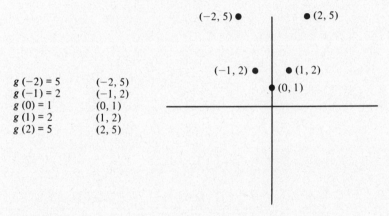

$g(-2) = 5$ $(-2, 5)$
$g(-1) = 2$ $(-1, 2)$
$g(0) = 1$ $(0, 1)$
$g(1) = 2$ $(1, 2)$
$g(2) = 5$ $(2, 5)$

Figure 6.5

The second column in Figure 6.5 is the solution set of the equation $y = x^2 + 1$ for the domain set $\{-2, -1, 0, 1, 2\}$.

The final and most interesting wrinkle in graphing functions occurs when we have a function defined by a formula on a domain which is either infinite or so large that it is impractical to deal with all the ordered pairs which would have to be plotted. In such a case we make a selection of a reasonable number of points from the domain and pretend that this set is representative of the whole domain.

Example 3

Consider the function $f(x) = x^2$ defined on the domain consisting of all numbers between -2 and 2. Graph this function.

Solution: Since there are an infinite number of numbers between -2 and 2 (we are not considering merely the whole numbers but all numbers, including fractions such as $\frac{1}{2}$ and $\frac{27}{18}$), we cannot list them all. We cannot plot all the ordered pairs that would be required if we were to follow the method as explained so far. Thus, we shall restrict ourselves to the sampling consisting of the numbers -2, -1, 0, 1, and 2. In Figure 6.6(a) we have

$f(-2) = 4 \qquad (-2, 4)$
$f(-1) = 1 \qquad (-1, 1)$
$f(0) = 0 \qquad (0, 0)$
$f(1) = 1 \qquad (1, 1)$
$f(2) = 4 \qquad (2, 4)$

$f(-3/2) = 9/4 \qquad (-3/2, 9/4)$
$f(-1/2) = 1/4 \qquad (-1/2, 1/4)$
$f(1/2) = 1/4 \qquad (1/2, 1/4)$
$f(3/2) = 9/4 \qquad (3/2, 9/4)$

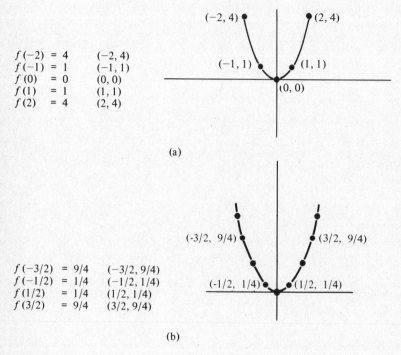

(a)

(b)

Figure 6.6

listed the values of the function at these points, computed the required ordered pairs, and plotted them. We have also drawn a smooth curve through these plotted points. This curve represents our prediction of where the remaining ordered pairs would fall if we had the time and patience to plot them. The general rule that is usually followed in "connecting the dots" is to do it so that the arc which connects adjacent points is a smooth curve. In Figure 6.6(b) we have taken a slightly larger sampling, adding the numbers $-\frac{3}{2}$, $-\frac{1}{2}$, $\frac{1}{2}$, and $\frac{3}{2}$ to our sampling. Only the additional ordered pairs that arise from these additional numbers are listed in the Figure 6.6(b) but all nine points are plotted.

Note that as x increases from an initial value of 0, the corresponding y values increase. As x decreases from an initial value of 0, the corresponding y values increase.

There are certain dangers associated with this process of taking a sampling of points to use in plotting ordered pairs. It may turn out that the sample may not consist of representative points so we may get a poor diagram, as shown by the following example.

Example 4

An ecologist makes a theoretical calculation to compute the freshwater supply of a certain town under a proposed change in ecology management policies. The freshwater supply, $W(t)$, t years after the beginning of the new policy, is given by the following functional model: $W(t) = t^3 - 5t^2 + 4t + 7$. [$W(t)$ is measured in billions of gallons.] In order to determine whether this policy is acceptable or not, the ecologist decides to plot a graph of the function. He first takes the following sampling of numbers: 0, 1, 4. The corresponding ordered pairs are $(0, 7)$, $(1, 7)$, $(4, 7)$; these points are shown plotted in Figure 6.7(a). After connecting the dots it seems that there will be no change in the water supply. If we add to our sampling the numbers 2 and 3, however, we get the graph in Figure 6.7(b), which shows a very differ-

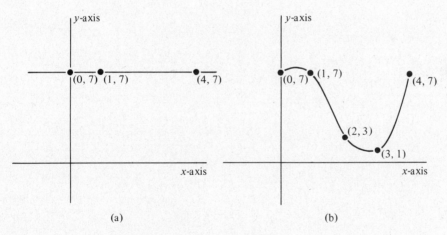

(a) (b)

Figure 6.7

ent picture. This graph shows that 3 years after the new policy takes effect, the water supply will be one-seventh of its initial value.

In this example, we first ignored the numbers 2 and 3 and our picture [Figure 6.7(a)] showed the water supply as constant. It turned out that these numbers were not of minor importance; when we took them into account, the diagram told a different story. The problem of choosing a good sampling to use in plotting a function is not an easy one. We shall have to be content with pointing out that it is a problem and that more sophisticated methods can be used to overcome the problem.

What is the value of making the graph of a function? The answer seems to be that the human mind finds the visual geometric patterns of the graph easier to interpret than formulas or functional notation. Would the cartoon of Figure 6.8 have more or less impact if the graph on the wall were

(*Copyright Ranaan Lurie, Life Magazine, © 1973 Time Inc.*)

"Looks better this way"

Figure 6.8

replaced with a formula? Another good example is the electrocardiogram. As the heart passes through its cycle of contraction and relaxation, electrical currents are generated that have different strengths at different times in the cycle. If we associate the strength of the current with each time in the cycle, this is a function. An electrocardiograph is a machine for plotting this function and displaying it visually. The electrocardiograph does not attempt to describe this function with a formula or by functional notation but directly constructs a graph called an *electrocardiogram* (see Figure 6.9).

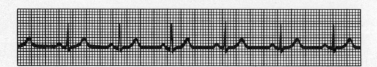

Figure 6.9

EXERCISES 6.2

1. Plot the points $(3, 4)$, $(-4, 5)$, $(-3, -2)$, and $(-1, 4)$.

2. Plot the points $(\frac{1}{2}, -1)$ and $(-\frac{3}{4}, \frac{3}{2})$.

3. Explain why no portion of the graph of the function $y = x^2 + 1$ can appear below the X axis.

4. Explain why no portion of the graph of the function $y = \sqrt{x^2 + 3}$ can appear below the X axis.

5. Draw a sketch of the graphs of the following functions. (Use the convention that the domain of the function is the largest collection of real numbers for which the formula is defined.)

(a)	$y = x$	(e)	$y = 2^{-x}$
(b)	$y = 2x$	(f)	$y = x^2 - 1$
(c)	$y = 3x + 1$	(g)	$y = 1/x$
(d)	$y = 2^x$	(h)	$y = 1 + x^3$

6. A graph of the function $y = f(x)$ is said to be symmetric with respect to the Y axis if, for every pair of numbers a and $-a$ in the domain of the function, $f(a) = f(-a)$. For each of the functions below, determine if the graph of the function is symmetric with respect to the Y axis. For example, consider $y = x^2$. For $x = 1$ and $x = -1$, the corresponding y value is 1. Similarly, for $x = a$ and $x = -a$, the corresponding value of y is a^2. Hence, $y = x^2$ is symmetric with respect to Y axis.

(a)	$y = x^4$	(d)	$y = x + 4$
(b)	$y = x^3 - x$	(e)	$y = x - 3$
(c)	$y = x^2 - x^4$	(f)	$y = -x^2 + 10$

7. Suppose that for any real number x, $[x]$ denotes the greatest integer less than or equal to x. (Thus, $[1.4] = 1$ and $[1.56] = 1$.) Draw a graph of the function $y = [x]$.

8. Find the domain that the convention given in Exercise 5 imposes on the following functions:

(a)	$y = 1/(x - 7)(x - 3)$	(e)	$y = \sqrt{(x - 1)/(x + 3)}$
(b)	$y = 1/x^2$	(f)	$y = 3^x$
(c)	$y = \sqrt{x - 4}$	(g)	$y = \sqrt{x^2 - 1}$
(d)	$y = \sqrt{x + 5}$		

9. Consider the function $y = (x - 1)(x - 2)(x - 3)(x - 4)$. Explain why plotting the graph of this function using as x values the numbers 1, 2, 3, 4 would give a misleading result. Can you generalize this example?

6.3 the linear function

There is a very special type of function called a *linear function* that appears quite often in applied mathematics. In this section we shall give a number of examples where linear functions provide good models for real-world phenomena.

Examples of linear functions are $f(x) = 2x$, $g(x) = -x + 7$, and $h(x) = \frac{1}{2}x - \frac{3}{4}$. On the other hand, $f(x) = x^3$ is not linear. Here is the definition.

DEFINITION 2

The function f is called *linear* if f has the form $f(x) = ax + b$ where a and b are constants. (Instead of x, the variable may be y, z, or anything else and this doesn't change the linearity.)

Example 1

The normal systolic blood pressure for an adult varies with age. As a rule of thumb, to compute the normal pressure, take one-third of the age in years and add 113. In the language of functions, if we let x stand for the age and f for the function that associates the corresponding normal blood pressure with each age x, then

$$f(x) = \tfrac{1}{3}x + 113$$

Note that f is a linear function.

Example 2

John owns an automobile and wants to use it on weekends as a taxi to earn money. He wants to charge a flat rate per mile but to determine this he needs to know what it costs him to run his car. He figures that gasoline, depreciation, and maintenance all together average out to 12¢/mi. In addition there is $200 insurance per year, which is independent of mileage. If he goes *m* miles in a year and we denote the total dollar cost by *c*(*m*), then

$$c(m) = (0.12)m + 200$$

Once again, this is a linear function.

Suppose John expects to be able to use his car as a cab for 20,000 mi and wants a clear profit of $2000. What should he charge?

The cost of going 20,000 miles is

$$c(20,000) = (0.12)(20,000) + 200 = 2400 + 200$$
$$= \$2600$$

If he profits $2000, he must take in

$$\$2000 + \$2600 = \$4600$$

His charge per mile should therefore be

$$\frac{\$4600}{20,000} = \frac{\$23}{100}$$
$$= \$0.23$$

Example 3

A ball is thrown straight upward with an initial velocity of 608 ft/sec. If Newton's laws govern the velocity of the ball, determine how long it takes the ball to reach its maximum height.

Solution: According to Newton's laws the velocity of an object projected straight up will lose 32 ft/sec in velocity for each second of flight. Hence, we can describe the velocity *v* of the ball at any instant in time *t* by

$$v = f(t) = -32t + 608 \qquad (t \geq 0)$$

Notice that we have used the fact that $f(0) = 608$; that is, there is an initial velocity of 608 ft/sec. The ball reaches its maximum height when its velocity is 0. Hence, setting $-32t + 608 = 0$ and solving the resulting equation

$$-32t + 608 = 0$$

we discover that 19 sec after the ball is thrown up, it reaches its maximum height.

EXERCISES 6.3

1. For each of the linear functions below, compute the required function values:
    (a) $f(t) = 3t + 8$
         Compute $f(4)$, $f(-3)$, $f(0)$.
    (b) $g(x) = -8x - 4$
         Compute $g(-3)$, $g(-7)$, $g(2)$.
    (c) $h(y) = 3y - 9$
         Compute $h(-4)$, $h(\frac{1}{3})$, $h(0)$.

2. A company is producing spun aluminum wire. Regardless of the length of the wire produced, the initial costs are $15,000. Furthermore, the cost of the materials needed to produce 100 ft of wire is $40.
    (a) Construct a function to describe the cost of producing aluminum wire in terms of the length of wire produced.
    (b) What is the cost of producing 8000 ft of wire? What is the cost of producing 10,500 ft of wire?
    (c) If the company has $135,000 available, how much wire can it produce?

3. The company of Exercise 2 makes a profit of $30 for every 1000 ft of wire it produces. Construct a function that relates the profit of the company to the amount of wire it produces.

4. Current (in amperes) is related to voltage (in volts) and resistance (in ohms) by Ohm's law:

$$I = \frac{E}{R}$$

    where I = current, E = voltage, and R = resistance. Write a function that relates the current to the voltage, assuming that the resistance is $\frac{1}{10}$ ohm.
    (a) When the voltage is 3 volts, find the current.
    (b) When the current is 120 amp, find the voltage.

5. Draw a graph of the function

$$f(t) = 3t + 8 \qquad (1 \leq t \leq 3)$$

6. Draw a graph of the function

$$f(t) = -t + 2 \qquad (-1 \leq t \leq 1)$$

7. The circumference of a circle is related to the size of its radius by a linear function.
    (a) Write the function relating the circumference of a circle to its radius.

    (b)   What is the length of the circumference when the radius is 8?
          When the radius is $\frac{6}{5}$?

8.   The length (L) of a rod and its temperature (t) are related by the function

$$L = L_0(1 + \alpha t)$$

where L_0 and α are fixed numbers. Suppose $L_0 = 6$ and $\alpha = \frac{1}{2}$. Find the function that gives the temperature of the rod in terms of its length. When the rod is at 80°C, what is its length?

9.   A bank is giving 6% annual interest on deposits. Write a function that describes the amount of interest at the end of a 1-year period in terms of the size of a deposit.

10.* Suppose that x denotes the amount of sales in dollars for 1 month of company Z. Production cost to the company is given by a fixed yearly cost of $9936 for rental of factory space and by a variable cost that is 10% of the amount of sales. The break-even point for the company is defined to be the value of sales for which the net profit (before taxes) is 0. Net profit is equal to the difference between sales and cost. Find the break-even point for company Z.

6.4 *linear equations*

We have seen that linear functions are often descriptive models for real-world situations; in addition, they are algebraically neat and have simple graphs. With this in mind, it is desirable to become familiar with linear equations, which are closely related to linear functions. Here and in Section 6.5, we shall develop the necessary algebraic machinery to discuss an important class of applied problems involving linear equations.

DEFINITION 3

If A, B, and C are constants, and A and B are not both 0, then $Ax + By + C = 0$ is called a *linear equation*.

Example 1

(1)   Choosing $A = 1$, $B = 0$, and $C = 0$, we obtain the linear equation $x = 0$.

(2)   Choosing $A = 0$, $B = 1$, and $C = 0$, we obtain the linear equation $y = 0$.

(3)   Choosing $A = 1$, $B = 2$, and $C = -1$, we obtain the linear equation $x + 2y - 1 = 0$.

Notice that except when $B = 0$, $Ax + By + C = 0$ defines a linear function since we can write $y = (-A/B)x - C/A$. For any linear equation we can plot the ordered pairs (or some of them) (x, y) that satisfy the equation and thereby form the graph of the equation. Figure 6.10 shows the graphs of the equations listed in Example 1 above. The graphs are shown in boldface.

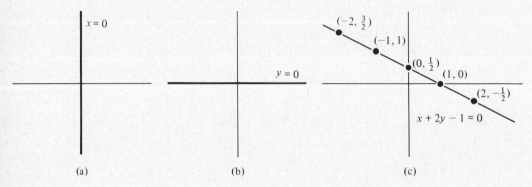

(a)   (b)   (c)

Figure 6.10

To graph $x + 2y - 1 = 0$, we pick a few values of x and calculate the corresponding values of y. Taking for x the values -2, -1, 0, 1, and 2, we determine that the corresponding y's are $\frac{3}{2}$, 1, $\frac{1}{2}$, 0, and $-\frac{1}{2}$ so we plot these ordered pairs: $(-2, \frac{3}{2})$, $(-1, 1)$, $(0, \frac{1}{2})$, $(1, 0)$, $(2, -\frac{1}{2})$. After connecting the points we appear to have a straight line.

For the skeptic we shall eventually give a proof that any linear equation has a graph that is a straight line. We shall begin backward, however, and first study how to find the equation of a given straight line.

Suppose L is a straight line.

Case 1

Line L parallel to the Y axis. All points on such a line have the same distance, say a units ($a \geq 0$), from the Y axis and are on the same side of the Y axis. Consequently, these points either all satisfy the equation $x = a$ or all satisfy the equation $x = -a$, depending on which side of the Y axis contains the line (see Figure 6.11).

Case 2

Line L parallel to the X axis. All points on such a line have a constant distance, say, b units ($b \geq 0$), from the X axis and are on the same side of the X axis. Consequently, these points all satisfy the equations $y = b$ or

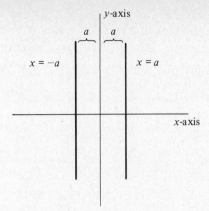

Figure 6.11

$y = -b$, depending on whether the line is above or below the X axis (see Figure 6.12).

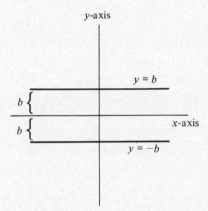

Figure 6.12

Case 3

Line L not parallel to either axis. Let $P_1(x_1, y_1)$ and $P_2(x_2, y_2)$ be two distinct fixed points on L and let $P(x, y)$ be a variable point on L. We drop perpendiculars from P and P_2 to the line $y = y_1$, touching the line at points Q and R, respectively. These constructions and the coordinates of the points Q and R are shown in Figure 6.13.

Triangles P_1PQ and P_1P_2R are similar, so

$$\frac{PQ}{P_1Q} = \frac{P_2R}{P_1R} \tag{6.1}$$

However, the length $PQ = y - y_1$, the length $P_1Q = x - x_1$, the length

185

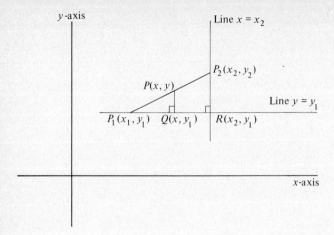

y-axis

Line $x = x_2$

$P_2(x_2, y_2)$

$P(x, y)$

Line $y = y_1$

$P_1(x_1, y_1)$ $Q(x, y_1)$ $R(x_2, y_1)$

x-axis

Figure 6.13

$P_2R = y_2 - y_1$, and the length $P_1R = x_2 - x_1$. Hence (6.1) can be written

$$\frac{y - y_1}{x - x_1} = \frac{y_2 - y_1}{x_2 - x_1} \qquad (6.2)$$

An algebraic simplification gives the following equation, which is clearly linear:

$$y = \left(\frac{y_2 - y_1}{x_2 - x_1}\right)x + y_1 - x_1\left(\frac{y_2 - y_1}{x_2 - x_1}\right)$$

This Shows that any point $P(x, y)$ which lies on the line determined by $P_1(x_1, y_1)$ and $P_2(x_2, y_2)$ satisfies a linear equation. Conversely, it can be shown that any point (x, y) that satisfies (6.2) must lie on a line with $P_1(x_1, y_1)$ and $P_2(x_2, y_2)$.

In all three cases, the equation that we found for a straight line was linear.

The formula (6.2) is called the *two-point formula for the equation of the line* and can be used to find the equation of any line when one knows two points that lie on the line, provided the two points do not determine a vertical line. (This happens only when the two given points have the same x coordinate.)

Example 2

Find the equation of the line connecting the points $(-1, 2)$ and $(2, 4)$.

Solution: Since $(x_1, y_1) = (-1, 2)$ and $(x_2, y_2) = (2, 4)$, substituting

186

in (6.2) we obtain

$$\frac{y-2}{x-(-1)} = \frac{4-2}{2-(-1)}$$

or

$$y - 2 = \frac{2}{3}(x+1)$$

or

$$y = \frac{2}{3}x + \frac{8}{3}$$

Finally this can be written $-2x + 3y = 8$, which is a linear equation.

Example 3

Find the equation of the line connecting the points (3, 2) and (3, 579).

Solution: Here the two-point formula is not applicable since the two points have the same x coordinate. The equation of the line is simply $x = 3$ and the line is a vertical one, three units to the right of the Y axis.

The constant $(y_2 - y_1)/(x_2 - x_1)$ that appears in (6.2) has a special significance. It is called the *slope* of the line L and is denoted by m. Reference to Figure 6.13 shows that m is the tangent of the angle the line L makes with the horizontal. The slope m is, therefore, a measure of the inclination of the line. The larger the slope of a line, the steeper its inclination is to the horizontal X axis.

In Figure 6.14, line L_1 has slope $m_1 = 3$, line L_2 has slope $m_2 = 1$,

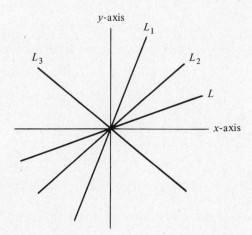

Figure 6.14

and line L_3 has slope $m_3 = -1$. Using the symbol m, we can rewrite (6.2) as

$$y - y_1 = m(x - x_1) \tag{6.3}$$

and in this form the equation is called the *point-slope form* of the line. It is most suitable when one is given the slope and a single point on the line and is asked to find the equation of the line. Another simple algebraic transformation gives $y = mx + (y_1 - mx_1)$. If we rename the constant $y_1 - mx_1$ as b, we obtain the so-called *slope-intercept form*:

$$y = mx + b \tag{6.4}$$

If we substitute 0 for x in (6.4), we find the value b for y. This shows that $(0, b)$ is the point on the Y axis at which the line crosses this axis. The constant b is, therefore, called the y *intercept*. Equation (6.4) is the most convenient one to use when one wishes to calculate the equation of a line for which the slope and y intercept are given (see Figure 6.15).

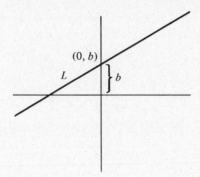

Figure 6.15

Example 4

(1)   Find the equation of the line passing through $(1, -3)$ and having slope -2.

(2)   Find the equation of the line with slope 3 and y intercept -2.

Solution:

(1)   We use (6.3) and find $y + 3 = (-2)(x - 1)$;   i.e. $y = -2x - 1$

(2)   We use (6.4) and find $y = 3x - 2$.

Now let us turn things around and start with a linear equation $Ax + By + C = 0$ in which A and B are not both 0. If $B = 0$, then $A \neq 0$ and the equation can be rewritten $x = -C/A$. We have seen earlier that if we take a vertical line whose distance from the Y axis is $-C/A$ (to the right of the axis if $-C/A$ is positive; to the left otherwise), then this line has as its equation $x = -C/A$. Thus, if $B = 0$, the equation represents a straight line. If $B \neq 0$, we can rewrite this equation as $y = (-A/B)x - C/B$. If we take a line with y intercept $-C/B$ and slope $-A/B$, this line will have the given equation as its equation. Consequently, *a linear equation always represents a straight line.* As a result of this fact, drawing the graph of a linear equation is especially easy: We know the graph is a straight line so it is necessary to plot only two points and then to connect them with a straight line.

Example 5

Draw the graph of the equation $2x + y - 7 = 0$.

Solution: We choose two values for x, say, $x = 0$ and $x = 1$, and calculate the corresponding y values by substitution: When $x = 0$, $y - 7 = 0$, and $y = 7$; when $x = 1$, $2 + y - 7 = 0$, and $y = 5$. The two ordered pairs are $(0, +7)$ and $(1, +5)$. These points and their connecting line are plotted in Figure 6.16.

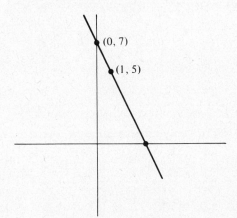

Figure 6.16

The two points that are especially convenient to choose for drawing a line are the points where the line cuts the X and Y axes. Here is an example.

Example 6

Graph the line with equation $3x - 4y = 12$.

Solution: When $x = 0$, $y = -3$ and we can locate the point $A = (0, -3)$ where the line cuts the Y axis. When $y = 0$, $x = 4$, and we can locate the point $(4, 0)$ where the line cuts the X axis (see Figure 6.17).

EXERCISES 6.4

1. For each of the following pairs of points, find the equation of the line they determine.

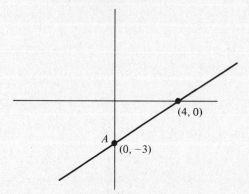

(4, 0)

A

(0, −3)

Figure 6.17

(a) (1, 5), (−2, −3)
(b) (0, 0), (3, 7)
(c) (2, 1), (4, −3)
(d) (8, −13), (4, 9)

2. For each of the following, determine the equation of the line with the given slope that passes through the given point.
(a) $m = 2$, $(-2, 3)$
(b) $m = -1$, $(17, 4)$
(c) $m = 0$, $(-4, -5)$
(d) $m = 9$, $(6, 7)$

3. For each of the following, determine the equation of the line with the given slope and y intercept.
(a) $m = 17$, $b = 0$
(b) $m = -7$, $b = 5$
(c) $m = 0$, $b = 9$
(d) $m = 4$, $b = -18$

4. Plot the graphs of each of the following linear equations.
(a) $3x + 2y - 7 = 0$
(b) $-5x + 4y + 8 = 0$
(c) $y - 9 = 0$
(d) $x + 12 = 0$

5. Temperatures on the Fahrenheit and centigrade scales are related by the equation

$$5F - 9C = 160 \qquad (*)$$

(a) When the temperature is 100°C, what is the temperature in Fahrenheit?

    (b)   When the temperature is 0°C, what is the temperature in Fahrenheit?

    (c)   When the temperature is 80°F, what is the temperature in centigrade?

    (d)   At what temperature will a thermometer read the same number on both scales?

    (e)   Draw a graph of the equation (*) above.

6.* Figure 6.13 is drawn with P between P_1 and P_2. Would the proof need changing to take into account the possibility that P is not so situated?

7.* Show that if (x, y) is any point satisfying Equation (6.1), then this point lies on the line determined by P_1 and P_2.

8.* Suppose A and B both equal 0 in the general form of the linear equation. What can you say about the set of all ordered pairs that satisfies the equation? (*Hint*: Does it make a difference whether or not $C = 0$?)

6.5 simultaneous linear equations

In this section we study the problem of finding the intersection of two lines. Of course, we don't mean this in the most naive sense: We don't mean to use a ruler and find where the lines cross but rather to find the coordinates of the point of intersection by manipulating the equations of the two lines. Having learned this technique, we shall apply it to a special kind of modeling problem.

    Consider Equations (6.5) and (6.6) in the system below; each represents a straight line in the plane. If we multiply (6.5) by $\frac{1}{2}$, we obtain another equation, which has the same solution set. Now our new equation and Equation (6.6) are identical, however, showing that we were dealing with two representations for the same line all along. Consequently, it makes no sense to ask for the intersection of the two lines. Algebraically, any ordered pair of real numbers that satisfies one equation also satisfies the other. This situation, which we call *dual representation of the same line*, can be detected as follows: Put both equations in the form where either the y coefficients or the x coefficients are identical. If the equations are now identical, we have the case of dual representation.

$$x + 2y = 8 \tag{6.5}$$

$$\tfrac{1}{2}x + y = 4 \tag{6.6}$$

Whenever one is confronted with a pair of linear equations, before looking for a common solution one should check whether or not there is a dual representation of the same line.

Now let us consider a case where we do not have dual representation. Within this case we shall subdivide our discussion into two parts. First, we consider the possibility that one of the equations does not contain one of the letters:

$$3x + 2y = 7$$

$$4y = 2$$

In such a case, we can use the equation with one variable missing ($4y = 2$ is missing the variable x in the example above) to solve for the other variable. Since $y = \frac{1}{2}$ in our example, this value is substituted in the other equation to solve for the remaining variable. After substitution we have $3x + 1 = 7$, whence we determine that $x = 2$. The one exception to these instructions occurs when both equations are missing the same variable as in the system:

$$2y = 7$$

$$4y = 2$$

As in the last system there is a variable missing in the second equation so we determine that $y = \frac{1}{2}$ from this equation. If we substitute in the first equation, however, we arrive at the nonsensical statement that $1 = 7$. The explanation is that the two equations in this system can never be simultaneously true. Geometrically we have two parallel lines (horizontal in this case) that have no point in common (see Figure 6.18).

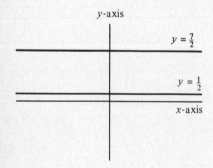

Figure 6.18

The second case to consider is that in which we have no dual representation and, in addition, both equations contain both variables. For example:

$$3x + 6y = 9$$

$$6x + 4y = 14 \tag{6.7}$$

Our goal here is to reduce this to the previous case in which one equation is missing a variable and then to proceed with the previous method.

We shall use the fact that if (a, b) is a solution of each equation of a system, then it is also a solution of any equation obtained by adding (or subtracting) the equations of the system. For example, since $(1, 3)$ is a solution of $x - 3y = -8$ and $2x + 4y = 14$, it is also a solution of $3x + y = 6$, obtained by adding the original equations together.

To solve the system (6.7), we can eliminate the variable x by multiplying the second equation by 1 and subtracting this new equation from the first, obtaining $4y = 2$. Taking this equation together with either of the originals, we have the sort of system described in the previous example.

The solution of the system is $x = 2$, $y = \frac{1}{2}$.

Example 1

Solve the system

$$2x - 4y = 2 \tag{6.8}$$

$$5x - 3y = 12 \tag{6.9}$$

Solution: We choose to eliminate x. Multiplying Equation (6.8) by 5 and Equation (6.9) by -2, we find

$$10x - 20y = 10$$

$$-10x + 6y = -24$$

Adding these equations, we obtain $-14y = -14$, or $y = 1$. Substituting $y = 1$ in Equation (6.8) gives $x = 3$.

The flow diagram in Figure 6.19 shows the process.

We shall now apply the foregoing algebraic techniques to modeling a certain class of practical problems. The type of modeling will sometimes be crude and impractical but will involve simple mathematics. In Sections 6.6 and 6.7 we shall refine the model and eliminate some of the defects but at the price of using somewhat more sophisticated mathematics.

Example 2

Espionage reports estimate that an enemy weapon factory received shipments of 870 tons of steel and 500 tons of aluminum in the last year. This

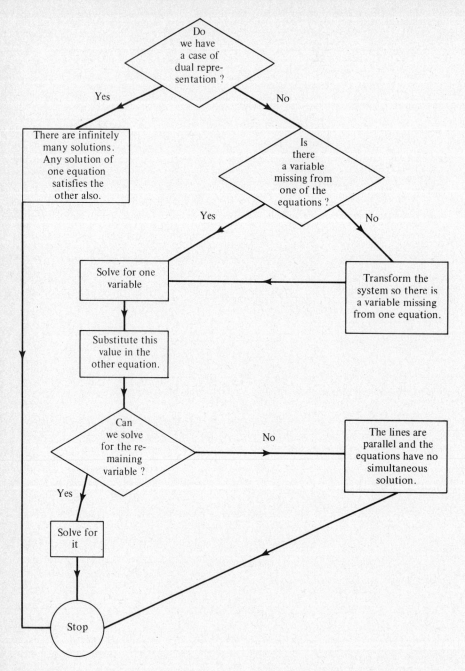

Figure 6.19

194

factory makes two kinds of weapons, the zinger and the zonker. A zinger uses 1 ton of steel and $\frac{1}{2}$ ton of aluminum while a zonker uses $\frac{3}{4}$ ton of steel and 2 tons of aluminum. Intelligence agents have been unable to determine how many of each type of weapon have been produced but they assume that all the aluminum and steel have been incorporated into the weapons. How many zingers and how many zonkers were produced?

Solution: Let x stand for the number of zingers produced and let y stand for the number of zonkers produced. Since each zinger needs 1 ton of steel, the amount of steel occurring in the zingers totals $(1)(x)$ tons. Since each zonker needs $\frac{3}{4}$ ton of steel, the total amount of steel used in zonkers is $(\frac{3}{4})(y)$ tons. Assuming that the total of the steel in the zingers and in the zonkers is equal to the amount of steel delivered, namely 870 tons, we can write the equation $870 = x + \frac{3}{4}y$. We total the aluminum in the same way: x zingers, each using $\frac{1}{2}$ ton of aluminum, account for $\frac{1}{2}x$ tons, while the y zonkers, each using 2 tons of aluminum, use a total of $2y$ tons. Consequently, $500 = \frac{1}{2}x + 2y$. We now have the following system of simultaneous linear equations, which we solve as previously described:

$$\tfrac{1}{2}x + 2y = 500$$

$$x + \tfrac{3}{4}y = 870$$

Multiplying the first equation by 2 and substracting the second from it gives

$$\tfrac{13}{4}y = 130$$

or

$$y = 40$$

Substituting in the equation $x + \frac{3}{4}y = 870$ gives

$$x + (\tfrac{3}{4})(40) = 870$$

or

$$x = 840$$

Thus, we conclude that 840 zingers and 40 zonkers were produced.

An important part of the modeling here was the modeling assumption that all the steel and aluminum was used. Using this assumption we were able to convert our practical problem to the mathematical problem of solving a system of two simultaneous equations. In this example this seemed reasonable. The next example makes a similar assumption but perhaps with less justification.

Example 3

A paper recycling company uses two materials, scrap paper and scrap cloth, to make two different grades of recycled paper. A single batch of grade A recycled paper is made from 4 tons of cloth and 18 tons of paper, while one batch of grade B needs 1 ton of cloth and 15 tons of paper. The company has 10 tons of scrap cloth and 66 tons of scrap paper on hand. If it wishes to use all its supplies of scrap cloth and scrap paper, how many batches of each grade should be produced?

Solution: If we let x stand for the number of grade A batches and y for the number of grade B batches, then $4x$ is the total number of tons of cloth in all the grade A batches, while y is the number of tons of cloth used altogether in the grade B batches. Since all the available cloth is used, $4x + y = 10$. We make a similar analysis with respect to scrap paper. The number of tons of scrap paper used altogether in all the grade A batches is $18x$, while $15y$ is the total number of tons of scrap paper used altogether in all the grade B batches. Since all the scrap paper is to be used, we must have $18x + 15y = 66$. Writing these equations in a system,

$$4x + \quad y = 10$$

$$18x + 15y = 66$$

To solve this system, multiply the first equation by 15 and subtract the second to obtain

$$42x = 84$$

or

$$x = 2$$

Substituting in the first equation gives $8 + y = 10$, whence $y = 2$.

The last two examples illustrate the kind of problem we have in mind to model here. They are problems in which the following assumptions are reasonable:

ASSUMPTION 1

Fixed amounts of two resources are at hand (aluminum and steel in Example 1 and scrap cloth and scrap paper in Example 2).

ASSUMPTION 2

Two different kinds of things can be produced from these resources by using them in different proportions (zingers and zonkers in Example 1, grade A and grade B paper in Example 2).

ASSUMPTION 3

All the resources available are used.

Assumption 3 will be called the *no slack* assumption.

ASSUMPTION 4

It is desired to determine how many of each kind of product must be made in order to meet Assumption 3, that is, to avoid having any resources left over.

Assumption 3 is a modeling assumption that may not always make sense. In the last problem, for example, the solution we derived based on Assumption 3 would not be practical if the selling price for grade B paper were very low. In that case it might be better to have some of the scrap cloth and/or scrap paper left over but to make as many grade A batches as possible. To be specific, suppose a grade A batch sells for $10,000, while a grade B batch sells for $2000. If we follow the solution just derived, we take in $24,000. Suppose, instead, that we make $2\frac{1}{2}$ batches of grade A paper and none of grade B; then we take in $25,000, which is better. Of course, you should check that it is possible to make as many as $2\frac{1}{2}$ batches of grade A, i.e., that there is enough scrap cloth and scrap paper to go around. How much scrap cloth and scrap paper would be left over? The sort of problem we raise here as a criticism of modeling assumption 3 will be discussed more fully in Section 6.5.

The procedure for solving problems modeled in this way is this:

(1) Let x be the number of units of one type of product produced and let y be the number of units of the second product produced.

(2) Corresponding to each resource there is a linear equation involving x and y. This equation expresses algebraically the fact that the total amount of that resource available equals the amount used in the entire batch of the first product plus the amount used in the entire batch of the second product.

(3) Solve the equations simultaneously.

Our final example illustrates this procedure once more but this time with two additional twists. The first twist concerns the resources. The word *resources* usually connotes some kind of commodity such as agricultural produce or something mined from the earth. In reality we mean the word a bit more generally. In Example 4 resources will be money and manpower. We shall discuss the second twist at the end of the example.

Example 4

The Mogul Film Company has 45 million dollars and 200,000 man-hours of staff time with which to make films this year. To make a grade A film requires 3 million dollars and 15,000 man-hours of time, while a grade B film requires 1 million dollars and 10,000 man-hours. How many films of each type should be made?

Solution: The example fits Assumptions 1 and 2. We shall make the additional assumption (Assumption 3) that the most efficient and profitable operating procedure would be to use all the money and manpower. Let x denote the number of grade A films and y the number of grade B films. The resource equations are the following:

Money equation: $45,000,000 = 3,000,000x + 1,000,000y$

Manpower equation: $200,000 = 15,000x + 10,000y$

Simplifying these equations gives

$$45 = 3x + y$$

$$40 = 3x + 2y$$

Subtracting from the second to the first gives

$$5 = -y \quad \text{or} \quad -5 = y$$

Substituting -5 for y in the first equation gives $50 = 3x$. Hence, $x = 16\frac{2}{3}$. Thus, the solution is to make $16\frac{2}{3}$ grade A films and -5 grade B films, which is absurd. For one thing, one can't make two-thirds of a film. Moreover, it is impossible to make a negative number of films. Thus, the second additional twist in this example is that after the theoretical solution is determined on the basis of the model, it is necessary to examine the solution to see if it makes sense. In this case we must conclude that our model for determining the company's policy was a bad one. A better model will be developed in Section 6.7.

EXERCISES 6.5

1. In each of the following cases, find the intersection point of the lines represented by the two equations if the lines have an intersection point. If there is no intersection, state whether we have a case of dual representation or whether the lines are parallel.

(a) $x + 3y = 1$ (d) $5x + 2y = 9$
             $3x - 4y = 4$ $-x - 3y = -7$

(b) $x - 2y = 7$ (e) $3x - 5y = 0$
            $-x + 2y = -7$ $x + y = 8$

(c) $x - 2y = 7$ (f) $125x + 500y = 1500$
            $-x + 2y = 8$ $50x - 200y = 200$

2. For each of the cases in Exercise 1, draw the graphs of the two lines whose intersection is sought and determine the point of intersection as closely as possible by visual inspection of the graph.

3. Do the following three lines have a single point in common?

$$x + y = 5$$

$$2x - 3y = 10$$

$$4x - 7y = 20$$

4. The Pony Express Shipping Company ships goods between Chicago, Ill., and St. Louis, Mo., using trucks and the railroad. The manpower requirements for sending off one truck are 1 man-hour of dispatcher's time and 3 man-hours of loader's time. The manpower requirements for sending off one railroad car are 1 man-hour of dispatcher's time and 5 man-hours of loader's time. The company has 70 man-hours of dispatcher's time next week and 300 man-hours of loader's time. It wishes to make use of all the man-hours available to it in each category. How many truck shipments and train shipments will accomplish this?

5. A dye company makes two shades of purple for the Easter season. A packet of deep purple is made by mixing 1 oz of red dye powder with 2 oz of blue. Light purple requires 2 oz of red and 3 oz of blue. The company has on hand 5000 oz of red and 7000 oz of blue. How much of each shade should be produced if the company wishes to use all its red and blue powder?

6. The antitrust division of the Justice Department has 28 tax lawyers and 48 antitrust lawyers. It decides to focus in the coming year on two types of situations: mergers and companies with a virtual monopoly of the market. A team of lawyers is set up for each situation. A team to investigate a merger requires 2 tax lawyers and 2 antitrust lawyers, while a team to investigate a quasi-monopoly needs one tax lawyer and 3 antitrust lawyers. If the division wishes to utilize all its lawyers, how many teams of each type should it create?

7. A college has 3200 man-hour units of professorial time and 400,000 units of floor space suitable for either classrooms or laboratories and other kinds of research facilities. The college wishes to determine how many professorial man-hours should be devoted to research and how many to teaching. For each man-hour devoted to teaching, 75 units of floor space are needed. For each man-hour devoted to research, 50 units of floor space are necessary. If the college wishes to use all man-hours and all floor space, how many man-hours should be devoted to research and how many to teaching?

8. A furniture firm makes two kinds of coffee tables using two basic machines, a planer and a molder. A fancy table requires 15 minutes

on the planer and 24 minutes on the molder, while a plain table needs 10 minutes on the planer and 18 minutes on the molder. In each day there are 600 minutes available on the planer and 500 on the molder. If all the machine time is to be used, how many tables of each type should be made?

9. The Little Varmint Undergarment Company makes two products, children's undershirts and children's underpants. Making a shirt requires just $\frac{1}{2}$ yd of cloth, while a pair of underpants needs $\frac{1}{4}$ yd of cloth and 18 in. of elastic. If the company has 200 yds of cloth on hand and 1440 in. of elastic, and if it wishes to use all the cloth and elastic, how many of each item should the company plan to make?

10. Camp Carraway has 30 counselors and 400 children. Two kinds of activities are planned to celebrate Watersports Day, a canoe trip down the river and a sailboat regatta. Camp tradition requires that a canoe contain 4 children and 2 counselors and that a sailboat contain 3 counselors and 30 children. If the camp wishes to have everyone participate in one of these activities, how many canoes and how many sailboats will be needed?

6.6   general systems of linear equations

The problems that we solved using a system of two equations in two letters lacked realism for several reasons. Problems that arise in the real world, like the construction of an armaments system of zonkers and zingers, might require hundreds of letters rather than just two letters to represent the raw materials needed to construct them. Furthermore, there might be a need to construct more than two types of armaments (i.e., zingers, zonkers, zappers, ...). As a step toward further realism, in this section we shall show an efficient method for solving systems of equations that involve many unknowns. To keep the discussion simple, we shall restrict the examples to easy cases but attempt to make clear how one proceeds in the general case.

Example 1

An armaments system consists of three types of weapons—zingers, zonkers, and zappers. A zinger requires 1 ton of aluminum and 2 tons of titanium. Each zonker requires 3 tons of aluminum and 3 tons of titanium. One zapper requires 2 tons of aluminum and 4 tons of titanium. If a munitions factory has 31 tons of aluminum and 41 tons of titanium, how many zingers, zonkers, and zappers can be produced if all the metallic materials available are used?

Solution (Started):   Let x stand for the number of zingers, y for the number of zonkers, and z for the number of zappers produced. Reasoning as in Example 2 on p. 193, we are led to the equations

$$x + 3y + 2z = 31$$
$$2x + 3y + 4z = 41$$

(6.10)

We see that the problem described above leads us to a system of two equations in three letters. Before proceding to discuss an algorithmic method for the solution of such a system, let us point out that these equations can be given a geometric interpretation. Suppose we consider three mutually perpendicular lines. These three lines (Figure 6.20) are known as the x, y, and z axes. The arrow on the line indicates the direction along the line that is positive. Each point in three-dimensional space can be located by giving a triple of numbers (x, y, z), which measures the perpendicular distance of that point from the plane determined by two of the lines. In Figure 6.20, the points

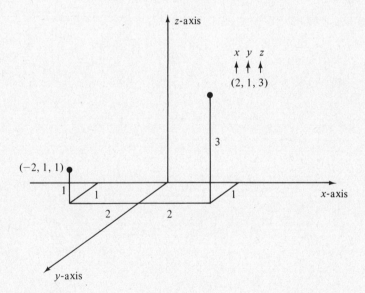

Figure 6.20

$(2, 1, 3)$ and $(2, -1, 1)$ can be located as shown. Just as an equation such as $x + y = 3$ turned out to be a line when plotted in a two-dimensional system, it can be shown using an analysis similar to that in Section 6.4, that an equation of the form $Ax + By + Cz = D$ when plotted in a three-dimensional coordinate system represents a plane.

Figure 6.21 shows a graph of the plane $k + 2y + 3z = 12$.

Unlike the situation for two equations in two letters, where the geometric patterns which can arise for the lines which represent the system are relatively simple, the situation for equations in three letters is much more complicated. We could go on to examine the geometry of linear equations in

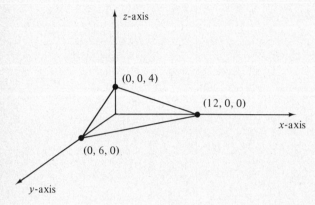

Figure 6.21

as many letters as we like. The equations we would get can be thought of as representing "hyperplanes" in higher dimensional spaces. Since we cannot draw any useful pictures of such hyperplanes, however, we shall return to the main discussion—the algebraic solution of systems of equations.

I. THE IDEA OF REDUCED ECHELON FORM

Basically, our methods rely on the same principles of manipulating an equation system that we have already used in the case of two equations and two unknowns. The three basic acceptable manipulations are:

Type 1

Multiply or divide an equation by a nonzero constant.

Type 2

Interchange two equations.

Type 3

Multiply or divide one equation by a nonzero constant and add to another equation.

What we mean by calling these manipulations acceptable is that they do not alter the solution set of the system. Therefore, if the transformed system is easier to solve than the original, we have made progress. As it turns out, it is possible to do these manipulations so as to find a system whose solution is obvious by inspection. The following example illustrates this.

Example 2

Solve the system

$$x_1 + x_2 - x_3 = 0 \tag{6.11}$$

$$3x_1 + x_2 - x_3 = 2 \tag{6.12}$$

$$2x_1 - x_2 + 2x_3 = 6 \tag{6.13}$$

Solution: First we add -3 times Equation (6.11) to Equation (6.12) and then we add -2 times Equation (6.11) to Equation (6.13). These are type 3 operations, which produce the following simpler system.

$$x_1 + x_2 - x_3 = 0 \tag{6.14}$$

$$-2x_2 + 2x_3 = 2 \tag{6.15}$$

$$-3x_2 + 4x_3 = 6 \tag{6.16}$$

Now we divide Equation (6.15) by -2, a type 1 operation.

$$x_1 + x_2 - x_3 = 0 \tag{6.17}$$

$$x_2 - x_3 = -1 \tag{6.18}$$

$$-3x_2 + 4x_3 = 6 \tag{6.19}$$

Now we add -1 times Equation (6.18) to Equation (6.17) and then add 3 times Equation (6.18) to Equation (6.19) to simplify further to

$$x_1 \qquad = 1 \tag{6.20}$$

$$x_2 - x_3 = -1 \tag{6.21}$$

$$x_3 = 3 \tag{6.22}$$

Finally, let us add Equation (6.22) to Equation (6.21) to produce a system that is its own solution.

$$x_1 \qquad = 1 \qquad\qquad (6.23)$$

$$x_2 \quad = 2 \qquad\qquad (6.24)$$

$$x_3 = 3 \qquad\qquad (6.25)$$

You may have noticed that the coefficients of the system are all that matters. The choice of multipliers in type 3 manipulations was made so that certain unknowns would drop out of certain equations, and this all depends on the coefficients. It is conventional to take advantage of this by "detaching" the coefficients and keeping track of them in a matrix containing no letters. We also put the constants from the right sides of the equations into the matrix but we separate them from the coefficients by a vertical line. The part of the matrix to the left of the line is called the *coefficient matrix*. The whole matrix is called the *augmented matrix*.

In the case of the system in Example 2, we have:

$$\left[\begin{array}{ccc|c} 1 & 1 & -1 & 0 \\ 3 & 1 & -1 & 2 \\ 2 & -1 & 2 & 6 \end{array}\right]$$

Now we can perform the same kinds of manipulations on the matrix as we did on the equations.

Type 1

Multiply or divide a row by a nonzero constant.

Type 2

Interchange two rows.

Type 3

Multiply or divide a row by a nonzero constant and add the result to another row.

Thus, corresponding to the manipulations of our system in Example 2, we have the following sequence of matrices. Since the original matrix for the system is

$$\left[\begin{array}{ccc|c} 1 & 1 & -1 & 0 \\ 3 & 1 & -1 & 2 \\ 2 & -1 & 2 & 6 \end{array}\right]$$

after adding -3 times row 1 to row 2 and adding -2 times row 1 to row 3, we find

$$\begin{bmatrix} 1 & 1 & -1 & \bigm| & 0 \\ 0 & -2 & 2 & \bigm| & 2 \\ 0 & -3 & 4 & \bigm| & 6 \end{bmatrix}$$

Now we divide row 2 by -2.

$$\begin{bmatrix} 1 & 1 & -1 & \bigm| & 0 \\ 0 & 1 & -1 & \bigm| & -1 \\ 0 & -3 & 4 & \bigm| & 6 \end{bmatrix}$$

Next we add -1 times row 2 to row 1 and we add 3 times row 3 to row 3 and find

$$\begin{bmatrix} 1 & 0 & 0 & \bigm| & 1 \\ 0 & 1 & -1 & \bigm| & -1 \\ 0 & 0 & 1 & \bigm| & 3 \end{bmatrix}$$

Finally, we add row 3 to row 2:

$$S = \begin{bmatrix} 1 & 0 & 0 & \bigm| & 1 \\ 0 & 1 & 0 & \bigm| & 2 \\ 0 & 0 & 1 & \bigm| & 3 \end{bmatrix}$$

Our general goal is to use manipulations of types 1, 2, and 3 to transform a matrix into a format somewhat like that of S. Unfortunately, since we can't always get the exact format of S, we may have to settle for formats like those of A, B, E, and F in Example 3. All of these can be subsumed under one general category called *reduced echelon form* (REF), which can be described as follows.

DEFINITION 4

An augmented matrix is in reduced echelon form (REF) provided the coefficient matrix part of it has these properties.

(1)   The rows of the coefficient matrix that are all 0 occur consecutively at the end of the matrix.

(2)   The leading nonzero coefficient of each row of the coefficient matrix is 1 unless all coefficients in that row are 0.

(3)   These "leading 1's" slant down to the right. More precisely, in any row the leading 1 occurs to the right of the leading 1 of any preceding row.

(4)   Each leading 1 is the only nonzero entry in its column.

Example 3

Of the matrices below, A, B, E, and F are in REF, but C, D, and G are not. Matrix C violates rule 2 above; matrix D violates rules 3 and 4, and matrix G violates rule 1.

$$A = \left[\begin{array}{ccc|c} 1 & 0 & 0 & 2 \\ 0 & 1 & 1 & 4 \\ 0 & 0 & 0 & 6 \\ 0 & 0 & 0 & 0 \end{array}\right] \qquad B = \left[\begin{array}{ccc|c} 1 & 0 & 0 & 1 \\ 0 & 1 & 0 & 2 \\ 0 & 0 & 1 & 4 \\ 0 & 0 & 0 & 3 \end{array}\right]$$

$$C = \left[\begin{array}{cccc|c} 4 & 0 & 0 & 2 & -1 \\ 0 & 0 & 1 & 0 & 2 \\ 0 & 1 & 0 & 3 & 3 \end{array}\right] \qquad D = \left[\begin{array}{ccc|c} 1 & 0 & 0 & 2 \\ 0 & 1 & 0 & -3 \\ 0 & 1 & 2 & 1 \\ 0 & 0 & 1 & 4 \end{array}\right]$$

$$E = \left[\begin{array}{cccc|c} 1 & 0 & 2 & 3 & 4 \\ 0 & 1 & -1 & 0 & 1 \end{array}\right] \qquad F = \left[\begin{array}{cc|c} 1 & 0 & 4 \\ 0 & 1 & 3 \\ 0 & 0 & 0 \end{array}\right]$$

$$G = \left[\begin{array}{ccc|c} 1 & 0 & 0 & 5 \\ 0 & 0 & 0 & 4 \\ 0 & 1 & 3 & 0 \end{array}\right]$$

There remain two important questions:

(1)   How do we transform a matrix to REF?

(2)   How do we interpret an REF matrix to find the solution to the original system?

The answer to question 1 is either to develop a "knack" for it or to study subsection III where an algorithm is given. Our next concern is question 2.

II. INTERPRETING AN REF MATRIX

There are four kinds of rows that appear in an REF matrix. Each kind is represented by one of the rows of matrix A of Example 3. The interpretations of these show how to proceed in general.

Row 1: $x_1 = 2$.

Row 2: $x_2 + x_3 = 4$; alternatively, $x_2 = 4 - x_3$. This may seem ambiguous since x_2 is not specified directly but defined in terms of x_3, which has no definite value. The explanation is that x_3 can be given any value whatever. For any such value, a corresponding value of x_2 that is part of a solution of the original system is determined by the formula $x_2 = 4 - x_3$.

Row 3: $0 = 0x_1 + 0x_2 + 0x_3 = 6$. This is absurd and indicates that the system has no solution. A single row of this type is enough to guarantee that there is no solution to the system.

Row 4: $0x_1 + 0x_2 + 0x_3 = 0$. This is neither absurd nor informative. It is totally irrelevant. Rows like this, consisting entirely of 0's, can always be disregarded.

Notice that for a matrix in REF form, if a column has no leading one among its entries then the letter corresponding to that column can be assigned any arbitrary value.

Example 4

Interpret the REF matrices B, E, and F of Example 3.

Solution:

Matrix B: There is no solution.

Matrix E: $x_1 = 4 - 2x_3 - 3x_4$

$x_2 = 1 + x_3$

x_3 and x_4 can take any values.

Matrix F: $x_1 = 4$, $x_2 = 3$.

III. AN ALGORITHM FOR REF

The achievement of the reduced echelon form depends entirely on the manipulations of certain elements of the matrix called *pivot elements*. These elements are the ones that eventually become the leading 1's of the rows in the REF matrix. In Example 2 the pivots were the entries on the main diagonal. In general, it is a little more difficult to determine the pivots than merely to pick the numbers on the main diagonal. Here is a description of how we pick the pivots and what we do with them. The algorithm is called *Gauss-Jordan elimination.*

THE FIRST PIVOT: The first pivot is the entry in the upper left position of the matrix. We can use whatever number originally occupies this position as long as it is not zero. If it is zero, we use a type 2 operation (interchange of rows, p. 204) to fill the upper left position with a nonzero element and use this number as a pivot.

WHAT TO DO WITH A PIVOT:

(1)   If the pivot is not 1, make it so by multiplying or dividing the row of the pivot by a suitable constant.
(2)   Use type 3 operations to make all other entries 0 in the pivot column. This is called *clearing the column.*

FINDING THE NEXT PIVOT: Suppose the previous pivot is in the ith row and jth column. Look for the first column after the jth column where there is a nonzero entry p in a row after the ith row. This entry p is the next pivot. However, it may not be in the $i + 1$ row. If necessary, use a type 2 operation (interchange of rows) to bring p into the $i + 1$ row.

Example 5

Reduce the following matrix to REF by Gauss-Jordan elimination.

$$M = \begin{bmatrix} 0 & 2 & 1 & \bigg| & 4 \\ -1 & 3 & -2 & \bigg| & -4 \\ 4 & 1 & 1 & \bigg| & 15 \end{bmatrix}$$

Solution: Since M has upper right entry 0, which is unsuitable as a pivot, we interchange the first two rows to produce a suitable pivot (circled).

$$\begin{bmatrix} \boxed{-1} & 3 & -2 & \bigg| & -4 \\ 0 & 2 & 1 & \bigg| & 4 \\ 4 & 1 & 1 & \bigg| & 15 \end{bmatrix}$$

Multiply the first row by -1 to make the pivot 1.

$$\begin{bmatrix} 1 & -3 & 2 & \bigg| & 4 \\ 0 & 2 & 1 & \bigg| & 4 \\ 4 & 1 & 1 & \bigg| & 15 \end{bmatrix}$$

Clear the column by adding -4 times the first row to the third row.

$$\begin{bmatrix} 1 & -3 & 2 & \vline & 4 \\ 0 & 2 & 1 & \vline & 4 \\ 0 & 13 & -7 & \vline & -1 \end{bmatrix}$$

The next pivot can be either 2 or 13. The 2 can be used in its present position, whereas the 13 would require an interchange, so we use the 2. We divide its row by 2 to make the pivot (circled) a 1.

$$\begin{bmatrix} 1 & -3 & 2 & \vline & 4 \\ 0 & ① & \frac{1}{2} & \vline & 2 \\ 0 & 13 & -7 & \vline & -1 \end{bmatrix}$$

Clear the column using the current pivot.

$$\begin{bmatrix} 1 & 0 & \frac{7}{2} & \vline & 10 \\ 0 & 1 & \frac{1}{2} & \vline & 2 \\ 0 & 0 & -\frac{27}{2} & \vline & -27 \end{bmatrix}$$

The next pivot can only be $-\frac{27}{2}$. Divide its row by $-\frac{27}{2}$ (equivalently, multiply by $-\frac{2}{27}$). Clearing the column gives

$$\begin{bmatrix} 1 & 0 & 0 & \vline & 3 \\ 0 & 1 & 0 & \vline & 1 \\ 0 & 0 & 1 & \vline & 2 \end{bmatrix}$$

From this REF matrix we can immediately read off the solution $x_1 = 3$, $x_2 = 1$, $x_3 = 2$.

Example 1 (continued)

We introduced this section with an armaments problem that led to solving the system

$$x + 3y + 2z = 31$$

$$2x + 3y + 4z = 41$$

Converting to matrix form and applying Gauss-Jordan elimination, we find the sequence of matrices

$$\begin{bmatrix} 1 & 3 & 2 & \vline & 31 \\ 2 & 3 & 4 & \vline & 41 \end{bmatrix}$$

$$\begin{bmatrix} 1 & 3 & 2 & \bigm| & 31 \\ 0 & -3 & 0 & \bigm| & -21 \end{bmatrix}$$

$$\begin{bmatrix} 1 & 3 & 2 & \bigm| & 31 \\ 0 & 1 & 0 & \bigm| & 7 \end{bmatrix}$$

$$\begin{bmatrix} 1 & 0 & 2 & \bigm| & 10 \\ 0 & 1 & 0 & \bigm| & 7 \end{bmatrix}$$

Thus

$$x = 10 - 2z$$

$$y = 7$$

$$z = \text{any value}$$

This system has an infinite number of solutions but since constructing fractions of zappers or negative numbers of zappers makes no sense, there are, in fact, only six realistic solutions. These six solutions are

$$x = 10 \qquad y = 7 \qquad z = 0$$

$$x = 8 \qquad y = 7 \qquad z = 1$$

$$x = 6 \qquad y = 7 \qquad z = 2$$

$$x = 4 \qquad y = 7 \qquad z = 3$$

$$x = 2 \qquad y = 7 \qquad z = 4$$

$$x = 0 \qquad y = 7 \qquad z = 5$$

The examples we have considered so far have only three letters but the method we described can be used in more general situations.

Example 6

Solve the system

$$x + y - z + w = 3$$

$$2x + 3y + z - 4w = 1$$

$$x + y + 2z + w = 3$$

$$3x - y + 4z + w = 3$$

$$5x + 2y + 5z - 3w = 4$$

Solution: Writing the system as a matrix and using Gauss-Jordan elimination, we find

$$
\begin{bmatrix}
① & 1 & -1 & 1 & | & 3 \\
2 & 3 & 1 & -4 & | & 1 \\
1 & 1 & 2 & 1 & | & 3 \\
3 & -1 & 4 & 1 & | & 3 \\
5 & 2 & 5 & -3 & | & 4
\end{bmatrix}
\longrightarrow
\begin{bmatrix}
1 & 1 & -1 & 1 & | & 3 \\
0 & ① & 3 & -6 & | & -5 \\
0 & 0 & 3 & 0 & | & 0 \\
0 & -4 & 7 & -2 & | & -6 \\
0 & -3 & 10 & -8 & | & -11
\end{bmatrix}
$$

(the pivot is circled)

$$
\longrightarrow
\begin{bmatrix}
1 & 0 & -4 & 7 & | & 8 \\
0 & 1 & 3 & -6 & | & -5 \\
0 & 0 & ③ & 0 & | & 0 \\
0 & 0 & 19 & -26 & | & -26 \\
0 & 0 & 19 & -26 & | & -26
\end{bmatrix}
\longrightarrow
\begin{bmatrix}
1 & 0 & -4 & 7 & | & 8 \\
0 & 1 & 13 & -6 & | & -5 \\
0 & 0 & ① & 0 & | & 0 \\
0 & 0 & 19 & -26 & | & -26 \\
0 & 0 & 19 & -26 & | & -26
\end{bmatrix}
$$

$$
\longrightarrow
\begin{bmatrix}
1 & 0 & 0 & 7 & | & 8 \\
0 & 1 & 0 & -6 & | & -5 \\
0 & 0 & 1 & 0 & | & 0 \\
0 & 0 & 0 & -26 & | & -26 \\
0 & 0 & 0 & -26 & | & -26
\end{bmatrix}
\longrightarrow
\begin{bmatrix}
1 & 0 & 0 & 7 & | & 8 \\
0 & 1 & 0 & -6 & | & -5 \\
0 & 0 & 1 & 0 & | & 0 \\
0 & 0 & 0 & ① & | & 1 \\
0 & 0 & 0 & -26 & | & -26
\end{bmatrix}
$$

$$
\longrightarrow
\begin{bmatrix}
1 & 0 & 0 & 0 & | & 1 \\
0 & 1 & 0 & 0 & | & 1 \\
0 & 0 & 1 & 0 & | & 0 \\
0 & 0 & 0 & 1 & | & 1 \\
0 & 0 & 0 & 0 & | & 0
\end{bmatrix}
$$

Hence the solution is $x = 1$, $y = 1$, $z = 0$, and $w = 1$.

EXERCISES 6.6

1. Solve the following system of equations:
   (a) $x + y - z = 1$
   $x - 3y + 2z = 0$
   $2x + y + z = 4$

(b) $2x + y - 3z = 3$
$x + y - z = 2$
$x + 2z = 1$

(c) $x - y = 5$
$x + z = 7$
$y - z = 0$

(d) $x + y - z + w = 2$
$x - y + 3z - w = 0$
$x - y + z + w = 0$
$x + 2y + z - w = 3$

2. Each of the systems of equations below has either
   (a)   no solution or
   (b)   infinitely many solutions.

Use your knowledge of Gauss-Jordan elimination to determine if part (a) or (b) holds.

(i) $x + y - z = 6$
$x - 3y + 4z = 8$
$2x - 2y + 3z = 14$

(ii) $x - y + 3z = 5$
$x + 2y - z = 6$
$2x + y + 2z = 10$

(iii) $x + y = 7$
$x - y + z = 8$
$x + z = 15$

(iv) $x - y + 3z = 2$
$3x - y + 4z = 8$
$5x - 3y + 10z = 12$

(v) $x - y + z = 6$
$x - 3y + 2z = 3$
$3x - 5y + 4z = 0$

(vi) $x - y = 4$
$x + z = 3$
$2x - y + z = 7$

3. A dietitian is attempting to meet the health needs of an army camp. Each soldier needs at least 12 units of vitamins, 8 units of proteins, and 8 units of carbohydrates for lunch. Suppose one glass of beverage X provides 1 unit of vitamins, 2 units of proteins, and 3 units of carbohydrates; one Zano sandwich provides 4 units of vitamins, 12 units of proteins, and 1 unit of carbohydrates; and one piece of cake provides 3 units of vitamins, 2 units of proteins, and 3 units of carbohydrates. What should a lunch consist of if the health needs are to be met exactly?

4. There are 52 tables in three rooms. The number of tables in the second room is one-half the number in the first. The first room has 48 sq ft of

floor space per table. The corresponding figures for the second and third rooms are 46 and 45 sq ft, respectively. If there are 2,424 sq ft in the three rooms, how many tables are there in each room?

5. A salesman is allowed 10 ¢/mi for the use of his car, $10/day for meals, and $15/night for a motel room. On a certain trip he averages 120 mi/day and his motel bill was $10 more than he spent for meals. His total bill was $160. Find the number of days and the number of miles he traveled and the number of nights he spent in a motel.

6.* Show that if (x_0, y_0, z_0) is a solution of a system, and if type 1, 2, or 3 operations are performed on the system, then the resulting new system has (x_0, y_0, z_0) as a solution.

6.7 inequalities and feasible regions

We have studied the problem of combining several kinds of resources into several kinds of products and we provided a method, based on the *no slack* assumption, which sometimes yields a reasonable solution and sometimes does not. It is desirable to have a method that always gives reasonable results. In this section we erect part of the scaffolding for an improved method that is one of the mainstays of modern applied mathematics, the technique of *linear programming*.

Our first step is to study linear inequalities and their solution sets. We are all familiar with the use of the sign "$=$" to indicate that two expressions are equal. We now recall the following inequality signs.

		Examples
$a < b$	means a is less than b.	$2 < 3$
$a \leq b$	means a is less than or equal to b.	$2 \leq 3$, or $3 \leq 3$
$a > b$	means a is greater than b.	$5 > 4$
$a \geq b$	means a is greater than or equal to b.	$5 \geq 4$, or $5 \geq 5$

Given an inequality such as $3x - 2y < 5$, we shall be interested in determining the solution set of the inequality. We say that a point (a, b) belongs to the solution set of $3x - 2y < 5$ if $3a - 2b < 5$. Thus $(0, 0)$ belongs to the solution set of $3x - 2y < 5$, while $(5, 1)$ does not belong to the solution set of $3x - 2y < 5$ since $3(5) - 2(1) = 13$, which is not less than 5.

We are accustomed to being able to perform certain operations on equations, such as multiplying (or dividing) both sides of the equation by the same constant. For example, the equation $-x + y = -7$ can be multiplied by -1 to produce $x - y = 7$, which is an equation with the same solution set. We may ask whether some similar manipulation would allow us to change the form of an inequality like $-x + y < 7$ into another inequality with the same solution set. It is also possible to transform an equation into an equivalent equation by adding the same constant to both sides of the equation. Again we may ask whether corresponding manipulations are valid for inequalities. It turns out that similar rules do exist for inequalities but they are not quite identical. The proper values are listed in the chart below with examples. To prove these rules would require a careful investigation of the number system, which we wish to avoid. We shall accept them on a commonsense basis.

Table 6.5   Rules for Transforming an Inequality into Another Inequality with the Same Solution Set.

	Examples
1. Multiply (or divide) both sides by a positive constant.	Multiply $2 < 3$ by 4 to obtain $8 < 12$. Divide $3x > 9$ by 3 to obtain $x > 3$.
2. Multiply (or divide) both sides by a negative constant and change the sense of the inequality (from $>$ to $<$ and vice versa or from $\geq$ to $\leq$ and vice versa.)	Multiply $-x + y \leq 7$ by -1 to obtain $x - y \geq -7$. Divide $-5x + 25 > 50$ by -5 to obtain $x - 5 < -10$.
3. Add (or subtract) the same constant to (from) both sides of the inequality.	Add 5 to both sides of $x - 5 < 10$ to obtain $x < 15$. Subtract 2 from both sides of $x + y > 2$ to obtain $x + y - 2 > 0$.
4. Add (or subtract) a multiple of a variable to (from) both sides of the inequality.	Subtract $2x$ from both sides of the inequality $3x > 2x + 2$ to obtain $x > 2$. Add y to both sides of the inequality $3x - y \geq y + 4$ to obtain $3x \geq 2y + 4$.

Example 1

Simplify these inequalities:

(1) $2x < 10$
(2) $2x - 3 > 7$
(3) $-3x + 7 < 2x + 4$
(4) $2x - y > 4 + y - 3x$

Solution:

(1) $2x < 10$
    $x < 5$ by rule 1
(2) $2x - 3 \geq 7$
    $2x \geq 10$ by rule 3
    $x \geq 5$ by rule 1

(3) $-3x + 7 < 2x + 4$

$-3x < 2x - 3$ by rule 3

$-5x < -3$ by rule 4

$x > \frac{3}{5}$ by rule 2

(4) $2x - y > 4 + y - 3x$

$5x - 2y > 4$ by rule 4

$5x - 2y - 4 > 0$ by rule 3.

We are particularly interested in linear inequalities in two variables, that is, inequalities of the form $Ax + By + C \leq 0$ where A, B, and C are constants and where A and B are not both 0. Such an expression resembles the equation $Ax + By + C = 0$. Indeed, when we plot the points (x, y) that satisfy the inequality $Ax + By + C \leq 0$, we shall discover that this solution set is closely related to the line that is the solution set of $Ax + By + C = 0$.

Let us take some particular examples.

(1) Choosing $A = -1$, $B = C = 0$, we obtain the inequality $-x \leq 0$, which can be transformed by rule 2 to $x \geq 0$. Clearly, the points (x, y) that satisfy $x \geq 0$ are those on and to the right of the (vertical) Y axis [see Figure 6.22(a)]. Note that the line with equation $x = 0$ is part of the solution set.

(2) Choosing $A = 0$, $B = -1$, $C = 0$, we obtain the inequality $-y \leq 0$, which can be transformed to $y \geq 0$. Clearly, the points (x, y) that satisfy $y \geq 0$ are all those on and above the (horizontal) X axis [see Figure 6.22(b)]. Note that the line with equation $y = 0$ is a part of the solution set.

(3) For a more complicated example, consider the inequality $4x + y \leq 10$. First notice that any point that satisfies $4x + y = 10$ also satisfies

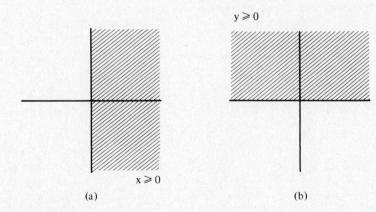

(a) $x \geq 0$

(b) $y \geq 0$

Figure 6.22

$4x + y \leq 10$. Thus, part of the solution set of $4x + y \leq 10$ is the line whose equation is $4x + y = 10$. But there is more, for suppose (x_0, y_0) is any point on the line $4x + y = 10$. That means it satisfies the equation of the line, that is, $4x_0 + y_0 = 10$. If we decrease y_0 by any *positive* amount e to a new value $y_0 - e$, the point $(x_0, y_0 - e)$ still satisfies the inequality since $4x_0 + (y_0 - e) = (4x_0 + y_0) - e = 10 - e \leq 10$. Paraphrasing this in words, any amount we move in a vertical direction downward from any point on the line $4x + y = 10$ keeps us in the solution set of $4x + y \leq 10$. Similarly, it is easy to check that if we move any amount upward to a point $(x_0, y_0 + e)$, we are no longer in the solution set of $4x + y \leq 10$ since $4x_0 + (y_0 + e) = (4x_0 + y_0) + e = 10 + e > 10$. Thus, our solution set consists precisely of those points on or below the line $4x + y = 10$. As this kind of analysis can be provided for any line, we always come to this conclusion:

The solution set of $Ax + By + C \leq 0$ (or $Ax + By + C \geq 0$) consists of all the points on the line $Ax + By + C = 0$ together with all points on one side of the line, which side needs to be determined by further analysis.

Because of this principle we refer to the solution set of the inequality $Ax + By \leq C$ as a *closed half plane*. The word *closed* signifies that the line $Ax + By = C$ is part of the solution set.

Given the inequality $Ax + By \leq C$, the best procedure for determining which side of the line $Ax + By = C$ constitutes the solution set is to take any point not on the line and test whether or not it satisfies the inequality. If it does, shade the side of the line containing that test point. Otherwise, shade the side of the line that does not contain the test point. [The point $(0, 0)$ makes a good test point if it does not lie on the line.] If we apply this test in the case of $4x + y \leq 10$, we find the same result as in our earlier analysis involving movement upward and downward from the line, namely, that the portion of the plane under the line $4x + y = 10$ constitutes the remainder of the solution set (see Figure 6.23). To be specific, using $(0, 0)$ as a test point (since it doesn't lie on the line) and substituting it into the inequality, we obtain $0 \leq 10$. The inequality is satisfied and we shade the side of

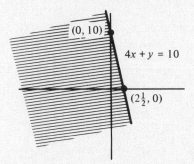

Figure 6.23

the line containing (0, 0). It is preferable to use the test-point method rather than the downward and upward movement method because it is faster and simpler.

Example 2

Plot the solution set of $18x + 15y \leq 66$. Our first step is to simplify the inequality by dividing by 3 to give $6x + 5y \leq 22$. The second step is to plot the equation of the line $6x + 5y = 22$ (see Figure 6.24).

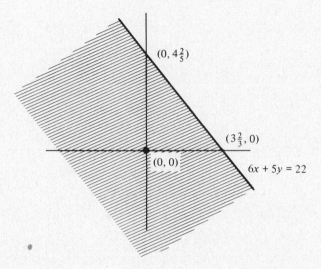

Figure 6.24

Next we look for a convenient test point not on the line. We choose (0, 0) and see whether it satisfies the inequality $6x + 5y \leq 22$ when substituted into it. Indeed, $6(0) + 5(0) \leq 22$ so the test point satisfies the inequality and we shade the side of the line containing (0, 0).

Now let us consider a system of linear inequalities and attempt to find the solution set of the system; that is, given a collection of inequalities, we seek the set of points that satisfy each of the inequalities in the collection. Take, for example, the system consisting of all the inequalities that we have separately analyzed in this section:

$$x \geq 0$$

$$y \geq 0$$

$$4x + y \leq 10 \qquad (6.36)$$

$$6x + 5y \leq 22$$

The fundamental principle in dealing with a system is this: *The solution set of the system is the intersection of the solution sets of the individual inequalities.*

Superimposing the various solution sets found for the individual members of the system above produces the shaded region in Figure 6.25 as the solution set for the system.

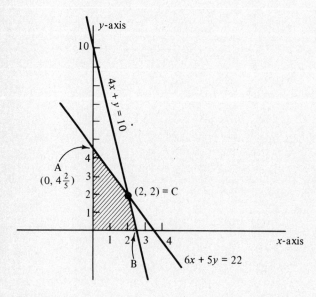

Figure 6.25

We shall now apply these techniques of dealing with inequalities to the resource allocation problems discussed earlier. We illustrate the method by reviewing the paper recycling problem and analyzing it from a new point of view.

Example 3

A paper recycling company uses two materials, scrap paper and scrap cloth, to make two different grades of recycled paper. A single batch of grade A recycled paper is made from 4 tons of cloth and 18 tons of paper, while one batch of grade B needs 1 ton of cloth and 15 tons of scrap paper. The company has 10 tons of scrap cloth and 66 tons of scrap paper on hand. What possible quantities of the two grades can the company produce?

Solution: As before, we let x stand for the number of grade A batches to be produced and y stand for the number of grade B batches. As before, the number of tons of cloth needed will be $4x + y$, while the amount of paper needed will be $18x + 15y$. Our first deviation from the *no slack* assumption is that instead of assuming that the amount of scrap cloth or paper used is equal to what is available, we assume that it *does not exceed* what is available. In other words, instead of dealing with resource equations, we deal with resource inequalities:

$$4x + y \leq 10$$

$$18x + 15y \leq 66$$

The first inequality states that the amount of scrap cloth to be used is not greater than the amount of scrap cloth available. The second inequality says that the amount of scrap paper to be used is not greater than the amount of scrap paper available.

In addition, we wish to rule out negative numbers of batches so we include these two inequalities:

$$x \geq 0$$

$$y \geq 0$$

We shall refer to these last inequalities as the *physical reality* inequalities. Taking these four inequalities together gives the system [Equation (6.36)], which we have already discussed and whose solution set is graphed in Figure 6.25. We call the shaded region of Figure 6.25 the *feasible region* for this allocation problem because the points of this region represent production goals for the two grades of paper that are feasible, in the sense that no negative quantities are called for and the amount of the two resources available are sufficient to fulfill the production goals. To be specific, the point (1, 2) is a member of the feasible region. To verify this, note that if we produce one batch of grade A and two batches of grade B, we can calculate that we shall need 6 tons of scrap cloth and 48 tons of scrap paper. Since we have 10 tons of scrap cloth and 66 tons of scrap paper, we have enough to fulfill the production goals and still have some left over (the so-called *slack*). It may be intelligently objected that (1, 2) is feasible but it is not optimal since there is so much left over (4 tons of scrap cloth, and 18 tons of scrap paper). In fact, at this point the feasible region presents us with an embarrassment of riches since it contains an infinity of feasible production goals and we have not found any way to pick out a best one. We shall work on this in the next section. For the moment, however, it is worth noting that if we wish at this point to readopt the no slack assumption (Assumption 3 of Section 6.5) in this problem, this would be a means of picking one production goal out of the feasible region. The point labeled C [= (2, 2)] in Figure 6.25 is the no slack solution and it does lie in the feasible region.

Example 4

The Mogul Film Company (see Section 6.5) has 45 million dollars and 200,000 man-hours of staff time with which to make films this year. To make a grade A film requires 3 million dollars and 15,000 man-hours of time, while a grade B film requires 1 million dollars and 10,000 man-hours of time. How many films of each type should be made?

Solution: If x is the number of grade A films to be made and y is the number of grade B films, then the total amount of money spent will be $(3,000,000)x + (1,000,000)y$ and so our first resource inequality will be $(3,000,000)x + (1,000,000)y \leq 45,000,000$. The man-hours needed will be $(15,000)x + (10,000)y$; this gives our second resource inequality, $(15,000)x + (10,000)y \leq 200,000$. Since one cannot make a negative number of films, we shall add the physical reality inequalities: $x \geq 0$ and $y \geq 0$. Figure 6.26 shows the feasible region for this allocation problem.

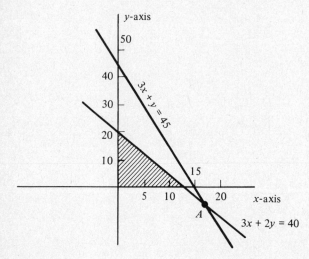

Figure 6.26

As in the last example, the feasible region consists of an infinite number of points and we have no way of choosing the best. In this case, however, if we fall back upon the no slack assumption to try to narrow the choice down to one point, we are unsuccessful. This is because the point determined by the no slack assumption is A, which is not in the feasible region since it involves a negative number of films. Also, it is worth noting that many points in the feasible region correspond to a fractional number of films, another reason why choosing a solution from the feasible region is not so simple.

EXERCISES 6.7

1. Simplify and then graph each of the following inequalities:
   (a) $x + y < -7 + 2x - 4y$
   (b) $12x - 4 > 11 - 3x$
   (c) $-7x + 12 > 13$
   (d) $-x - y - 3 < 10$
   (e) $x - 3(x + y) - 4(x - 3y) < x + y$
   (f) $x + 3(x - 4) < 8(x - 3)$
   (g) $4(x - 3) < 2[x - 3(x - 2)]$

2. (a) Find the feasible region for Exercise 4 of Section 6.5 if the no slack assumption is not made.
   (b) Do the same for Exercise 5 of Section 6.5.
   (c) Do the same for Exercise 6 of Section 6.5.
   (d) Do the same for Exercise 7 of Section 6.5.
   (e) Do the same for Exercise 8 of Section 6.5.
   (f) Do the same for Exercise 9 of Section 6.5.
   (g) Do the same for Exercise 10 of Section 6.5.

3. For each part of Exercise 2, determine if the no slack solution belongs to the feasible region.

6.8   linear programming

One fact that we have not discussed extensively in connection with resource allocation problems is the matter of the profits which can be gained from the two kinds of products which can be produced. Not surprisingly, it can be useful to take this into account. We shall, in fact, use this as the criterion for choosing a point out of the feasible region.

For definiteness, let us return to the paper recycling example whose feasible region is shown in Figure 6.25. We shall consider three different profit assumptions and see how each leads us to select a point from the feasible region as the best production goal.

PROFIT ASSUMPTION 1

Suppose there is no profit on grade B batches but $500 profit on each grade A batch of recycled paper.

We naturally wish to make as many grade A batches as we can. In terms of the feasible region, we want to select a point with the largest possible x coordinate. This point would be $B = (2\frac{1}{2}, 0)$. The profit would total $1250. Notice that we would use all the scrap cloth but only 45 tons of the scrap paper, leaving 21 ($=66 - 45$) tons of scrap paper unused. In estimating the fit of this solution, we would need to know whether it is possible to have a fraction of a batch of paper.

George Dantzig is one of the pioneering creaters of linear programming, which is perhaps the most important development in applied mathematics in the last half-century. (*Photo courtesy of George Dantzig.*)

PROFIT ASSUMPTION 2

Suppose that the grade A batches are the ones with no profit but that there is $1000 profit on each grade B batch.

In this case, it is apparent that we should simply make as much grade B as possible; that is, we want to choose a point (x, y) from the feasible region that has as large a y coordinate as possible. This would be the point A, where $y = 4\frac{2}{5}$ and $x = 0$. The profit here is $4400. Can you calculate how much scrap paper or cloth is left as slack if these production goals are instituted?

Before leaving this simple discussion, let us suppose that for some reason we chose the point $(0, 3)$ for our production goal. Now our profit would be $3000. It turns out that there are other points in the feasible region where we can obtain the same profit, for example, $(1, 3)$. In fact, any point (x, y) that is in the feasible region and satisfies the equation $y = 3$ gives the same $3000 profit. This set of points is just the intersection of the feasible region and the line $y = 3$.

PROFIT ASSUMPTION 3

Suppose now that there is $1000 profit on grade A batches and $500 on grade B.

In this case, it seems likely that some combination of the two types would yield the most profit. But which combination? If x and y denote, as usual, the number of batches of grade A and B, respectively, then our profit, denoted P, will be expressed by the equation

$$P = 1000x + 500y \qquad (6.37)$$

Let us for a moment put the cart before the horse; we shall decide what profit we would like and then see if we can get it. If we aim for $P = 5000$, then we have $5000 = 1000x + 500y$, which can be simplified to

$$10 = 2x + y$$

The graph of this equation is a straight line shown in Figure 6.27. It is called the profit line corresponding to the value 5000. The points on this line represent production goals that yield a profit of exactly \$5000. Unfortunately, none of these points is in the feasible region so the profit of \$5000 is unattainable with the resources at hand. Let us be more modest and ask for a profit of \$2000. Our profit equation is now $2000 = 1000x + 500y$, which simplifies to $4 = 2x + y$. This profit line is also shown on Figure 6.27 and we note that it intersects the feasible region in a line segment. Any point on this

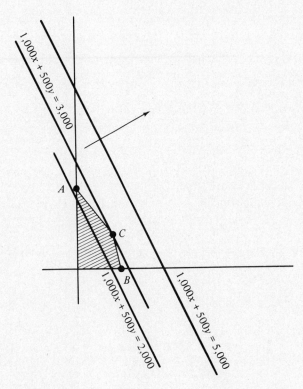

Figure 6.27

segment represents a production goal that is feasible and gives $2000 profit. Thus, we see that while $5000 was too much to hope for, $2000 can be achieved. We can do even better however. As we increase our desired profit, our profit lines move in the direction of the arrow in Figure 6.27. These profit lines will continue to intersect the feasible region until we reach a profit of $3000 at which point the profit line whose equation is $1000x + 500y = 3000$ touches the feasible region at the corner point C $(2, 2)$. If the desired profit is increased further, the profit line will no longer intersect the feasible region. Thus $3000 is the largest profit that can be achieved and it can be achieved only by using the production goals corresponding to the point C $(2, 2)$, namely, two batches of grade A and two batches of grade B.

Let us abstract from this example those important features that make it a linear programming problem. We have a practical problem of deciding how much of each of two types of products to produce with fixed resources at hand. The possible production goals (x, y), which are thought of as points in the plane, form a feasible region that we determine from inequalities that are derived from the statement of the problem. This feasible region will be an intersection of closed half planes and may be either bounded (see Figure 6.28) or unbounded (see Figure 6.29).

In the latter case, when the region "goes off to infinity," our analysis won't apply. All problems in this book deal with the bounded case. Now we have a linear profit function of the form $P = ax + by$, which we wish to

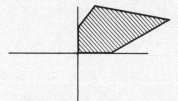

Figure 6.28

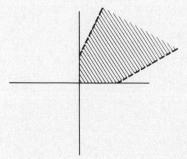

Figure 6.29

maximize. We can rely on the following principle to produce a point where the profit is the largest it can be in the feasible region:

THEOREM 1

A linear function $ax + by$ (a and b constant) achieves its maximum value over a bounded region that is the intersection of a finite number of closed half planes at one or more corner points of the region. (A *corner point* is a point that lies on two of the boundary lines of the region.)

We shall omit the proof of this theorem. It follows the ideas in the last example very closely, where we vary our profit lines in the direction of increasing profit until we reach a point where further increase of desired profit produces a profit line that completely misses the feasible region. Then we show that where further increase of desired profit is impossible, our profit line touches the region in a corner point or a boundary edge with corner points as end points.

This theorem suggests the following method of solving linear programming problems:

(1)   Draw the feasible region (as described in the previous section).
(2)   Determine the corner points of the feasible region. Since each corner point lies on two boundary lines, its coordinates can be found by solving the equations for these two lines simultaneously.
(3)   Calculate the profit at each of the corner points and select as the answer a point where the profit is largest.

In step 3 you may have noticed that we directed you to select "a" corner point with largest profit rather than "the" corner point with largest profit. The reason for this is that it can happen that there will be a whole boundary segment where the largest profit can be obtained. In that case there will be two corner points with largest profit, one at each end of the segment. This phenomenon can be observed in the recycling problem if we take as the profit function $P = 600x + 500y$. Here is the result of evaluating this profit function at the various corner points.

$$(0, 0) \qquad P = (600)0 + (500)0 = 0$$

$$(0, 4\tfrac{2}{3}) \qquad P = (600)0 + (500)(4\tfrac{2}{3}) = 2200$$

$$(2\tfrac{1}{2}, 0) \qquad P = (600)(2\tfrac{1}{2}) + (500)0 = 1500$$

$$(2, 2) \qquad P = (600)(2) + (500)(2) = 2200$$

The profit is the same at $(0, 4\tfrac{2}{3})$ as at $(2, 2)$ so either production goal can be chosen; in either case the profit will be 2200. The profit line for 2200 is shown in Figure 6.30.

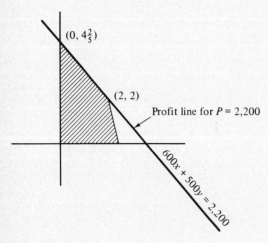

Figure 6.30

Example 1

On a certain day the sanitation department of a city finds itself with 180 collection trucks in operating condition and 480 men reporting for duty. Two types of collection teams can be formed from the two resources of men and trucks: a full-strength team consists of 1 truck and 3 men; a half-strength team consists of 1 truck and 2 men. It is desired to find out how many teams of each type to form if we assume that a full-strength team collects 10 tons of garbage in a day, while a half-strength team collects 5 tons, and if our goal is to maximize the garbage collected.

Solution: Let x denote the number of full-strength teams and let y denote the number of half-strength teams. Then we get resource inequalities corresponding to the resources men and trucks.

$$3x + 2y \leq 480 \qquad \text{(Men)}$$

$$x + y \leq 180 \qquad \text{(Trucks)}$$

$$x \geq 0 \qquad \text{(Physical reality inequalities)}$$

$$y \geq 0$$

When we go to find the profit function, we may be temporarily dismayed to note that the problem says nothing about profit. In place of profit, we are interested in maximizing something else, namely, tons of garbage collected, which we shall call G. We can work with this exactly as if it were a dollars and cents profit and write the equation

$$G = 10x + 5y$$

The feasible region is shown in Figure 6.31.

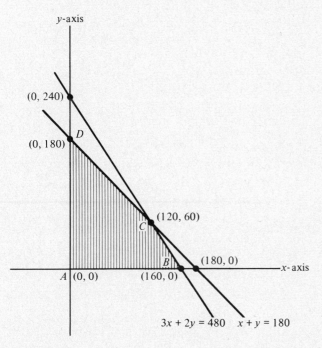

Figure 6.31

The corner points are computed as follows:

A: No computation needed. A is $(0, 0)$

B: Solve the following simultaneously. $\left\{\begin{matrix} y = 0 \\ 3x + 2y = 480 \end{matrix}\right\}$

C: Solve the following simultaneously. $\left\{\begin{matrix} 3x + 2y = 480 \\ x + y = 180 \end{matrix}\right\}$

D: Solve the following simultaneously. $\left\{\begin{matrix} x = 0 \\ x + y = 180 \end{matrix}\right\}$

These calculations produce the coordinates of the points as listed below. We then evaluate G at each of these points.

A (0, 0) $G = (10)0 + (5)0 = 0$

B (160, 0) $G = (10)(160) + (5)0 = 1600$

C (120, 60) $G = (10)(120) + (5)(60) = 1200 + 300 = 1500$

D (0, 180) $G = (10)0 + (5)(180) = 900$

The largest amount of garbage collected occurs at B, where we form 160 full-strength teams and no half-strength teams and collect 1600 tons of garbage. In this problem we were lucky that the optimal solution also turned out to be an integer solution. When the optimal solution is not integral but for physical reasons should be so, different methods must be employed. These methods belong to the area known as *integer programming*, which we shall not discuss.

In our next example we have more than two resource inequalities and, consequently, we have more corner points to calculate than in our examples thus far. In addition, the corner points are a bit harder to identify.

Example 2

Consider once again the paper recycling company, except that this time let us suppose that a batch of grade A paper requires, in addition to the 4 tons of scrap cloth and 18 tons of scrap paper, 3 tons of wood pulp. Similarly we suppose that a batch of grade B requires, in addition to 1 ton of scrap cloth and 15 tons of scrap paper, 9 tons of wood pulp. As before, we assume that we have 10 tons of scrap cloth and 66 tons of scrap paper available but we also assume that there are 27 tons of wood pulp. Assuming a profit of $700 on each grade A batch and a profit of $500 on each grade B batch, how many batches of each grade should be produced in order to maximize profit?

Solution: We have the following inequalities to deal with:

$$x \geq 0$$
$$y \geq 0$$
$$4x + y \leq 10$$
$$6x + 5y \leq 22$$

(These inequalities are the same as in Example 3 on p. 219.)

$$3x + 9y \leq 27 \qquad \text{(Wood pulp inequality)}$$

The function for the profit is $P = 700x + 500y$. As usual, the first task is to draw the feasible region. We have already found the feasible region corresponding to the first four inequalities so all we have to do, apparently, is graph the line $3x + 9y = 27$, pick out the closed half plane determined by it that corresponds to the inequality $3x + 9y \le 27$, and then intersect this half plane with the feasible region corresponding to the first four inequalities. After this we can find the corner points of the resulting feasible region. This is certainly what ought to be done but there is a difficulty not encountered before. It is this: Unless we are extremely careful about plotting the line $3x + 9y = 27$, we shall not be sure exactly how it cuts the other lines and so we shall not be exactly sure of the feasible region or its corner points. For example, each of Figures 6.32(a) and (b) looks as if it might represent the true

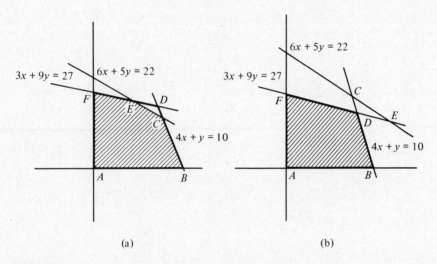

(a)   (b)

Figure 6.32

state of affairs. (The region enclosed in boldfaced line segments is the feasible region for the first four inequalities; the shaded region represents the feasible region for all the inequalities.) Which of these figures is correct? It makes a difference because in Figure 6.32(a), A, B, C, E, and F are all corner points, while in Figure 6.32(b) only A, B, D, and F are. Extremely careful drawing and measurement would overcome this difficulty but mathematicians dislike relying upon such methods and prefer the following approach. The question revolves around whether the intersection of lines $3x + 9y = 27$ and $4x + y = 10$, which is labeled D in both figures, is above or below the line $6x + 5y = 22$. Figure 6.32(a) corresponds to the case of D lying above

$6x + 5y = 22$, i.e., where the coordinates of D satisfy $6x + 5y > 22$. On the other hand, Figure 6.32(b) corresponds to the case where the coordinates of D satisfy $6x + 5y < 22$. This is a question that is easy to answer by explicit calculation. To find the coordinates of D, we solve the equation $3x + 9y = 27$ simultaneously with $4x + y = 10$ and arrive at the coordinate representation $(\frac{21}{11}, \frac{26}{11})$. We substitute into $6x + 5y$ and find $6(\frac{21}{11}) + 5(\frac{26}{11}) = \frac{256}{11} = 23.27$ (approximately). Since $23.27 > 22$, the point D lies above the line $6x + 5y = 22$ and Figure 6.32(a) is the correct one. Thus we know our corner points to be A, B, C, E, and F. Their values turn out to be

$A\ (0, 0)$ $P = (500)0 + (700)0 = 0$

$B\ (2\frac{1}{2}, 0)$ $P = (500)(2\frac{1}{2}) + (700)0 = 1250$

$C\ (2, 2)$ $P = (500)(2) + (700)(2) = 2400$

$E\ (\frac{21}{13}, \frac{32}{13})$ $P = (500)(\frac{21}{13}) + (700)(\frac{32}{13}) = 2531$ (approximately)

$F\ (0, 3)$ $P = (500)0 + (700)(3) = 2100$

The largest of these profits occurs at E and is approximately 2531.

EXERCISES 6.8

1. Solve Exercise 4 of Section 6.5 as a linear programming problem (do not make the no slack assumption), where the profit is $100 for each truck dispatched and $500 for each train dispatched.

2. Solve Exercise 5 of Section 6.5 as a linear programming problem (do not make the no slack assumption), where the profit is 5¢ for each packet of deep purple and 6¢ for each packet of light purple.

3. Solve Exercise 6 of Section 6.5 as a linear programming problem (do not make the no slack assumption), where each merger team will obtain five convictions and each monopoly team will obtain seven and where it is desired to maximize the total number of convictions obtained.

4. Solve Exercise 8 of Section 6.5 as a linear programming problem (do not make the no slack assumption), where each fancy table brings a profit of $20 and each plain table brings a profit of $15.

5. Solve Exercise 9 of Section 6.5 as a linear programming problem (do not make the no slack assumption), where each shirt brings 12¢ profit and each pair of underpants brings 10¢ profit.

6.9 other linear programming problems

The special nature of the problems we have been discussing (i.e., utilization of resources) may have obscured the fact that the principles of linear programming allow the solution of a very broad range of realistic problems. To illustrate this we give the solution to three problems that give the flavor of the many areas in which linear programming is useful. For simplicity, we restrict ourselves to the two-variable situations. Notice that one can solve either maximization or minimization problems using linear programming.

Example 1

A farmer has 400 acres on which he can grow corn or soybeans. Because of government regulation he cannot plant more than 300 acres of corn or more than 200 acres of soybeans. A further regulation requires that at least as much corn as soybeans be planted. If the profit from growing corn is $100 an acre and the profit from growing soybeans is $200 an acre, how much of each crop should he plant to maximize his profit?

Solution: Let x and y, respectively, denote the number of acres of corn and soybeans planted.

We have the following:

$$\left. \begin{aligned} x &\geq 0 \\ y &\geq 0 \end{aligned} \right\} \quad \text{(The physical reality inequalities)}$$

and

$$\left. \begin{aligned} x &\leq 300 \\ y &\leq 200 \\ y &\leq x \end{aligned} \right\} \quad \text{(By government regulation)}$$

$$x + y \leq 400 \qquad \text{(The farmer has a total of 400 acres)}$$

Figure 6.33 shows the region that results from plotting these inequalities.

The feasible region is bounded by four lines and has as its vertices $A = (0, 0)$, $B = (300, 0)$, $C = (300, 100)$, and $D = (200, 200)$.

The farmer's profit is given by

$$P = 100x + 200y$$

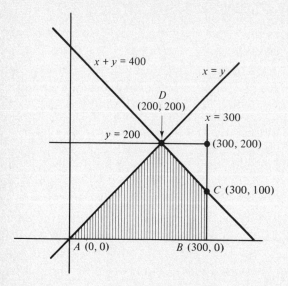

Figure 6.33

P evaluated at A is $100(0) + 200(0) = 0$

P evaluated at B is $100(300) + 200(0) = 30,000$

P evaluated at C is $100(300) + 200(100) = 30,000 + 20,000$
$$= 50,000$$

P evaluated at D is $100(200) + 200(200) = 20,000 + 40,000$
$$= 60,000$$

Hence the farmer should grow 200 acres of corn and 200 acres of soybeans.

Note that if the profit on corn was $100 and the profit on soybeans $50, then 300 acres of corn and 100 acres of soybeans should be grown.

Example 2

The dietitian of a summer camp for boys wishes to serve each camper a healthy lunch every day. She decides that for the entire summer she will serve each camper the same lunch, which consists of no more than 10 glasses of milk and no more than 10 Zamo sandwiches. Minimum health requirements for lunch for young boys are 9 units of vitamins, 22 units of proteins and 8 units of carbohydrates. A food supplier's manual shows that 1

232

glass of milk provides 3 units of vitamins, 4 units of proteins, and 1 unit of carbohydrates, while 1 Zamo sandwich provides 1 unit of vitamins, 3 units of proteins, and 2 units of carbohydrates. The dietitian knows that the cheaper a lunch is, the more likely it is that she will be hired again next year. How many sandwiches and glasses of milk should each lunch contain to satisfy the minimum health requirements above and to maximize the dietitian's reemployment prospects if a glass of milk costs 10¢ and a Zamo sandwich costs 25¢?

Solution: Let z and m denote the number of Zamo sandwiches and glasses of milk in a lunch, respectively. Then the amounts of vitamins, proteins, and carbohydrates provided by a lunch are

$$\text{Vitamins} = 3m + z$$

$$\text{Proteins} = 4m + 3z$$

$$\text{Carbohydrates} = m + 2z$$

In order for the lunch to be healthy we must have, therefore,

$$3m + z \geq 9 \qquad \text{(Vitamin requirement)}$$

$$4m + 3z \geq 22 \qquad \text{(Protein requirement)}$$

$$m + 2z \geq 8 \qquad \text{(Carbohydrate requirement)}$$

These inequalities, together with $0 \leq m \leq 10$ and $0 \leq z \leq 10$, give the feasible region shown in Figure 6.34. The cost C of a lunch of m glasses of milk and z sandwiches is

$$C = \$0.10m + \$0.25z$$

The corner points of the feasible region and their costs are

$$U = (0, 10) \qquad C = (\$0.10)(0) + (\$0.25)(10) = \$2.50$$

$$W = (0, 9) \qquad C = (\ 0.10)(0) + (\ 0.25)(9) = \$2.25$$

$$X = (1, 6) \qquad C = (\ 0.10)(1) + (\ 0.25)(6) = \$1.60$$

$$Y = (4, 2) \qquad C = (\ 0.10)(4) + (\ 0.25)(2) = \$0.90$$

$$Z = (8, 0) \qquad C = (\ 0.10)(8) + (\ 0.25)(0) = \$0.80$$

$$V = (10, 0) \qquad C = (\ 0.10)(10) + (\ 0.25)(0)\ = \$1.00$$

$$T = (10, 10) \qquad C = (\ 0.10)(10) + (\ 0.25)(10) = \$3.50$$

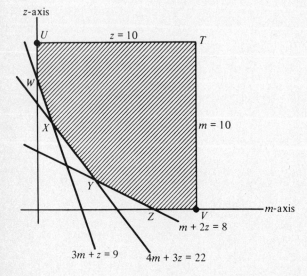

Figure 6.34

The lowest cost occurs at (8, 0). The dietitian followed this policy but to her great wonderment she was not rehired the next year.

Example 3

Mrs. Jones has a maximum of \$12,000 that she can invest in corporate bonds. Bond A returns 6% and bond B returns 10%. Since bond B is riskier, investment in bond B is restricted so that it cannot exceed 50% of the investment in bond A. How much should she invest in each type of bond to maximize her return?

Solution: Let x denote the amount of money invested in bond A and y the amount of money invested in bond B. The following inequalities must hold:

$$\left.\begin{array}{l} x \geq 0 \\ y \geq 0 \end{array}\right\} \qquad \text{(Physical reality inequalities)}$$

$$x + y \leq 12{,}000 \qquad \text{(Mrs. Jones has at most \$12,000 to invest)}$$

$$2y \leq x \qquad \text{(Restriction on investment of type B bond.)}$$

Figure 6.35 shows the graph of the feasible region.

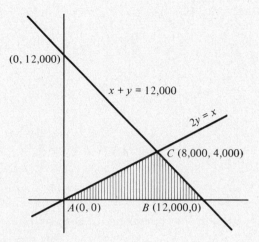

Figure 6.35

Mrs. Jones' return is given by

$$R = (.06)x + (.10)y$$

Evaluating R at the three vertices of the feasible region, $A = (0, 0)$, $B = (12,000, 0)$ and $C = (8000, 4000)$, we find

R evaluated at $A = 0$

R evaluated at $B = (.06)(12,000) + .10(0) = \$720$

R evaluated at $C = (.06)(8000) + .10(4000) = \$480 + \$400 = \$880$

Hence, Mrs. Jones should invest \$8000 in bond A and \$4000 in bond B to obtain her maximum return of \$880.

EXERCISES 6.9

1. A company that manufactures animal feed must produce 200-lb packages of a mixture of two grains, G_1 and G_2. G_1 costs \$4/lb and G_2 costs \$6/lb. Since too much G_1 is undesirable, no more than 80 lb of that

grain can be used in a package. Health requirements for the animals require that at least 150 lb of G_2 be contained in a package. What mixture should the company use to minimize cost?

2. The maximum daily production of a refinery is 1000 units. The refinery produces two kinds of gasoline—*A* and *B*. A minimum of 500 units of *A* and 300 units of *B* is required to meet the demands of a steady customer. The profit on *A* is $2.00 a unit and the profit on *B* is $4.00 a unit. What is the maximum profit that can be made?

3. A manufacturer can produce two types of stoves—*S* and *T*. Stove *S* can be sold for $50 and *T* can be sold for $80. For marketing, the stoves must first be shipped by rail and then by truck. Rail shipping rates are $3 for *S* and $2 for *T*. Truck shipping rates are $5 for *S* and $4 for *T*. Furthermore, total cost due to rail shipping *or* to truck shipping cannot exceed $40. How many of *S* and *T* should be manufactured so as to maximize potential gross sales?

4. The Pogal Company manufactures nipperkins and pipkins. Government contract requires at least 100 nipperkins and at least 300 pipkins per month. For each nipperkin produced at least two pipkins must be produced. A total of 200 man-hours is required to construct a nipperkin and 100 man-hours to construct a pipkin. The company has a minimum work force of 600 and a maximum work force of 1000. Each worker puts in 150 hr/month. The profit from a nipperkin is $1000 and the profit from a pipkin is $2000. How many nipperkins and pipkins must be produced each month for maximum profit?

5. The Custom Clothes Corporation makes dresses and blouses. The profit on a dress is $8 and on a blouse, $6. Both dresses and blouses require the labor of designers and tailors. There are 60 hr of designer time available and 48 hr of tailor time. Designers require 4 hr to process a dress and 2 hr to process a blouse. Tailors process a dress in 2 hr and a blouse in 4 hr. How many dresses and blouses should be produced to maximize profit?

SUGGESTED READING

GASS, S., *An Illustrated Guide to Linear Programming*, McGraw-Hill Book Co., New York, N.Y., 1970. A whimsical, nontechnical view of linear programming.

GLICKSMAN, A., *An Introduction to Linear Programming and the Theory of Games*, John Wiley & Sons, New York, N.Y., 1963. An elementary introduction to basic ideas and methods of linear programming.

Methods are developed for solving game theory problems using linear programming.

KAYSEN, C., "The Computer that Printed Out W*O*L*F*," in *Foreign Affairs*, **50**, no. 4, July 1972. A review of *The Limits To Growth* by a leading economist in a small but influential journal.

MEADOWS, D., ET AL., *The Limits To Growth*, Universe Books, New York, N.Y., 1972. Available in paperback. This book presents the gloomy opinion that we must drastically slow the growth of population and industrialization if we are to avoid catastrophe. Many important scholars disagreed; for example, see the readings by Kaysen and by Passel, et al.

PASSELL, P., M. ROBERTS, L. ROSS, Book review of *The Limits To Growth*, *New York Times*, Book Review Section, April 2, 1972. A review by three economists.

It has been suggested that the U. S. Constitution was a great influence on computer development in the United States because it calls for taking a national census. By 1890 the task of processing census data and calculating statistics from this data was so troublesome that Herman Hollerith (1860–1929), an engineer with the Census Office, was led to develop tabulating equipment to automate the data processing. Although born of German immigrants, Hollerith showed classic Yankee ingenuity by using data cards the size of dollar bills so that he could borrow processing equipment from the Mint. Hollerith later formed a commercial organization which ultimately became the present International Business Machines Corporation. (*Courtesy of IBM.*)

Machines such as this one were used by Herman Hollerith to tabulate census data. (*Courtesy of IBM.*)

7
facts and figures

7.1 *understanding data*

The American public seems to have an insatiable need for facts and figures. We are constantly being assaulted with the high, low, and average temperature of the day; cost of living indices; the Dow-Jones averages; unemployment figures; or the number of residents of Phoenix, Arizona who own cheetahs as pets. The Constitution of the United States calls for the taking of a national census. This operation, now performed every 10 years, provides us with a stream of numerical information about ourselves, which seems to dry up just prior to the start of the next census. The world of sports generates an almost unending torrent of figures and "averages." But even more important than the "informational" aspect of numerical facts and figures has been a dramatic increase in the use of numerical information in analyzing and developing policy in such areas as health, welfare, education, and war. In light of this phenomenon it seems worthwhile to discuss how numerical information is gathered, classified, and used in policy making.

We begin with that increasingly omnipresent phenomenon of the current scene, the survey. Imagine that a survey of the students in the senior class in a small school has been taken. Two questions are asked on the survey:

QUESTION 1

How many children do you wish to have?

QUESTION 2

How much does your father earn from all his jobs?

The information obtained from the survey is recorded below.

RESPONSES TO QUESTION 1

4, 2, 0, 5, 2, 1, 1, 2, 1, 1, 3, 2, 0, 7, 1, 1, 2, 2, 0, 1, 0, 2, 1, 1, 2, 0, 3, 1, 4, 2

RESPONSES TO QUESTION 2

$10,350, $8470, $9050, $9680, $8480, $10,600, $8900, $8340, $8160, $8050, $9220, $8400, $8080, $8550, $8490, $8450, $10,950, $8360, $9970, $9400, $8440, $8670, $8600, $10,200, $8180, $9600, $8250, $8780, $9340, $8740

Clearly the data listed above is not in a form to be of much use. What we have in each case is a list of 30 numbers in no particular order. From such a disarray of numbers it is not possible to obtain any useful information. As a step in the direction of obtaining order and information from this disorder of numbers, construct a table for question 1 with the number of children desired listed in increasing order from top to bottom, and tally the number of the people who wanted the various numbers of children (see Table 7.1). Thus, for example, the indication of 2 in line 4 of the frequency column means that two students desired families with three children.

Table 7.1   Table of Responses to Question 1 (Number of Children Desired) in a Class of 30 Students

Number of Children Desired	Tally	Frequency	Relative Frequency
0	|||	5	$\frac{5}{30} = 0.17$
1	|||| ||||	10	$\frac{10}{30} = 0.33$
2	|||| ||||	9	$\frac{9}{30} = 0.30$
3	||	2	$\frac{2}{30} = 0.07$
4	||	2	$\frac{2}{30} = 0.07$
5			1	$\frac{1}{30} = 0.03$
6		0	$\frac{0}{30} = 0.00$
7			1	$\frac{1}{30} = 0.03$
Total	30	30	1.00

This procedure goes a long way toward organizing the information contained in the raw data above. For example, one can state that more students desired one child than any other number. Furthermore, one can state that 17% of the students wanted no children. This number, the *relative frequency* of the response "no children," was obtained by taking the number of students who wanted no children and dividing it by the total number of students surveyed. Numerous other conclusions could be drawn from the table. It is often convenient, however, to have some visual display of the information contained in such a frequency table. There are three methods commonly used for displaying the information in a frequency table visually: the frequency polygon, the histogram, and the pie chart. We shall discuss each of these methods briefly in turn, using for convenience the example above. The general procedure should be apparent from the example.

FREQUENCY POLYGON: To draw a frequency polygon for the data given in Table 7.1 we draw two axes at right angles, one vertical and the other horizontal. Equally spaced along the horizontal axis, we list the entries of the left-hand column of the frequency table, starting with its top row. In this case, we list the various numbers of children. Along the vertical axis we list at equally spaced intervals, starting with zero, the frequencies that can occur. The largest number appearing would correspond to the largest frequency. For each of the numbers along the horizontal axis we plot a point at the proper vertical height, as required by the table and according to the scale along the vertical axis (see Figure 7.1). To obtain the frequency polygon, we join up these plotted points by straight lines.

Warning: A fairly common error in the reading of frequency polygons can be illustrated from this example. Even though the only significant points in the frequency polygon are the end points of the straight line segments, some persons reading the chart might conclude that $4\frac{1}{2}$ people wanted $1\frac{1}{2}$ children by interpolating information between the end points of the

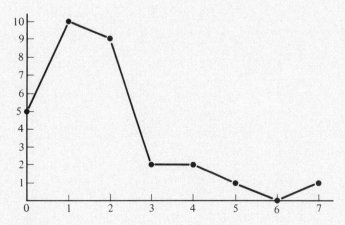

Figure 7.1

segments. The meaninglessness of this procedure may be less obvious when the data being analyzed doesn't clearly rule out the "information" obtained.

HISTOGRAM: The procedure for constructing a histogram from a frequency table is similar to that of constructing a frequency polygon (see Figure 7.2). Note that the center of each "bar" of the histogram lies above the number on the horizontal line.

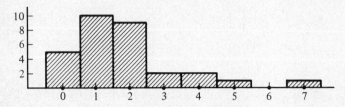

Figure 7.2

PIE CHART: To construct a pie chart, we proceed as follows: First, we compute the relative frequency of each of the responses to the survey. For example, the relative frequency of the response "three children" is $\frac{1}{15}$. (See the relative frequency column in Table 7.1.) This number is obtained by observing that out of a total of 30 students there were 2 students who want three children. We now divide a circle into sectors, apportioned according to the relative frequencies (see Figure 7.3). Since a whole circle is made up of 360 degrees, to find the angle of any particular slice of the pie diagram we

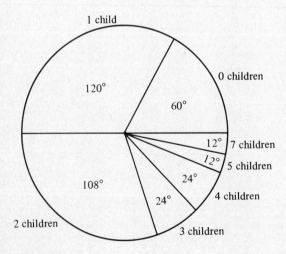

Figure 7.3

multiply the relative frequency by 360 degrees. Thus, for example, the fact that 9 students wanted two children (relative frequency of .3) is indicated by apportioning .3 times 360 degrees, or 108 degrees, of the circle to represent this situation.

Pie charts have certain disadvantages and are used less commonly than histograms or frequency polygons. Nevertheless, they are superior to these other schemes for showing proportions of a whole taken up by various categories. For instance, a glance at Figure 7.3 shows that half of all persons surveyed want no more than one child and three-quarters of all persons surveyed want no more than two children. These facts are much less apparent in Figures 7.1 and 7.2.

Now let us attempt to carry out the same procedure as above for the data concerning incomes (i.e., the answers to question 2) as we did for the data about numbers of children desired. If we arrange the items in a tally chart as before, we discover (Table 7.2) that each income yields only one tally and, except for having rearranged the data in increasing order, we have not made much progress toward analyzing it.

Table 7.2   Table of Father's Income for a Class of 30 Students

Salary	Tally	Frequency	Salary	Tally	Frequency
$8,050	|	1	$8,670	|	1
8,080	|	1	8,740	|	1
8,160	|	1	8,780	|	1
8,180	|	1	8,900	|	1
8,250	|	1	9,050	|	1
8,340	|	1	9,220	|	1
8,360	|	1	9,340	|	1
8,400	|	1	9,400	|	1
8,440	|	1	9,600	|	1
8,450	|	1	9,680	|	1
8,470	|	1	9,970	|	1
8,480	|	1	10,200	|	1
8,490	|	1	10,350	|	1
8,550	|	1	10,600	|	1
8,600	|	1	10,950	|	1

Consider the first two lines of the table containing the entries $8050 and $8080. Since these entries differ by only $30, while the first and last entries differ by $2900, we might agree that for many purposes we could "lump together" the first two. This suggests that we agree to group the incomes in classes, where people with approximately the same income would be put in the same class. This would give us a means of getting more information from the data. Noting that the difference between the largest and smallest incomes is approximately $3000, let us agree to form six classes, each with a $500

244

facts and figures

range of incomes. When the data is grouped into classes by this procedure, we obtain the new table in Table 7.3. On the basis of this table, we can proceed to draw a frequency polygon, a bar graph, and a pie chart for the grouped data, as shown in Figures 7.4, 7.5, and 7.6.

Table 7.3   Grouped Income Table

Class Limits*	Number in this Class
$ 8050–8550	13
8550–9050	6
9050–9550	4
9550–10,050	3
10,050–10,550	2
10,550–11,050	2

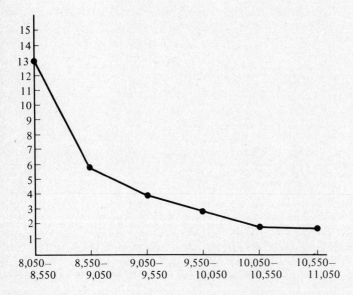

Figure 7.4

A few more remarks on the question of how data can be grouped into classes are in order. Recall that we made the somewhat arbitrary decision to use 6 classes of $500 "width" for the data in Table7.2. Could we have used 2, 18, or 9 classes? To answer this question, let us first begin with the following definition.

*We make the convention that if a number falls on the borderline between two classes, we put it into the higher class.

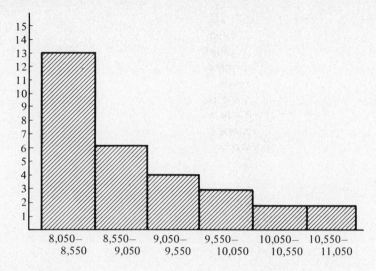

Figure 7.5

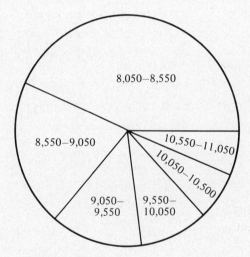

Figure 7.6

DEFINITION 1

Given a collection of numbers that are the outcomes of a poll, survey, or experiment, the *range* of the data is the difference between the largest and the smallest number in the collection.

This use of the word *range* corresponds with our ordinary use of the word *range* in such expressions as "the temperature range was 40°F," meaning that the difference between the high and low temperatures was 40°F.

Example 1

The weather bureau gives the temperatures for a 12-hr period as 75, 75, 74, 76, 76, 77, 77, 78, 79, 79, 81, and 83°F. The lowest of the temperatures is 74° and the highest is 83°. Hence, the range of temperatures is 9°, the difference between 74 and 83.

When we decide on a procedure for grouping the data into classes, we are choosing a model for our data. Here are two possible alternatives.

MODELING PROCEDURE 1

Choose a convenient or desired width for a class, and then use as many classes as are needed to cover the whole range of measurements.

MODELING PROCEDURE 2

Choose a convenient number of classes, and then determine a convenient width for each class by dividing the range by the number of classes and rounding off to a convenient number.

Example 2

George is attempting to analyze how many miles per gallon he has been getting from his new car. Each week he calculates the number of miles per gallon he has gotten, and after 18 weeks he has 18 numbers ranging from 11.4 to 20.7 mi/gal. One possible set of classes would be 11.4 to 14.4, 14.4 to 17.4, 17.4 to 20.4 and 20.4 to 23.4. To avoid decimals, he decides on the alternative set of classes 11 to 14, 14 to 17, and 17 to 20, and 20 to 23. We use the convention that if a number falls on the borderline between two classes, say at 14, we put it into the higher class.

Suppose George wanted to model his mileage data using five classes. He divides the range, which is 9.3, by 5 to get 1.86. Now he can mark off classes, beginning at 11.4. The arithmetic is a little tedious but here is how one set of classes might look: 11.4 to 13.26, 13.26 to 15.12, 15.12 to 16.98, 16.98 to 18.84, 18.84 to 20.70. Recall, however, that George dislikes decimals. Consequently, he decides to round off the class size to two units and start with 11. Thus he gets 11 to 13, 13 to 15, 15 to 17, 17 to 19, and 19 to 21.

The question now arises as to which of the two modeling procedures to use. No hard and fast answer can be given. Loosely speaking, the choice is governed by whether convenience of class width, or having a prechosen number of classes is paramount. Notice that if too few classes are used, then much information contained in the data will be destroyed. On the other hand, when there are too many classes, there is no gain in information over what was present in the ungrouped data itself.

EXERCISES 7.1

1. The following numbers are the ages of the signers of the Declaration of Independence at the time that document was signed. Group the data in 5-year classes and construct a frequency polygon.

    40, 53, 46, 39, 38, 35, 40, 37, 48, 41, 70, 31, 41, 52, 39, 50, 65, 46, 29, 34, 71, 38, 45, 33, 44, 41, 63, 60, 26, 42, 34, 50, 42, 41, 37, 45, 36, 42, 48, 46, 30, 26, 55, 57, 46, 33, 60, 62, 35, 46, 33, 53, 49, 50

2. The following numbers are the ages of the first 36 United States presidents at the time of taking office. Group the data and construct a frequency polygon.

    57, 61, 57, 57, 58, 57, 61, 54, 68, 51, 49, 64, 50, 48, 65, 52, 56, 46, 54, 49, 50, 47, 55, 54, 42, 51, 56, 55, 51, 54, 51, 60, 62, 43, 55, 55

3. The distributions of ages in Exercises 1 and 2 are significantly different. Prepare groupings of data and frequency polygons designed to highlight the difference and facilitate comparison. Try to explain historically and politically why the difference occurs.

4. The following are per capita income figures for the nations of South and Central America as of 1970. Group the data and make a frequency polygon.

    $280, 510, 215, 121, 308, 568, 709, 190, 200, 325, 227, 350, 622, 850, 248, 426, 284

5. Take any book, choose 40 words from it at random, and write the number of letters in each word as it is chosen. With these word lengths as data, make a frequency polygon after suitable grouping.

6. Here are some test scores for a class of sixth graders. Use 10 data classes to group the data and then make a frequency polygon.

    89, 37, 93, 51, 92, 76, 65, 77, 79, 71, 82, 85, 63, 75, 69, 80, 65, 75, 87, 38, 65, 68, 85, 98, 59, 60, 80, 78, 74, 71, 71, 49, 91, 65, 80, 68, 91, 78

7. The following is the list of crimes entered on the docket of a criminal court for a certain time period. Make a frequency polygon (note that no grouping is needed since the data are not numbers.)

    A, A, B, R, A, B, B, F, L, R, R, L, R, B, A, R, F, L, R, B, B, L, L, A, R, F, R

    A = assault; B = burglary; L = larceny; F = fraud.

8–14. For Exercises 1 to 6, draw a histogram.

15–22. For Exercises 1 to 7, draw a pie chart.

23. For which of Exercises 1 to 7 is a pie chart especially useful?

7.2   uses and abuses of statistics

> *The words figure and fictitious both derive from the same Latin root, fingere, Beware!*
>
> —*M. J. MORONEY**

One of the most important aspects of the modeling process is to be able to use the model, if possible, to predict the course of future events. We shall give an example of how a seemingly naive model can be of great assistance to city planners.

Suppose the school board of the town of Candlesville is examining census data about its population, with a view toward determining future needs for school buildings and teachers. Imagine that in 1975 Candlesville had bulk data consisting of the birthdays of its residents. From these very exact ages can be calculated (e.g., 2 years and 47 days). Of course this is a little more detail than is useful so the data about ages is grouped into classes of width equal to 1 year. In addition, the numbers of students have been rounded off to the nearest hundred. The result is Table 7.4. It may be worth noting in passing that grouping and rounding off, while perfectly natural and necessary, does involve destroying or overlooking some of the data. This is an example of the fact that sometimes detail stands in the way of insight.

The numbers in the second column of Table 7.4 can be used to predict

*Courtesy of M. J. Moroney, *Facts from Figures,* Middlesex, England: Penguin Book Ltd., 1951.

Table 7.4

Ages in 1975 (years)	Number of Children	Year of Entrance to First Grade
0–1	1800	1981
1–2	1800	1980
2–3	1900	1979
3–4	2100	1978
4–5	2000	1977
5–6	1900	1976
6–7	1800	1975
7–8	1900	1974
8–9	1800	1973

future school populations because for each age group one can determine in what year the members of that group (or most of them) will enter first grade. These figures are shown in the third column. From similar projections one can determine the school population in each grade for any year in the near future. For example, in 1980, 5 years after the census, the children who will be entering first grade are those in the 1- to 2-year age bracket in 1975; the children who will be entering seventh grade are in the 7- to 8-year age bracket in 1975; and so on. Therefore, in 1980 classrooms and teachers for 1900 pupils in the seventh grade will be needed.

It is especially interesting to note that the largest number of children lie in the 3- to 4-year age bracket. Hence, in 1978 the first grade will have to absorb the largest number of students and this "bulge" will then go through the remaining grades, a grade higher each year. The school board and the town will have to decide how to meet the problems associated with the bulge. Will new teachers be hired and new classroom space be constructed or can this be avoided?

Suppose that the school board decides to present the information in Table 7.4 to the voters in the form of a frequency polygon. Which of the two frequency polygons in Figure 7.7 would be better suited to the purpose? Perhaps surprisingly, both present the same information; however, Figure 7.7(b) seems much more dramatic. It makes the population bulge seem quite drastic and is likely to suggest to people strong remedies such as building another school and hiring more teachers. Figure 7.7(a) is more likely to suggest to people a temporary increase in the size of classes. The dramatic effect of Figure 7.7(b) is produced in two ways: The scale for the vertical axis has been expanded [in comparison with the vertical scale in Figure 7.7(a)] and has been started with the number 17 instead of 0.

Are both graphs equally good models? They make such different impressions that it would be disquieting to think so, even though they both

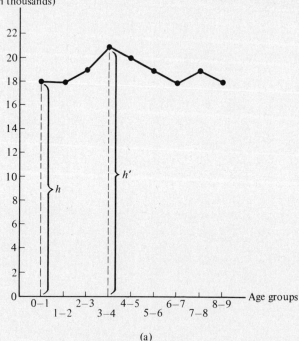

(a)

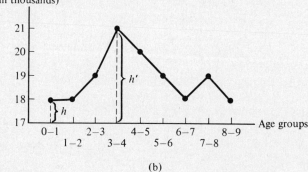

(b)

Figure 7.7

present the same data. Perhaps the best way of expressing our disquiet is to ask "Which graph is more honest?" People knowledgeable about statistics agree that while the scale can be expanded or contracted in good faith, it is usually best to start at 0. Thus, Figure 7.7(b) represents an abuse of statistics. The reasoning behind this verdict is as follows. In Figure 7.7(b) the number of

250

students in the 0- to 1-year class is represented by height h while the number in the 3- to 4-year class is represented by height h'. Since h' is about four times the size of h, this suggests to the casual glance that the 3- to 4-year class is four times as large as the 0- to 1-year class. Of course this is wrong. We are being misled because h and h' are not exactly proportional to the populations of the age groups, due to the fact that our scale starts at 17 and not at 0. In Figure 7.7(a), where we have started our scale with 0, the heights h and h' are in proportion to the sizes of the age groups they represent; hence, the fact that h' is 16% higher than h reflects accurately the fact that the 3- to 4-year class is 16% larger than the 0- to 1-year class.

Figure 7.8 shows another example of an abuse in the presentation of

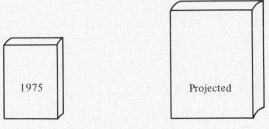

Figure 7.8

data. The school board wants to present the projected increase in the expense of educating the children of the bulge whether or not new buildings or teachers are to be added. These expenses would be for books and supplies, school lunches, and other items that are necessary whether or not new buildings and teachers are added. The school board calculates that these expenses during the bulge will be one and a half times what they were in 1975. It decides to represent the 1975 expenses using a single textbook symbol and the projected expenses by a book whose dimensions are both one and a half times those of the first book.

The reason why such a diagram, often called a *pictogram*, is misleading stems from the fact that the eye responds to the area of the books. If the length and the width of a book are multiplied by a factor of k, the area of the book increases by a factor of k^2. Hence, although costs have gone up by a factor of 1.5, the area of the second book is $(1.5)^2 = 2.25$ times as large as that of the first book.

EXERCISES 7.2

1. Using the data from Table 7.4, find how many students will be entering the fifth grade in 1978?

2. Again using Table7.4, find how many students will enter the ninth grade in 1979?

3. If the number of students in a class is to be kept at 17 in grade 8, how many classrooms would be necessary in 1978 in grade 8? In 1979? In 1980?

4. Here is some data about percentages of men who wear various shoe sizes. Make an honest frequency polygon exhibiting this. Now make a dishonest one that exaggerates the discrepancy between the popular and the less frequent sizes. Size 8, 7%; size $8\frac{1}{2}$, 10%; size 9, 15%; size $9\frac{1}{2}$, 20%; size 10, 15%; size $10\frac{1}{2}$, 10%; size 11, 7%.

5. A new bridge is to be opened shortly in Metropolis and the city planning commission wishes to estimate in advance its effect on traffic patterns. Thirty-five citizens who cross a bridge on their way to work are asked which of the bridges *A, B, C, D*, and *E* (the new one) they expect to use to reach work. The responses, in the order in which they were received, are *A, B, B, C, A, D, E, C, C, A, E, C, D, A, B, C, D, B, C, D, C, A, D, D, C, B, D, E, E, A, B, E, B, C, E.* Draw an honest frequency polygon and then a dishonest one designed to exaggerate discrepancies.

6.* Suppose that 5% of the students in a given grade fail each year and must repeat that grade. How many students would there be in the various grades from 1 to 6 in the year 1979?

7.* If each year, starting in 1975, 300 additional children enter town due to new immigration and we assume that there is no departure of families, what is the maximal number of students that might enter the fourth grade in 1978? Assume that each year one-sixth of the children who have moved into town would be fourth graders in 1978.

7.3 averages of various sorts

The word *average* is probably the most abused word in the whole field of statistics. The source of the confusion is that there are various kinds of averages. When one computes an average, the purpose is to find a single number that gives a "summary" of the information contained in a collection of data. Depending on which average is chosen, one can obtain totally different pictures of the same data. We shall examine a few sets of data to show how these different types of averages provide models for sets of data.

Example 1

An income survey of shopkeepers in a small town produces the numbers in Figure 7.9 (this is a modified histogram in which we use dots in place of the bars—for example, the three dots over the 11 represent three shopkeepers whose incomes are about $11,000). If we wish to find a single representative number to model this data, perhaps the most natural procedure would be to calculate the *mean*, that is, to add all the incomes and then divide by the number of people. The result is shown in Figure 7.9.

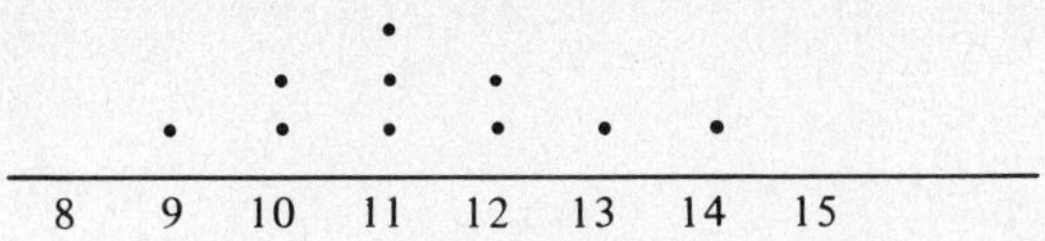

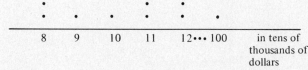

```
       •
   •   •   •
 • • • • • • •
─────────────────────────
 8  9  10  11  12  13  14  15
```

$$\text{Mean} = \frac{9 + 10 + 10 + 11 + 11 + 11 + 12 + 12 + 13 + 14}{10} = 11.3$$

Figure 7.9

DEFINITION 2

Given a collection of measurements $x_1, x_2, \ldots, x_n$ (these need not all be distinct numbers), their *mean* is

$$\frac{x_1 + x_2 + \cdots + x_n}{n}$$

This type of calculation is what usually comes to mind when the word *average* is mentioned, say, in a newspaper or on television. People familiar with statistics, however, use the word *mean* so as to avoid confusion with other averages.

Example 2

Suppose our income data gives rise to the dot diagram of Figure 7.10. If we calculate the mean, we find the results shown in Figure 7.10. In this case, however, the mean does not produce a realistic picture of the data because one very large income skews (shifts) the mean. The mean income turns out to be almost twice what most shopkeepers are earning. To overcome this problem, one can use a different type of average, the median.

DEFINITION 3

Given a collection of data arranged in increasing order, if the number of measurements in the data set is odd, the *median* is the number for which equal

$$\text{Mean} = \frac{8 + 8 + 9 + 10 + 11 + 11 + 12 + 12 + 12 + 100}{10} = 19.3 \text{ thousand}$$

```
                   •
 •         •   •
 • • • • • •
─────────────────────────
 8   9   10  11  12••• 100    in tens of
                             thousands of
                             dollars
```

Figure 7.10

253

numbers of data lie above and below it. If the number of measurements in the data set is even, the *median* is the mean of the two middle measurements.

For the data of Figure 7.10 the median is the mean of the fifth and sixth measurements, namely, 11. This comes closer to the income that most shopkeepers enjoy but it doesn't indicate the existence of one large income.

The Zero Growth Paycheck

To the Editor:

A letter by Prof. Bertram Gross, in your issue of March 1, suggests that if a zero-economic-growth policy were instituted, American families could look forward to an average income of over $13,000, instead of $10,000 as I had suggested earlier. I wish Professor Gross were right, but unfortunately even my own figure probably was much too high if, as I had also suggested, a zero-growth policy should compel us to share part of our income with the poor countries.

That demands for such equalization would be strong becomes evident from Robert Reinhold's report in your business issue of March 3 of the Smithsonian Meeting on Zero Economic Growth. Even partial justice would probably compel us to reduce our family standard well below the $10,000 currently constituting the median income (which, Professor Gross notwithstanding, is as defensible an average as the mean which he prefers).

Professor Henry C. Wallich
Yale University
New Haven, March 6, 1972

A dispute about the relative merits of the mean and the median lurks in the letter to the New York Times.
(*Courtesy of Henry Wallich and The New York Times.* © *1972 by The New York Times Company. Reprinted by permission.*)

Here is the final kind of average that we shall consider.

DEFINITION 4

The *mode* of a set of data is that number that occurs most frequently. If there is no such number, that is, if there are two or more numbers that are tied for first place, then there is no mode.

For example, in Figure 7.9, 11 is the mode, while in Figure 7.10 the mode is 12. Our next example (Figure 7.11) shows a case where the mode does not exist.

Example 3

We calculate the mean, median, and mode for the data shown in Figure 7.11.

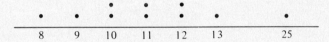

| 8 | 9 | 10 | 11 | 12 | 13 | 25 |

Mean = 12.1

Median = 11

Mode does not exist

Figure 7.11

Lest the mode concept seem a useless one, recall that in Section 7.2 the future planning of the school board to meet the "bulge" of students depended essentially on knowing the mode.

Table 7.5 shows how the three types of averages come out for the three

Table 7.5

	Mean	Median	Mode
Figure 7.9	11.3	11	11
Figure 7.10	19.3	11	12
Figure 7.11	12.1	11	does not exist

sets of income data we have considered. It underscores again the different impressions that may be conveyed by the various averages.

Since averages are often used in formulating policy, these differences are more than mere curiosities. If, for example, a governmental agency making loans to small businesses decides to earmark money for different towns according to one of these averages, it makes an important difference which average is used. If the mean is used, the town corresponding to the data of Figure 7.9 will receive the most money, whereas this town would receive the same amount as the other towns if the median is used. (See Table 7.5.)

The problem of representing a mass of data with a single number can be thought of as a modeling problem. The mean, median, and mode are different solutions to this problem of modeling a set of data with a single number—all of them far from perfect solutions. In each case, a good deal of criticism of their "goodness of fit"—the extent to which they really represent the data—can be offered. We have already illustrated the fact that they give different pictures of the same data. Here are some other observations about the *averages* we have discussed.

First, regardless of which of these averages is used, very different sets of data can have the same average.

Example 4

The sets of data in Figures 7.12(a) and (b) have the same mode, namely, 3. Notice that in this example the mode is the largest number in the set of data in one case and the smallest in the other.

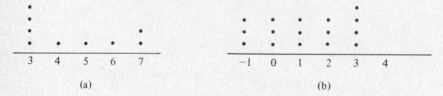

(a)   (b)

Figure 7.12

Example 5

The data collections in Figure 7.13(a) and (b) have the same median of 3.

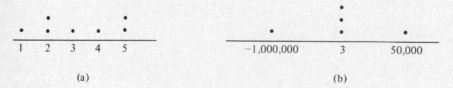

(a)   (b)

Figure 7.13

Example 6

The data collections in Figures 7.14(a) and (b) have the same mean.

| 0 | 1000 | | 500 |
| (a) | | | (b) |

Figure 7.14

A second problem with the mean, the median, and the mode is that they give no information about the distribution of the data. Suppose, for example, the proprietor of a store that sells nurses' uniforms reads in a trade journal that the mean uniform size for nurses is 12. Should he order all his uniforms of this size? Clearly not. But how many should he order of the various possible sizes? If he knew the median instead, he would know that half the nurses had sizes above the median and half below the median. On the other hand, if he knew the mode was 12, he would at least know that he should order more size 12's than any other size but this is still not good enough information on which to base an order.

Averages are particularly inadequate in cases where many measurements that lie at the extremes are deemed particularly significant. Thus, if the mean of environmental radiation levels for a group of nuclear power plants is low, this does not preclude the possibility that the level may be high enough to be dangerous at some plants.

EXERCISES 7.3

1. Calculate the mean, median, and mode of the data in Exercise 1 on p. 247. Which do you think gives the best model of the data?
2. Calculate the mean, median, and mode of the data in Exercise 2 on p. 247. Which do you think gives the best model of the data?
3. Calculate the mean, median, and mode of the data in Exercise 4 on p. 247. Which do you think gives the best model of the data?
4. Calculate the mean, median, and mode of the data in Exercise 5 on p. 247. Which do you think gives the best model of the data?
5. Calculate the mean, median, and mode of the data in Exercise 6 on p. 247. Which do you think gives the best model of the data?
6. One often reads statements like "The average American family has 2.4 children." In this statement, is 2.4 a mean, a median, or a mode?
7. Calculate the means, medians, and modes for each of the following distributions in Figure 7.15.

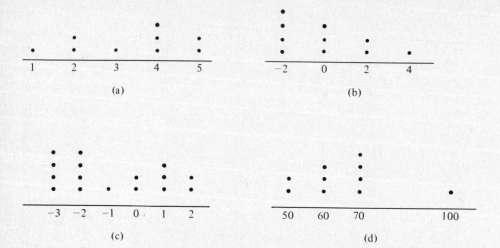

Figure 7.15

8.* What is the law of averages, and what, if anything, does it have to do with the mean, median, or mode?

9.* Is the batting average, as used in baseball, an average of a set of data in the sense of being a median, mean, or mode? If so, which of the three is it, and how would you go about calculating the other two types of averages for a player's batting record?

10.* A modification of the concept of the mean is often used in situations where data are to be assigned weights to emphasize their relative importance. For example, suppose a teacher wishes to weight the final exam twice as much as the midterm exam. If John received 80 on the final and 50 on the midterm, his weighted mean grade would be

$$\frac{2(80) + 1(50)}{2 + 1} = \frac{210}{3} = 70$$

Use the idea from the example to solve the following problems:
(a) June bought 14 gal of gas at 30¢/gal and 20 gal of gas at 34¢/gal. Compute the weighted mean cost of the gas she bought.
(b) April got 100, 90, 100, and 70 on hour exams; a 50 on the midterm exam; and 80 on the final exam. If the final counts twice as much as the midterm, and the midterm counts twice as much as an hour exam, and if the hour exams count equally, compute her weighted mean grade.

7.4* estimates of fit for various averages

In the last section we discussed three types of averages as models for a collection of data but we tried to avoid the question of which model was best. Actually there is no best model because each has its uses and each is good in its own way. In this section we want to perform some estimates of fit for the mean, median, and mode as models for a collection of data. We shall set up three reasonable definitions of what we might mean by a "good fit" and then we'll see how the mean, median, and mode stack up against these standards. What we shall do is to set up three formulas for the error that is incurred when we use a particular number to model a certain set of data; then we shall say that the lower the error is, the better the fit. We shall let the measurements be denoted by $x_1, x_2, x_3, \ldots, x_n$ and denote by μ the number being used to model this set of data.

ERROR DEFINITION 1

Count the number of measurements x_i that differ from μ. This is the error.

Example 1

Let the measurements be 1, 1, 1, 5, 5, 6, 6, 7. We can easily calculate that

$$\text{Mean} = 4$$

$$\text{Mode} = 1$$

$$\text{Median} = 5$$

The number of measurements different from the mean is 8, the number of measurements different from the mode is 5, and the number of measurements different from the median is 6. Thus, in this case the mode wins by virtue of having the least error.

The outcome of Example 1 is not a quirk dependent on the particular data set chosen. In general, for any collection of data, the mode (if it exists) does at least as well by this definition of error as either of the other two averages. The reason, of course, is that the mode is by its very definition the most frequently occurring number in the set of measurements. This means all other numbers, in particular the mean or median, occur less often than the mode among the data. In turn this means that both the mean and the median, except in the case where either coincides with the mode, differ from more of the measurements and so have a higher error.

ERROR DEFINITION 2

Subtract μ from each measurement in the data set and then add up all these differences and take the absolute value. (The absolute value of a number is the magnitude of the number without the sign. Thus the absolute value of a nonnegative number is the number itself but the absolute value of a negative number is the negative of that negative number. The absolute value of x is denoted $|x|$. Thus, $|7| = 7$, while $|-4| = 4$.)

Example 2

Using the same data as in Example 1, we compute the errors for the mean, median, and mode.

Mean	Median	Mode
$1 - 4 = -3$	$1 - 5 = -4$	$1 - 1 = 0$
$1 - 4 = -3$	$1 - 5 = -4$	$1 - 1 = 0$
$1 - 4 = -3$	$1 - 5 = -4$	$1 - 1 = 0$
$5 - 4 = 1$	$5 - 5 = 0$	$5 - 1 = 4$
$5 - 4 = 1$	$5 - 5 = 0$	$5 - 1 = 4$
$6 - 4 = 2$	$6 - 5 = 1$	$6 - 1 = 5$
$6 - 4 = 2$	$6 - 5 = 1$	$6 - 1 = 5$
$7 - 4 = 3$	$7 - 5 = 2$	$7 - 1 = 6$
$\overline{0}$	$\overline{-8}$	$\overline{24}$

$$\text{Error} = |0| = 0 \quad \text{Error} = |-8| = 8 \quad \text{Error} = |24| = 24$$

We observe that for this second type of error, the mean scores best. Again, this is not accidental. The mean by the present definition would be best for any set of data. This can be demonstrated by an easy algebraic calculation. If the data are $x_1, x_2, \ldots, x_n$, the mean is, by definition,

$$\frac{x_1 + x_2 + \cdots + x_n}{n}$$

Therefore,

$$n \,(\text{mean}) = x_1 + x_2 + \cdots + x_n$$

Now the calculation of the error goes as follows:

$$
\begin{aligned}
\text{Error} &= |(x_1 - \text{mean}) + (x_2 - \text{mean}) + \cdots + (x_n - \text{mean})| \\
&= |(x_1 + x_2 + \cdots + x_n) - n \,(\text{mean})| \\
&= |n \,(\text{mean}) - n \,(\text{mean})| \\
&= 0
\end{aligned}
$$

Since 0 is the smallest nonnegative number, no other error can be smaller. Consequently, this shows that neither the median nor the mode can give a smaller type 2 error. We see that the error is least for the mean.

ERROR DEFINITION 3

Compute the difference between each measurement and μ, and then take the absolute value of all these differences. Now add all the nonnegative numbers just calculated to find the error.

Example 3

Using the same data as in Example 1, we compute the errors for the mean, median, and mode.

Mean	Median	Mode
$\lvert 1 - 4 \rvert = 3$	$\lvert 1 - 5 \rvert = 4$	$\lvert 1 - 1 \rvert = 0$
$\lvert 1 - 4 \rvert = 3$	$\lvert 1 - 5 \rvert = 4$	$\lvert 1 - 1 \rvert = 0$
$\lvert 1 - 4 \rvert = 3$	$\lvert 1 - 5 \rvert = 4$	$\lvert 1 - 1 \rvert = 0$
$\lvert 5 - 4 \rvert = 1$	$\lvert 5 - 5 \rvert = 0$	$\lvert 5 - 1 \rvert = 4$
$\lvert 5 - 4 \rvert = 1$	$\lvert 5 - 5 \rvert = 0$	$\lvert 5 - 1 \rvert = 4$
$\lvert 6 - 4 \rvert = 2$	$\lvert 6 - 5 \rvert = 1$	$\lvert 6 - 1 \rvert = 5$
$\lvert 6 - 4 \rvert = 2$	$\lvert 6 - 5 \rvert = 1$	$\lvert 6 - 1 \rvert = 5$
$\lvert 7 - 4 \rvert = 3$	$\lvert 7 - 5 \rvert = 2$	$\lvert 7 - 1 \rvert = 6$
18 = Error	16 = Error	24 = Error

We shall now show that by using Error Definition 3 we shall always obtain the least error by using the median. Instead of giving the proof in the greatest generality, we shall make a few simplifying assumptions to allow the essential ideas of the proof to stand out. First, we shall suppose that the measurements in the data set are all different (which was not the case in the set used in our examples) and that they are subscripted in the order of their magnitude. This means $x_1 < x_2 < x_3$ and so on. Second, we shall assume that the median is one of the measurements rather than the midpoint of two measurements. Thus, if x_k is the median measurement, there are as many measurements after it as measurements before it, namely, $k - 1$. Consequently $x_{k+1}, x_{k+2}, \ldots, x_{2k-1}$ are the measurements following x_k. With these simplifications and notations, we can justify our claim about the median giving the lowest type 3 error by proving the following theorem.

THEOREM 1

Let $x_1 < x_2 < x_3 < \cdots < x_{2k-1}$ be a set of measurements with x_k as median. The expression $\lvert x_1 - y \rvert + \lvert x_2 - y \rvert + \cdots + \lvert x_{2k-1} - y \rvert$ is a minimum when we take $y = x_k$ (i.e., the median).

Proof: Each term of the form $|x_i - y|$ can best be visualized as the distance from x_i to y along a number line (see Figure 7.16). If we think of it this way and write $d(x_i, y)$ instead of $|x_i - y|$, then we want to show that the

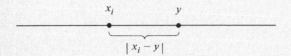

Figure 7.16

following expression is a minimum when $y = x_k$:

$$d(x_1, y) + d(x_2, y) + \cdots + d(x_{2k-1}, y) \tag{7.1}$$

We shall proceed by showing that if y moves from the value x_k to values greater than x_k, the size of the expression in (7.1) increases. A similar argument will handle the case where y moves to smaller values. First let's imagine a small rightward movement of y (see Figure 7.17), by an amount e that is

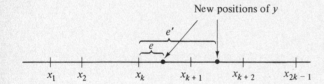

Figure 7.17

small enough so that the new position of y is not beyond x_{k+1} (although y may go as far as being exactly equal to x_{k+1}). Each of the k terms $d(x_1, y)$, $d(x_2, y), \ldots, d(x_k, y)$ will increase by e, while each of the $k - 1$ terms $d(x_{k+1}, y), d(x_{k+2}, y), \ldots, d(x_{2k-1}, y)$ will decrease by e. The total increase is therefore ke, while the total decrease is $(k - 1)e$, yielding a net increase of e in the value of (7.1).

This argument has taken care of movement to the right that does not go beyond the next measurement x_{k+1}. If we continue our movement to the right, beginning now at x_{k+1}, we increase (7.1) at an even faster rate because a further movement to the right from x_{k+1} of magnitude e' decreases the distances $d(x_{k+2}, y), \ldots, d(x_{2k-1}, y)$, while it increases the distances $d(x_1, y)$, $\ldots, d(x_{k+1}, y)$. The number of distances which increase is $k + 1$ while the number which decrease is $k - 2$. Consequently, there is a net increase of $(k + 1)e' - (k - 2)e' = 3e'$. This type of argument can be continued: No matter how many measurements we pass as y moves to the right from its initial value of x_k, the value of (7.1) always increases. Since a similar result holds if we move to the left, there is no other value of y that gives as small an error of type 3 as x_k.

EXERCISES 7.4

1. Calculate the minimum error according to each error definition for each of the following sets of data.
   (a)   1, 2, 2, 4, 5, 5, 6, 8, 8
   (b) $-2, -1, -1, 0, 0, 5, 6$
   (c)   3, 3, 3, 4, 5, 6

2. For each of the data sets of Exercise 1, calculate the error for the mean, median, and mode, according to this error definition:

$$\text{Error} = (x_1 - \mu)^2 + (x_2 - \mu)^2 + \cdots + (x_n - \mu)^2$$

3.* If $x_1, x_2, \ldots, x_n$ is a set of measurements and the function f is defined by $f(y) = |x_1 - y| + |x_2 - y| + \cdots + |x_n - y|$, plot the graph of $f(y)$ for values of y lying between -5 and $+5$ in each of the following cases:
   (a)   The measurements $x_1, \ldots, x_n$ are: $-5, -5, -1, -1, -1, 2, 2, 3$
   (b)   The measurements $x_1, \ldots, x_n$ are: $-5, -3, -1, 1, 3, 5$
   (c)   The measurements $x_1, \ldots, x_n$ are: $-1, 0, 0, 0, 1, 2, 2$

4.* Supply the proof of Theorem 1 in the case where the median is not itself a data point.

7.5 modeling data
with two numbers

We have seen that the idea of using a single number (i.e., average) to model a large collection of data has only limited success. The idea of using a single number as a representative of a large collection of data seems to run into difficulty because it cannot give any measure of *dispersion*. By dispersion we mean the extent to which the numbers are spread out. This suggests that a more accurate model of a large collection of data might be obtainable by modeling the data with *two* numbers. One of the two numbers would be one of the averages, while the other would measure dispersion of the data (with respect to the number used to indicate the average). Because of its important theoretical role we shall use the *mean* as our choice of the average.

The measure of dispersion that we shall use is called the *standard deviation* and is measured this way:

DEFINITION 5

The *standard deviation*, σ, of a set of data is calculated by subtracting each data number from the mean, squaring the result, taking the mean of these numbers, and taking the square root of the resulting number. In symbols, if the data numbers are $x_1, x_2, \ldots, x_n$ and if the mean is μ, then

$$\sigma = \sqrt{\frac{(x_1 - \mu)^2 + (x_2 - \mu)^2 + \cdots + (x_n - \mu)^2}{n}}$$

Example 1

Suppose the given data is

$$3, 3, 3, 4, 4, 5, 6, 20$$

$$\text{Mean} = 6$$

Computing the difference of the data from the mean, and squaring, we obtain

$3 - 6 = -3$	$(-3)^2 = 9$
$3 - 6 = -3$	$(-3)^2 = 9$
$3 - 6 = -3$	$(-3)^2 = 9$
$4 - 6 = -2$	$(-2)^2 = 4$
$4 - 6 = -2$	$(-2)^2 = 4$
$5 - 6 = -1$	$(-1)^2 = 1$
$6 - 6 = 0$	$(0)^2 = 0$
$20 - 6 = 14$	$(14)2 = 196$

Adding together the squares, $9 + 9 + 9 + 4 + 4 + 1 + 0 + 196 = 232$. Now, taking the number we find and dividing it by the total number of measurements, we obtain $\frac{232}{8} = 29$. This number, σ^2, is called the *variance*. The standard deviation is the square root of the variance. In our example the standard deviation $\sigma = \sqrt{29}$, which is approximately 5.4 (see square root table in Appendix 12.2).

Example 2

Find the standard deviation of

$$1, 2, 3, 1, 4, 1, 3, 1, 1, 3$$

Solution: The mean equals

$$\frac{1 + 2 + 3 + 1 + 4 + 1 + 3 + 1 + 1 + 3}{10} = 2$$

$$\begin{array}{cc}
\text{Score} - \text{mean} = & \text{Square of difference} \\
1 - 2 = -1 & 1 \\
2 - 2 = 0 & 0 \\
3 - 2 = 1 & 1 \\
1 - 2 = -1 & 1 \\
4 - 2 = 2 & 4 \\
1 - 2 = -1 & 1 \\
3 - 2 = 1 & 1 \\
1 - 2 = -1 & 1 \\
1 - 2 = -1 & 1 \\
3 - 2 = 1 & \underline{1} \\
& 12
\end{array}$$

$$\sigma^2 = \tfrac{12}{10} = 1.2$$

Hence,

$$\sigma^2 = \text{variance} = 1.2$$

$$\sigma = 1.10$$

Intuitively one feels that if two sets of data have the same mean, then if the standard deviation for one set is smaller than for the second, the first set is more tightly clustered about the mean than the second.

How much better is this two-number modeling process involving the mean and standard deviation than the various one-number modeling possibilities we dealt with in the last section? In other words, what can one say about the goodness of fit, the extent to which the mean and standard deviation give a good picture of the data? It turns out to be a good deal better than it was for our one-number models because we can draw conclusions about how much of our data lies close to the mean by looking at the standard deviation and doing a certain calculation.

To see how we do this, examine the chart in Figure 7.18. The data have

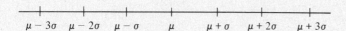

$$\mu - 3\sigma \quad \mu - 2\sigma \quad \mu - \sigma \quad \mu \quad \mu + \sigma \quad \mu + 2\sigma \quad \mu + 3\sigma$$

Figure 7.18

not been indicated on the chart, but the mean μ has been plotted. We have also plotted, above and below the mean of the missing data, several intervals of width equal to the standard deviation σ. Using only arithmetic we can obtain some information about how many of the data numbers must lie within these intervals. To be specific, we shall be able to determine that at least a certain percentage of the data must lie between $\mu - \sigma$ and $\mu + \sigma$; similarly, for the interval from $\mu - 2\sigma$ to $\mu + 2\sigma$, we shall be able to determine that a certain minimum percentage (and perhaps more) will lie in that interval. In general, for any number of standard deviations above and below the mean (not necessarily integral), we can determine a minimum percentage such that at least that percentage of the data lies within those limits.

The exact means of establishing these percentages is contained in the following theorem of the Russian mathematician Chebyshev.

THEOREM 2

Given a collection of measurements, then within h standard deviations of the mean (i.e., within the interval from $\mu - h\sigma$ to $\mu + h\sigma$) there are at least $[1 - 1/(h)^2] \times 100\%$ of the measurements.

Example 3

Suppose $\mu = 2$ and $\sigma = 10$ and we are interested in knowing how many measurements lie between the numbers -18 and 22. These limits are 2 standard deviations above and below the mean so we take $h = 2$ in Chebyshev's theorem. Thus, between -18 and 22 we have at least $[1 - 1/(2^2)] \times 100\% = 75\%$ of all measurements.

Example 4

Suppose $\mu = 2$ and $\sigma = 10$ and we want to set limits above and below the mean so that at least 40% of the data lies within those two limits. What should the limits be? Let h be the distance of these limits above and below the mean. We want to choose h so that

$$\left(1 - \frac{1}{h^2}\right) \times 100 = 40$$

Solving for h,

$$1 - \frac{1}{h^2} = 0.4$$

$$\frac{1}{h^2} = 0.6$$

$$h^2 = 1.66$$

The square root of 1.67 is approximately 1.3 (see table in Appendix 12.2) so if h is 1.3, we shall have at least approximately 40% of the measurements between $\mu - h\sigma$ and $\mu + h\sigma$. Since 1 standard deviation is 10 units, 1.3 of them amounts to 13 units. We conclude that at least 40% of the measurements lie between -11 and 15. Thus, the limits we want are -11 (i.e., $2 - 13$) and 15 (i.e., $2 + 13$).

There is one subtlety about Chebyshev's theorem that limits its usefulness, however. In the context of the last example we determined that at least 40% of the measurements lay in a certain interval. Chebyshev's theorem does not guarantee that precisely 40% of the measurements lie in the interval but rather that at least that many do. The true number could be a good deal more. For example, the data of Figure 7.19 has $\mu = 2$ and $\sigma = 10$, just as

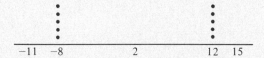

Figure 7.19

in the last example. Consequently, we can conclude by Chebyshev's theorem, as we did in the last example, that within 1.3 standard deviations of the mean we shall have at least 40% of the measurements. Since 1 standard deviation is 10 units, 1.3 of them amounts to 13 units, and we conclude that at least 40% of the measurements lie between -11 and 15. As you can see from the figure, this is not a very good estimate because all the measurements actually lie within the limits -11 and 15.

Example 5

A political party decides to appeal to middle-income voters by suggesting lower tax rates for them. The party concludes that to be effective the policy must favor at least three-quarters of all families in the United States. What income brackets should be given favorable treatment, assuming that mean family income is \$11,000 and that $\sigma = 2000?

Solution: As an initial trial, the party experiments with setting the limits for favored treatment from \$9000 to \$13,000. These limits are 1 standard deviation above and below the mean so we can use an h value of 1 in Chebyshev's theorem to see what percentage of families is guaranteed to lie within these limits. However, $1 - 1/(1)^2 = 0$ and so we are not guaranteed that anyone lies within these income brackets. Senator Marble, the party leader, snorts that this is ridiculous since there must be many families within these limits. His girl Friday, who is a statistical wonder, explains that he is right, of course, and that the problem is that the mean and standard deviation do not tell the whole story about the data. Unfortunately, Marble's knowledge doesn't extend to knowing how many families lie within the limits from $9000 to $13,000 so it is necessary to find some information from the admittedly inadequate mean and standard deviation model. Increasing the size of the interval so that its limits are 2 standard deviations from the mean will accomplish this. For with $h = 2$, $1 - 1/(h)^2 = \frac{3}{4}$ and so at least 75% of all families in the United States have incomes in the range from $7000 to $15,000 (assuming still that $\mu = \$11,000$ and $\sigma = \$2000$).

EXERCISES 7.5

1. For each of the following sets of data, calculate the mean μ and the standard deviation σ.
   (a)   1, 2, 8, 5, 5, 4, 6, 7, 8, 4
   (b)   1.1, 1.3, 1.5, 1.7
   (c)   3, 4, 4, 3, 2, 1, 3, 3, 4

2. For each of the sets of data in Exercise 7 on p. 257, calculate the mean μ and the standard deviation σ.

3. The week's low temperatures have been

$$40°, 30°, 40°, 40°, 50°, 20°, 60°$$

   Find the mean and the standard deviation.

4. Compute the mean μ and the standard deviation σ of the following set of numbers:

$$1, 3, 8, 4, 5, 4, 6, 7, 9, 3$$

   How many measurements actually lie between $\mu - 2\sigma$ and $\mu + 2\sigma$? What number is predicted by Chebyshev's theorem?

5. In a labor-management dispute, management claims that since the mean of the workers income is $12,000, they are doing quite well and don't need a wage increase. Labor counters by saying that the mean models the situation poorly because there are a great many workers earning a great deal less than the mean. Some research turns up the fact that the mean figure of $12,000 is correct and that the standard deviation is $500. Which side has the better argument?

6. A crutch manufacturer wants his product to accommodate 99% of the adult population and so he makes his crutches adjustable. What should the maximum and minimum height adjustments be to achieve the 99% goal? Assume that the mean height of adults is 66 in. and that the standard deviation is 2 in.

7. A computer file of unemployed job seekers is being set up. It is necessary to set a limit to the length of a surname the computer can handle.

Since people do not want to see their names mangled, it is decided that the limit should be large enough to handle 99.5% of all surnames without cutting them short. If mean surname length is seven letters and if the standard deviation is 1.5, what upper limit should be set?

8. A home developer decides to build homes to accommodate family sizes of up to 7 persons. What percentage of all families in the United States could be guaranteed to find one of his homes suitable if mean family size is 4.5 persons and if the standard deviation is 1.7?

9. Shelves are being put up for a library and it is necessary to determine what vertical distances should be maintained between shelves. The librarians want a minimum distance that will accommodate 95% of the books in the library. There is no time to measure all the books so a sampling is made that produces these numbers for the heights of the books in the sample: 9, $9\frac{1}{2}$, 10, 10, 9, 8, 10, 9, 8, $8\frac{1}{2}$, $8\frac{1}{2}$, $8\frac{1}{2}$. Assuming that the collection of all the books in the library would have the same mean and standard deviation as this sample, can you determine the required distance?

10.* In what situations might it occur that a person knew the mean and the standard deviation for a collection of data but didn't have access to the data itself?

SUGGESTED READING

FREUND, J., *Statistics, A First Course*, Prentice-Hall, Inc., Englewood Cliffs, N.J., 1970. A clear development of elementary statistical and probabilistic methods.

HUFF, D., *How To Lie With Statistics*, W. W. Norton & Co., Inc., New York, N.Y., 1954. A humorous view of the pitfalls of interpreting statistics.

MORONEY, M. J., *Facts from Figures*, Penguin Books Inc., Baltimore, Md., 1951. A fine exposition that starts at the beginning and reaches some fairly technical subjects almost painlessly.

8
probability

8.1 The meaning of probability

All of us have been faced with situations that involve chance and uncertainty. Here is a true-false test involving such questions. See how well you can do at answering them. The remainder of this chapter will be devoted to developing an appropriate mathematical framework for dealing with questions of this kind.

(1)   A newly married couple is planning to have four children. Suppose that at each birth the chances are even that they will have a boy or a girl. The chances are also even that they will have two boys and two girls altogether.

True   *False*

(2)   A doctor tells you that you have a 50-50 chance of surviving a delicate heart operation. This means that if you were to undergo the operation four times, you would only die twice.

True   *False*

(3)   A fair roulette wheel has had a run of 20 reds. There is less than 1 chance in 10,000 that the next color to appear will be red again.

True   *False*

(4)   Any given coin can be determined by an experiment to be fair (having even chances of coming up heads or tails) or not.

True   *False*

Extra Credit

(5)   There is a close to one-third chance of getting a score of 75% or better on the exam above (consisting of the first four questions) by merely guessing.

True   False

Let us consider the fourth question. A first approach might be to examine the physical construction of the coin to see whether it is well-balanced, whether the weight is evenly distributed, and so on. Assuming that no unusual irregularities or marked assymetries are discovered, the next step would undoubtedly be to flip the coin a number of times and keep track of the outcomes. After each flip you might compute the fraction of total flips that turned up heads, the so-called *relative frequency* of heads. In the table below, we list the sequence of heads and tails that was recorded in an actual experiment of this type. Under each entry of *H* (head) or *T* (tail) we list the relative frequency of heads up to that point. Figure 8.1 shows how the relative frequency of heads varies as the number of tosses increases.

H	T	H	H	T	H	H	T	T	H	T	H	T
$\frac{1}{1}$	$\frac{1}{2}$	$\frac{2}{3}$	$\frac{3}{4}$	$\frac{3}{5}$	$\frac{4}{6}$	$\frac{5}{7}$	$\frac{5}{8}$	$\frac{5}{9}$	$\frac{6}{10}$	$\frac{6}{11}$	$\frac{7}{12}$	$\frac{7}{13}$

H	H	H	T	T	T	T	T	H	T	T	T	H
$\frac{8}{14}$	$\frac{9}{15}$	$\frac{10}{16}$	$\frac{11}{17}$	$\frac{11}{18}$	$\frac{11}{19}$	$\frac{11}{20}$	$\frac{11}{21}$	$\frac{12}{22}$	$\frac{12}{23}$	$\frac{12}{24}$	$\frac{12}{25}$	$\frac{13}{26}$

We can draw two conclusions. First, we notice that the relative frequency is a bit erratic at the outset. For this reason, judging the fairness of the coin after just a few tosses would be unreliable. Second, we notice that in the long run the relative frequency stops jumping around and *stabilizes*—in this case at around 0.50, which suggests that the coin is fair or tolerably fair. This phenomenon of stabilization of the relative frequency is common in many other circumstances; for example,

(1)   The relative frequency of correct answers given while guessing one's way through a true-false exam.
(2)   The relative frequency of defective items produced by a machine under standard operating conditions.
(3)   The relative frequency of patients suffering from a certain disease who show improvement when treated with drug X.
(4)   The relative frequency of male babies born.

All the examples above where stabilization of relative frequency occurs can be conceptualized in a similar way. By the word *experiment*, we mean

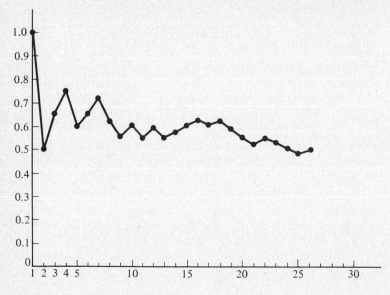

Figure 8.1

a situation which can be repeated many times but the results of which cannot be predicted in advance. Each result of an experiment is called an *outcome*. To clarify this use of terminology, we shall show how the terminology applies to several of the examples above. In the first example, the experiment consists of answering a question on a true-false examination. The experiment can have two outcomes—either a correct or an incorrect answer is given. Which result will occur is not predictable in advance. In the third example, the experiment consists of testing a person with a new drug. One cannot predict which of the two outcomes, "improvement" or "no improvement," will occur.

In light of the tendency for the relative frequency for many situations to stabilize, as in our coin tossing example (see Figure 8.1), we make the following empirical assumption:

EMPIRICAL ASSUMPTION

Let an experiment be repeated N times. If N_A denotes the number of times in these N repetitions that the outcome A occurs, then as N increases, N_A/N stabilizes at a definite number p_A.

The number p_A, which exists by our empirical assumption, is the *probability* that on any performance of the experiment we shall observe that A occurs. The number p_A represents approximately the fraction of the time that A occurs in a large number of repetitions of the experiment.

Example 1

John says that he has a probability of two-thirds of hitting the target every time he shoots at it with his bow and arrow. What does he mean?

Solution: If John were to repeat the experiment of shooting at the target many times, then one can compute the relative frequency of getting a hit. If the probability of his hitting the target is two-thirds, then the relative frequency of getting a hit must stabilize at two-thirds.

In practice, one can never perform an experiment an infinite number of times so we must *estimate* the probability by using a relative frequency for a large number of repetitions of the experiment. But how large? In our coin toss experiment, the relative frequency seems to stabilize after 25 tosses so we would calculate the relative frequency for 25 or more tosses and use that for the probability of heads. In doing this we are only getting an approximate result because we can never tell what the coin might do if we continued tossing it. This is a weak spot in the theory but as far as anyone can tell, it does not affect the practical usefulness of this concept of probability.

Another shortcoming of the *relative frequency* concept of probability is that there are many experiments that are not repeatable—and so no relative frequency can be computed. In such situations however, many people feel a strong intuitive urge to assign a *subjective probability*. For example, if a man considers whether to ask a woman to marry him, he will undoubtedly estimate the *probability* of her accepting the offer. The experiment of asking for her hand in marriage is rarely repeated, however, and so his *probability* has little scientific basis. Similarly, a businessman may estimate the *probability* that he will get a certain government contract, even though the situation will never recur. A related problem occurs in estimating the probability of surviving a certain operation, as in question 2 of the true-false examination. For a given patient, the operation will be performed only once. Since the surgeon may have performed the same operation on many patients in the same physical condition, however, he might calculate the relative frequency for survival on this basis.

EXERCISES 8.1

1. A poll is taken of 481 voters to see whom they favor for dog catcher in the upcoming election. If 322 persons polled favor Mr. Galp, estimate the probability that any one person will favor Mr. Galp.
2. An airline's records show that 643 of its 813 flights from New York to Dallas have arrived less than 5 min late. Estimate the probability that a given New York to Dallas flight will arrive less than 5 min late.

3. A department of motor vehicles has compiled records for the month of January indicating that 11,564 out of 23,465 motorists selected the upper deck rather than the lower deck of the Gaklic River Bridge. Estimate the probability that a motorist making this choice in January will select the *lower* deck.

4. The weather bureau in Lorainnesville has statistics indicating that it has snowed 60 times on the last 80 New Year's Days. Estimate the probability that it will snow in Lorainnesville on New Year's Day.

5. A department store makes a survey indicating that of 670 customers who entered the store, 340 made at least one purchase. Estimate the probability that a person entering the store will make at least one purchase.

6. Two common phrases used in situations involving chance are *odds for an outcome* and *odds against an outcome*. The definition of these phrases is as follows:

    If the probability of an outcome is p, then the odds for the outcome are p to $1 - p$. If the probability of an outcome is p, then the odds against the event are $1 - p$ to p.

    (a) Someone claims the odds of winning in the state lottery are 1000 to 1. What is the probability of winning in the lottery?

    (b) If John is willing to give 5 to 1 odds that he can beat Jake at chess, what probability would John attach to his being able to beat Jake? Give an interpretation of this probability in terms of relative frequencies.

    (c) The weatherman says that the odds against snow are 4 to 5. What are the odds for snow? What is the probability that it will snow? Do you think a weatherman's use of the word probability falls under the frequency definition? If so, explain why.

    (d) The odds against getting a parking ticket when one parks in front of a fire hydrant are 9 to 1. What is the probability of getting a parking ticket when one parks in front of a hydrant?

7. In 30 times at bat, Flashy Fred has gotten 18 hits as a pinch hitter. Estimate the probability that Flashy Fred will get a hit when he bats. Discuss the role of the concept of "hot streak" in making a decision such as whether or not to use Flashy Fred all the time.

8. Invent several examples where an experiment would arise only once. Are such examples tied to the idea that human beings are unique as individuals?

9. Criticize the legitimacy of using a relative frequency interpretation of the word *probability* in the context of surviving an operation. (*Hint:* How many people of the same age, race, weight, etc., are there who have undergone any given operation?)

10. Attempt to formulate a framework in which the weatherman's use of the word *probability* can be interpreted using a frequency definition of probability.

11. Can frequency interpretations be given to the following phrases that involve the use of the word probability?
    (a) I'll go to the movies on Friday with a probability of one-half.
    (b) The probability that the sun will rise tomorrow is one.
    (c) The probability that I pass Mathematics 150 is zero.
    (d) The probability that Senator X will win the election is seven-tenths.
    (e) The probability of power failures this summer is four-tenths.
    (f) The probability that I shall graduate from college is nine hundred ninety nine-thousandths.

12.* Try to construct an example of an experiment that can be repeated arbitrarily often but for which the relative frequency of an occurrence will not stabilize.

13.* Take an ordinary penny and toss it 25 times. Use the data that you get to construct a chart like the one in the table on p. 271. Does your chart tend to verify the empirical assumption? Continue your chart for another 25 tosses. Compare the probability of a head that you would estimate on the basis of 25 tosses to that probability that you would estimate on the basis of 50 tosses. Which estimate would you tend to prefer? Place a wad of gum on or attach a paper clip to one side of the coin and repeat the procedure above. How has the presence of the gum or paper clip affected the estimates for the probability of a head?

14.* For each of the situations below, determine empirically (by constructing a chart as in the table on p. 271) if the empirical assumption, p. 272, appears to be valid.
    (a) Pick two pages of a book and compute the relative frequency of the letter *e* to the total number of letters.
    (b) Toss an ordinary die, and compute the relative frequency that the number 6 will occur.
    (c) Using a telephone book, compute for one page of the book the relative frequency with which the number 4 appears (at least once) in a telephone number.

8.2 probability assignments

The determination of probabilities is not in the domain of probability theory but is a matter for an applied statistician. Our concern will be to draw conclusions from given assumptions about probabilities that can be obtained by manipulating these numbers. However, what is reasonable to assume about a probability? Is it reasonable to say that there is a minus one-third chance of winning a bet? Would you feel uneasy if someone said that the probability of surviving an operation was one-fourth and the probability of not surviving was also one-fourth?

Before treating these questions, let us first develop some terminology.

When an experiment has been conducted, the collection of all outcomes of the experiment is called the *sample space* for the experiment. The outcomes that make up the sample space are called *simple events*.

Example 1

A large number of students at Ellenville High School are asked their ages. Construct a sample space for the experiment of asking a student his age.

Solution: Assuming that the high school is an ordinary one, the students for the most part will range in age from 12 to 19. Thus, the sample space S will consist of the outcomes 12, 13, 14, 15, 16, 17, 18, and 19. Each of these numbers is referred to as a simple event.

Now suppose we have a sample space S which consist of n outcomes which we shall denote by $o_1, o_2, \ldots, o_n$. Suppose the probabilities assigned (by some method not necessarily known to us) are denoted by $p(o_1)$, $p(o_2)$, $\ldots, p(o_n)$, respectively. We can now deduce some restrictions on these numbers that must hold in order for them to "make sense" if we were to interpret them as stabilized relative frequencies.

Imagine we were to repeat our experiment (whose sample space is S) N times and that o_i occurred n_i times out of N. Then we would have $0 \leq n_i \leq N$. Hence, if we divide by N, we obtain $0 \leq n_i/N \leq 1$; that is, the relative frequency with which o_i occurs is a number that is greater than or equal to 0 but less than or equal to 1. Since as N becomes larger, we want n_i/N to stabilize at $p(o_i)$, it follows that we want $0 \leq p(o_i) \leq 1$. Therefore, we can reasonably make the following assumption about probability numbers:

ASSUMPTION 1

For each outcome o_i (i taking on values 1 up to n) of an experiment, the probability number $p(o_i)$ assigned to the outcome o_i must satisfy the inequality

$$0 \leq p(o_i) \leq 1$$

It may be useful to point out the meaning of saying $p(o_i) = 0$. Intuitively, it may seem to mean that o_i is impossible and that outcome o_i can never occur. If we interpret $p(o_i)$ as a stabilized relative frequency, however, then we can conclude that o_i is extremely unlikely and occurs so rarely that the stabilized relative frequency is 0. A similar analysis is given in suggesting that the meaning of $p(o_i) = 1$ is that o_i almost always happens.

Example 2

Consider the sample space S constructed from the outcomes of the experiment of asking a woman who was a member of the Zero Population Growth Movement how many children she wanted to have. The sample space S would consist of the numbers 0, 1, or 2; i.e., $S = \{0, 1, 2\}$. The following assignments of probabilities would all satisfy assumption 1.

(1) $P(0) = \frac{1}{4}$ $P(1) = \frac{1}{4}$ $P(2) = \frac{1}{2}$
(2) $P(0) = 1$ $P(1) = 0$ $P(2) = 0$
(3) $P(0) = \frac{1}{3}$ $P(1) = \frac{1}{3}$ $P(2) = \frac{1}{3}$
(4) $P(0) = \frac{1}{10}$ $P(1) = \frac{1}{7}$ $P(2) = \frac{52}{70}$
(5) $P(0) = \frac{1}{10}$ $P(1) = \frac{1}{10}$ $P(2) = \frac{1}{10}$

Note that in each case the number assigned to any outcome is between 0 and 1.

Once again, imagine that when our experiment is repeated N times, o_1 occurs n_1 times, o_2 occurs n_2 times, $\ldots$, o_n occurs n_n times. Thus, we have that $n_1 + n_2 + \cdots + n_n = N$. If we divide this equation by N, we obtain

$$\frac{n_1}{N} + \frac{n_2}{N} + \cdots + \frac{n_n}{N} = 1$$

The interpretation of this last equation is that the sum of the relative frequencies must add up to 1. Since as N, the number of times we repeat the experiment, becomes larger, we want n_i/N to stabilize near $p(o_i)$, the probability of o_i, the following equation should hold for probability numbers:

$$p(o_i) + p(o_2) + \cdots + p(o_n) = 1$$

This shows the need for the following assumption:

ASSUMPTION 2

Given the sample space $\{o_1, \ldots, o_n\}$, the probability numbers $p(o_1), \ldots, p(o_n)$ assigned to the outcome $o_1, \ldots, o_n$, respectively, must satisfy

$$p(o_1) + \cdots + p(o_n) = 1$$

Example 3

The assignments of probability given in parts 1, 2, and 3 of Example 2 satisfy Assumption 2 as well as Assumption 1. For the assignments in parts 4 and 5 however, although Assumption 1 is satisfied, we see that Assumption 2 is not satisfied since the probabilities do not sum to 1. A sample space *S*, together with an assignment of probability numbers that satisfies Assumptions 1 and 2, is called a *probability model*. Below are listed some examples of situations where one might be interested in constructing probability models. In each example various different models are considered.

Example 4

A subject in an experiment in behavioral psychology is asked which of the figures below he finds most aesthetically pleasing.

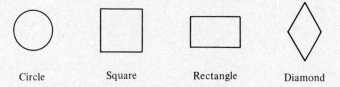

    Circle   Square   Rectangle   Diamond

We shall use the first letters of the words below the diagrams to indicate the outcomes of the experiment. Hence, the sample space is

$$S = \{c, s, r, d\}$$

PROBABILITY MODEL 1

Each of the simple events is assigned a probability of one-quarter.

$$P(c) = \tfrac{1}{4} \qquad P(s) = \tfrac{1}{4} \qquad P(r) = \tfrac{1}{4} \qquad P(d) = \tfrac{1}{4}$$

PROBABILITY MODEL 2

Probability numbers are assigned as shown:

$$P(c) = \tfrac{1}{8} \qquad P(s) = \tfrac{1}{8} \qquad P(r) = \tfrac{1}{2} \qquad P(d) = \tfrac{1}{4}$$

Which of these probability models is correct? The answer is that both are correct. Why? Because all that is required of a probability model is that it obey the requirements of Assumptions 1 and 2. For some particular collection of individuals one model might be more useful than others.

Example 5

An ordinary (six-sided) die is tossed. Construct a probability model for this experiment.

Solution: An ordinary die can show any of six numbers, 1, 2, 3, 4, 5, or 6, when it is rolled. The sample space thus consists of

$$S = \{1, 2, 3, 4, 5, 6\}$$

PROBABILITY MODEL 3:

$$P(1) = \tfrac{1}{6} \quad P(2) = \tfrac{1}{6} \quad P(3) = \tfrac{1}{6} \quad P(4) = \tfrac{1}{6} \quad P(5) = \tfrac{1}{6} \quad P(6) = \tfrac{1}{6}$$

In this model we have assigned each simple event the same probability number.

PROBABILITY MODEL 4: The following is an acceptable probability model since both Assumptions 1 and 2 have been met.

$$P(1) = \tfrac{1}{2} \quad P(2) = \tfrac{1}{4} \quad P(3) = \tfrac{1}{8} \quad P(4) = \tfrac{1}{16} \quad P(5) = \tfrac{1}{32} \quad P(6) = \tfrac{1}{32}$$

PROBABILITY MODEL 5: Since the requirements made by Assumptions 1 and 2 have been met, this is an acceptable probability model:

$$P(1) = \tfrac{1}{2} \quad P(2) = 0 \quad P(3) = 0 \quad P(4) = 0 \quad P(5) = \tfrac{1}{4} \quad P(6) = \tfrac{1}{4}$$

Models 4 and 5 correspond to a biased die, while Model 3 corresponds to a fair die.

Example 6

A drug called hexaglorifane is being tested to determine if it cures the common cold. A patient treated with hexaglorifane can improve, die, or show no change. Construct probability models for the experiment of treating a patient with hexaglorifane.

Solution: The sample space for the experiment has three outcomes, which we shall denote by I, D, N for improve, die, or show no change, respectively.

PROBABILITY MODEL 6:

$$P(I) = \tfrac{1}{3} \quad P(D) = \tfrac{1}{3} \quad P(N) = \tfrac{1}{3}$$

PROBABILITY MODEL 7:

$$P(I) = 0 \quad P(D) = 1 \quad P(N) = 0$$

Pierre-Simon, Marquis de Laplace (1749–1827), was the son of a poor farmer. His most famous work was in celestial mechanics. Using and extending Isaac Newton's work, he calculated the orbits for the moons of Jupiter, as well as the velocity of rotation for Saturn's rings. His work in probability, begun in 1774, culminated in the book *Théorie analytique des probabilités*, which appeared in 1812. This work deals with the foundations of probability theory, geometric probability, least squares (a method of fitting empirical data using curves), games of chance, and probability functions. Part of his writings on probability indicated applications to the theory of elections.
(Courtesy The Bettmann Archive.)

PROBABILITY MODEL 8:

$$P(I) = \tfrac{1}{4} \qquad P(D) = \tfrac{1}{4} \qquad P(N) = \tfrac{1}{2}$$

The choice of model would depend on experimental results concerning hexaglorifane's effectiveness.

In each of Examples 4 to 6, the first probability model given is the kind that is most often encountered. In this kind of probability model all the outcomes in the sample space are assumed to have the same probability. Thus, if the sample space has n outcomes in it, for each outcome o_i we would have $p(o_i) = 1/n$. One reason the *equiprobable model* arises so often is that it corresponds to outcomes to an experiment that *occur at random* or that *occur with equal likelihood*. This model is especially common in gambling situations.

Example 7

What is the probability of drawing the ace of spades from a well-shuffled deck of cards in one try?

Solution: As in all probability problems, the first thing to do is to describe the sample space. In this case, the sample space consists of all the 52 cards in a standard deck. Since the drawn card is selected from a well-shuffled deck, it is natural to assume that each card has the same probability of being drawn. Thus, the probability of drawing the ace of spades is $\frac{1}{52}$. In terms of relative frequency, this means that over a very long run of repetitions of this experiment the ace of spades will be drawn about 1 time out of 52.

Another argument in favor of the equiprobable model is that it is appealingly simple. It should be borne in mind, however, that in real problems relative frequencies are rarely equal. For example, even if a symmetrical die is rolled 600 times, it is not likely that each of the six possible outcomes will occur exactly 100 times.

As a further example, consider the experiment of asking a high school student his age. It seems safe to say that the outcomes of this experiment are the numbers from 12 to 19. It is also safe to say, however, that the probability assignment would not be the equiprobable one—the numbers 12 and 19 would be much less likely to occur than the others under normal circumstances. For the sake of our further discussion, suppose the probability assignment is the following:

$$P(12) = 0.05$$
$$P(13) = 0.16$$
$$P(14) = 0.18$$
$$P(15) = 0.19$$
$$P(16) = 0.19$$
$$P(17) = 0.13$$
$$P(18) = 0.06$$
$$P(19) = \underline{0.04}$$
$$1.00$$

Now suppose we change the experiment by selecting a student at random and asking him whether he is eligible to vote in the Presidential election that is occurring that year. What is the probability that the answer is yes? One approach would be to construct a new sample space $S = \{yes, no\}$. But what probability assignment would we use? The equiprobable one is certainly no good. One feels instinctively that the probability assignment for the age experiment ought to help us out.

The answer will be *yes* provided the student is 18 or 19, and we intuitively feel that the probability of a *yes* answer would be $P(18) + P(19) = 0.06 + 0.04 = 0.10$. We formalize this type of reasoning, without recourse to a new sample space, as follows.

DEFINITION 1

If S is a sample space, then any subset E of S is referred to as an *event*. The subset of S consisting of all elements of the sample space not in E is called the *complement of E* and is denoted by E'. The probability of any event is the sum of the probabilities of the outcomes constituting that event.

In the example concerning the high school students, the event *being able to vote in the Presidential election* is just a wordy way of describing the subset {18, 19}. Consequently, according to our definition, $P(18, 19) = P(18) + P(19) = 0.06 + 0.04 = 0.10$. The event {12, 13, 14, 15, 16, 17} is the complementary event to {18, 19} because it contains all the remaining outcomes of the sample space. Its probability is $0.05 + 0.16 + 0.18 + 0.19 + 0.19 + 0.13 = 0.90$. A simpler way to calculate the probability of the complementary event is $P(12, 13, 14, 15, 16, 17) = 1 - P(18, 19) = 1 - 0.10 = 0.90$. This example is a special case of the following:

THEOREM 1

If E has probability $P(E)$, then $P(E') = 1 - P(E)$.

Example 8

What is the probability of drawing a spade from a well-shuffled standard deck of cards? What is the probability of drawing either a spade or an ace? Describe the complementary events and their probabilities.

Solution: The sample space consists of the 52 cards. The event we are interested in, described in words as *drawing a spade*, consists of 13 of these 52 cards. Using the equiprobable assignment of probabilities, $P(\text{spade}) = \frac{13}{52} = \frac{1}{4}$. The event *spade or ace* consists of 16 cards (13 spades, together with 3 other aces) and so $P(\text{spade or ace}) = \frac{16}{52} = \frac{4}{13}$. The complementary event to *drawing a spade* is *drawing a heart, club, or diamond* and has probability $1 - \frac{1}{4} = \frac{3}{4}$. The complementary event to *drawing a spade or an ace* is *drawing a heart, club or, diamond that is not an ace*. Its probability is $\frac{9}{13}$.

Occasionally the construction of a suitable sample space and probability assignment is a bit less routine than in the foregoing examples. Here is an illustration.

Example 9

A married couple expects to have two children and wants to know the probability of the event *both children will be of the same sex*.

Solution 1: Take as our sample space B_0, B_1, B_2 (where B_i represents the outcome *i boys*). Assuming the equiprobable model, the event we are interested in has probability one-third.

Solution 2: Take as the sample space *BB, BG, GB, GG* (where *BG*, for example, means the first child is a boy and the second is a girl) and assume the equiprobable model. Then the event we are interested in is $\{BB,\ GG\}$ and has probability $\frac{1}{4} + \frac{1}{4} = \frac{1}{2}$.

Both solutions may seem reasonable but only one is useful to model the real-world situation. Solution 2 actually fits the real data while Solution 1 does not. (You can verify this with a coin tossing experiment—interpret heads as having a girl.) There are also theoretical reasons for arriving at Solution 2 that we shall discuss in Section 8.4. For the moment, we wish merely to make the following observations motivated by this example: Overly casual use of the equiprobable model may lead to errors.

EXERCISES 8.2

1. For the various assignments of probabilities to the sample space below, state which assignments would constitute a probability model. If the assignment is not a probability model, give a reason.

    John asks his date to select an ice cream flavor from a list of 5 on a chart at the restaurant. The sample space S of outcomes uses appropriate abbreviations for the flavors.

    $S = \{C$ (chocolate), V (vanilla), M (mint), P (pecan) S (strawberry)$\}$

    (a) $P(S) = \frac{1}{3}, P(C) = \frac{1}{3}, P(P) = \frac{1}{3}, P(V) = 0, P(M) = 0$
    (b) $P(S) = \frac{1}{3}, P(C) = \frac{1}{5}, P(P) = \frac{1}{5}, P(V) = \frac{1}{5}, P(M) = \frac{1}{5}$
    (c) $P(S) = -\frac{1}{2}, P(C) = 1, P(P) = \frac{3}{4}, P(V) = -\frac{1}{4}, P(M) = 0$
    (d) $P(S) = \frac{1}{8}, P(C) = \frac{1}{8}, P(P) = \frac{1}{8}, P(V) = \frac{1}{8}, P(M) = \frac{3}{2}$
    (e) $P(S) = \frac{1}{6}, P(C) = \frac{1}{3}, P(P) = \frac{1}{3}, P(V) = \frac{1}{6}, P(M) = \frac{1}{12}$
    (f) $P(S) = \frac{1}{6}, P(C) = \frac{1}{3}, P(P) = \frac{1}{6}, P(V) = \frac{1}{12}, P(M) = \frac{1}{6}$

2. The National Energy Commission is planning to permit the building of a nuclear power plant at one of these sites: Chicago, Ill.; Los Angles, Calif.; Big River, Mont.; Ogalala, Neb.; Superior, Wis; or Hornitos, Calif. At present it seems that all sites are equally likely to be chosen. Find the probabilities of the following events:
    (a) The site chosen is adjacent to a major population center.
    (b) The site chosen is near a major body of water.
    (c) The site chosen is in a major earthquake zone.

3. A student selects a book off the shelf in his library and records the first letter of the first word in Chapter 1. Construct a sample space for this experiment.

4. A single card is drawn from a well-shuffled deck. If we assume the equiprobable measure is used, what is
    (a) The probability a spade or a heart is drawn?
    (b) The probability a 4 or an 8 is drawn?
    (c) The probability of a 5 or a heart?

5. Referring to Example 2 and using Model 1, compute the probabilities of the following events:
   (a) The figure chosen has no right angle.
   (b) The figure chosen has curved lines.
   (c) The figure chosen is a quadrilateral.
   Compute the answers for the problem above for Model 2.

6. Suppose A and B are two events in sample space S. Show that $P(A \cup B)$ $= P(A) + P(B) - P(A \cap B)$. Consider Example 5, Model 4 and the events

$$A = \text{an odd number appears}$$

$$B = \text{a number greater than 4 appears}$$

$$C = 4 \text{ appears}$$

   Compute
   (a) $P(A')$
   (b) $P(B')$
   (c) $P(A \cup B)$
   (d) $P(A \cup C)$
   (e) $P(B \cup C)$
   (f) $P(C')$
   (g) $P(A' \cup B)$

7. A mathematics class has 20 girls and 10 boys. The names of the students are written on tags that are then placed in a jar and mixed. One tag is drawn. What probability model would you use for the sample space S $= \{B, G\}$ ($B =$ boy chosen, $G =$ girl chosen)?

8. Jerry has some socks in the dryer. There are two pairs of brown socks, one pair of blue socks, and two pairs of black socks. Suppose a single sock is drawn from the dryer, and we assume any individual sock out of the ten socks is as likely to be drawn as any other.
   (a) What is the probability a black sock is drawn?
   (b) What is the probability a black or a blue sock is drawn?
   (c) What is the probability a brown sock is not drawn?
   (d) What is the probability a blue sock is not drawn?

9. A single marble is drawn from a bag containing three red, four green, and two blue marbles. If we assume the marbles are all equally likely to be chosen.
   (a) The probability a blue or red marble is chosen.
   (b) The probability a green marble is not chosen.

10.* Using the four figures of Example 3, ask 40 individuals which shape they prefer. Use the relative frequencies you get to construct a probability model.

11.* Using your local phone book, record the last digit of the first 50 telephone numbers. Compute the relative frequencies you obtain. Does your data support the use of the equiprobable model for the experiment: *record the last digit of a number drawn from the telephone book*?

12.* For a period of 30 nights, count the number of phone calls your family gets between 7 and 10 P.M. Use the data you collect to estimate the probability of your getting k ($k \geq 0$) phone calls between 7 and 10 P.M.

13.* A person is chosen at random and is asked in what state he was born. A probability of $\frac{1}{50}$ is assigned to each possible outcome. Is the resulting probability model reasonable?

8.3   multistage experiments

Let us consider again the plight of the couple looking for a suitable sample space to describe the experiment of having two children. This experiment differs from our earlier examples in that the experiment is really a multistage experiment consisting of two repetitions of the simpler experiment of having one child. For this simpler experiment there is only one reasonable sample space, $\{B, G.\}$ The sample space for the multistage experiment can be constructed from the simpler experiment by drawing a certain kind of graph called a *tree*.

The outcomes of the multistage experiment will be built up, one stage at a time, from a single vertex called a *root*. Let us illustrate the procedure by the multistage experiment of the couple planning to have two children. Starting at the root of the tree (see Figure 8.2) we draw as many edges as there are

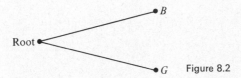

Figure 8.2

outcomes to the first stage of the multistage experiment. In the case of the birth of a child, there are two outcomes, a boy or a girl. Hence, as shown in Figure 8.2, we draw two edges from the root. We label the end points of these edges by B and G to indicate, respectively, that the first child born was a boy and that the first child born was a girl. We now imagine that the second stage of the experiment is carried out. If the result at the first stage was a girl, there are two possible outcomes for the second child—boy or girl. We indicate this by taking the 1-valent vertex G and drawing from it two edges, one for each outcome of the second stage of the experiment. Our tree would now appear as shown in Figure 8.3. The labels of the end points of the new edges are designed to accumulate the history of the previous stages. Thus, the label *GG* means the first child was a girl and the second child was a girl.

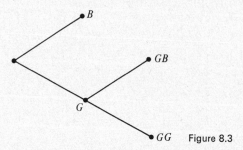

Figure 8.3

The label *GB* indicates that the first child was a girl and the second child was a boy. If the result at the first stage was a boy, we again have two possible outcomes for the second child, a boy or a girl. To indicate these outcomes in the second stage, we take the 1-valent vertex *B* (Figure 8.3) and draw two edges emanating from the vertex, one for each of the possible outcomes. The final result for the two stages, incorporating both cases from the first stage, is shown in Figure 8.4. Further stages of this experiment would be depicted

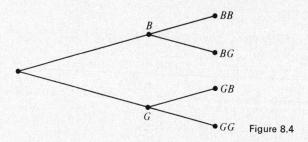

Figure 8.4

the same way. Thus, Figure 8.5 shows the tree for the experiment of having three children.

Using the procedure developed here, together with your knowledge of the equiprobable model, can you answer any of the questions of the true-false examination at the start of this chapter?

It may happen that the simple experiments that constitute the multi-stage experiment are not identical. For example, consider the two-stage experiment of determining environmental conditions at a certain weather station and suppose that this consists of two simpler experiments: First we determine whether or not it is raining—the sample space obviously being

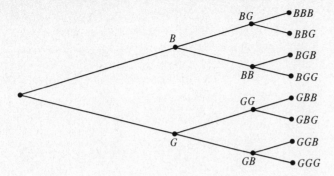

Figure 8.5

{Rain, Fair}—and then we check air pollution readings, with the sample space being {Acceptable, Unacceptable, Hazardous}. The tree for this two-stage experiment is shown in Figure 8.6. We could also do our classifications

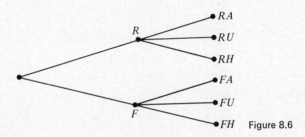

Figure 8.6

in the reverse order, checking for pollution before rain. Figure 8.7 shows the tree we would get in that case. The two trees look different but the outcomes at the right-hand end are the same.

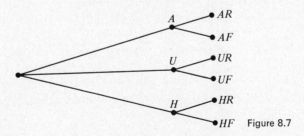

Figure 8.7

Sometimes it is useful to know the total number of outcomes in the sample space of a multistage experiment without listing all these outcomes. For example, how many outcomes are there if we flip a coin six times? The outcomes for the first three stages (flips) are shown in Figure 8.8 (H = heads; T = tails). If we imagine extending the tree of Figure 8.8 three more stages, in each extension doubling the number of outcomes, we obtain 64 possible outcomes after six stages. This reasoning is a particular case of the following general principle.

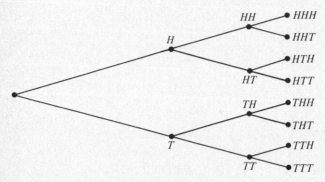

Figure 8.8

PRINCIPLE OF COUNTING

Given k experiments whose results do not depend on one another, if there are n_1 outcomes in one experiment, n_2 outcomes in a second experiment, ... , n_k outcomes in a kth experiment, then the number of different outcomes in the multistage experiment built up from the k experiments is $n_1 \times n_2 \times \cdots \times n_k$.

As an example of where this principle comes in handy, suppose in a group of 24 people we ask each person his birthday and we wish to determine the probability that there will be some pair of persons with the same birthday. This is best thought of as a 24-stage experiment where each stage is the simple experiment of asking one person his birthday. We shall ignore leap years and suppose that the sample space for each such simple experiment consists of the integers from 1 to 365, representing the various days of the year. If we were to draw a tree diagram for this 24-stage experiment (an impossible task actually), we would have, according to the Principle of Counting, 365^{24} different outcomes possible. We can think of these outcomes as being sequences with 24 entries, each entry being one of the numbers from 1 to 365. The event E we are interested in is the collection of all sequences where at least two entries are the same. We shall assume the equiprobable probability assignment on the sample space of 365^{24} outcomes, and so

$$p(E) = \frac{n(E)}{365^{24}}$$

where $n(E)$ is the number of elements in E. Alternatively,

$$p(E) = 1 - p(E') = 1 - \frac{n(E')}{365^{24}}$$

where E' is the complementary event which can be described as the set of all sequences in S which have no entries the same.

It turns out that we can easily count $n(E')$ by the Principle of Counting. Clearly, the first entry in a typical sequence in E' can occur in any one of 365 ways. The second entry, however, corresponding to the second person's birthday, must be different from the first entry if the sequence is in E'. Thus, the second entry can occur in only 364 different ways. Continuing this way and applying the Principle of Counting, we see that the total number of sequences in E' is $365 \times 364 \times \cdots \times 342$. Therefore,

$$p(E) = 1 - \frac{365 \times 364 \times \cdots \times 342}{365 \times 365 \times \cdots \times 365}$$

The computation of the value of $p(E)$ is tedious but not difficult. The value of $p(E)$ is 0.538. Does this surprise you? For some reason it does not seem intuitively acceptable that in a group of 24 persons the chances are better than even that there will be two or more people with the same birthday. It may be of interest to see how the value of $p(E)$ depends on the number of people in the group. This is shown in Table 8.1.

Table 8.1   The Probability $P(E)$ (to two decimals) for at Least Two Persons in a Group of n Having the Same Birthday

$n =$	$P(E) =$	$n =$	$P(E) =$
3	0.01	17	0.32
4	0.02	18	0.35
5	0.03	19	0.38
6	0.05	20	0.41
7	0.06	21	0.45
8	0.08	22	0.48
9	0.10	23	0.51
10	0.12	24	0.54
11	0.15	25	0.57
12	0.17	26	0.60
13	0.20	27	0.63
14	0.23	28	0.66
15	0.26	29	0.68
16	0.29	30	0.70
		31	0.73
		32	0.75

To return briefly to the subject of tree diagrams, it should be pointed out that there are cases where the Principle of Counting won't work but where a tree diagram can still be an aid in enumerating possibilities. The following examples illustrate this.

Example 1

A man has five coins in his pocket, one nickel (N), one dime (D) and three pennies (P). He reaches into his pocket and draws out one coin. Without replacing it, he then draws another. What outcomes can be obtained? How many outcomes are there?

Solution: The final result can be built up in two stages. First one coin and then another must be drawn from the pocket. However, the types of coins available on the second drawing depend on what coin was drawn the first time. Thus, if the first coin drawn was a nickel, the second coin drawn could not be a nickel but must be either a dime or a penny. On the other hand, if the first coin drawn was a penny, the next coin could be of three types, a penny, a nickel, or a dime. With these ideas in mind one obtains the tree shown in Figure 8.9. There is a total of seven outcomes. This could not have been calculated from the Principle of Counting. Do you see why?

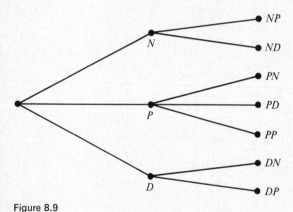

Figure 8.9

Example 2

A couple decides to have children until they have a boy and a girl or until they have a total of three children, whichever comes first. Describe and enumerate the possible outcomes of this multistage experiment.

Solution: Figure 8.10 shows a tree diagram for the outcomes. As can be seen, the six outcomes are *BBB, BBG, BG, GB, GGB, GGG.*

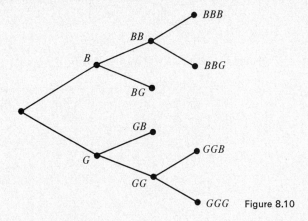

Figure 8.10

EXERCISES 8.3

1. Mary owns three skirts and four blouses. Use a tree diagram to construct the sample space for the experiment of picking a skirt-blouse outfit.

2. A coin is tossed four times. Use a tree graph to construct the collection of possible outcomes.

3. A student takes a five-question multiple-choice examination. Use a tree diagram to construct a sample space for the experiment of taking such an examination.

4. An amateur weatherman classifies December days according to whether or not they are clear or cloudy and according to whether they are cool, cold, or very cold. Draw a tree diagram to indicate the outcomes of the experiment of classifying a December day.

5. A couple is planning to have four children. Draw a tree diagram to indicate the possible outcomes in the sample space.
   (a) What outcomes make up the event that they have an equal number of boys and girls? Call the event E.
   (b) Construct the equiprobable model for the sample space.
   (c) What is the probability of event E? Does this result seem unintuitive?

6. In Chapter 9 we shall study the theory of games. A game can be classified according to whether it has two or more players, whether it is zero-sum or not, and whether it permits communication or not. Draw a tree diagram to show the different outcomes for classifying a game.

7. In anthropology a culture can be classified as matrilineal or patrilineal (i.e., whether or not you belong to your mother's clan or your father's clan). A culture can also be classified as matrilocal or patrilocal (i.e., whether a married couple lives at the home location of the husband

or wife). If you pick a culture to study, draw a tree diagram to show the possible outcomes.

8. A psychologist is interested in studying emotional stability in families with one child. The mother, the father, and the child can each be classified according to whether they are emotionally unstable or not. A one-child family is selected and its members tested for emotional stability. Draw a tree diagram for the possible outcomes.

9. A sociologist is making a survey concerning political views. He classifies the persons he interviews by income (low, middle, high), sex, and educational level (graduated college, did not graduate college). Draw a tree diagram to show the possible outcomes.

10. The main course of a meal consists of meat with two different kinds of vegetables. If there are four kinds of meat available and a choice of six possible vegetables, use the Principle of Counting to compute how many main courses can be served.

11. A drug is tested on four patients. The patient either improves (I) or does not improve (I'). Draw a tree diagram to construct the sample space for the experiment of testing four patients. Let A be the event that all the patients improve. Let B be the event that exactly three patients improve. If the equiprobable measure is used on the sample space, what is the probability of A, and what is the probability of B?

12. A coin is tossed five times or until a head appears. Draw a tree diagram for the outcomes of the multistage experiment. (*Hint:* The first toss has two outcomes, H and T. At the second stage only T is branched since the experiment would terminate at H, had an H occurred.)

13. A neighborhood health center classifies children in its outpatient clinic by age and sex and whether or not they have had vaccinations for diphtheria, smallpox, or polio. The center treats children from ages 3 to 14. A typical category might be 7 year old girls who have been vaccinated for diphtheria but not for smallpox or polio. What total number of categories is needed?

14. The Gruenbox Car Manufacturing Company advertises that its cars come in four colors; with or without automatic transmission; and with a choice of no radio, FM only radio, AM only radio, or FM and AM radio. How many models of cars does the company produce?

15.* Use the Principle of Counting to determine the number of anagrams (i.e., rearrangements using the same letters) of your name. (*Note:* If your name has repeated letters, you can't answer the question in a straightforward way. However, you can think of the repeated letters as being different at first and then decide how many anagrams you have overcounted when the repeated letters are not considered different.)

16.* Seven people are each asked their horoscope signs (of which there are

12 in all). What is the probability that some pair of people will have a common horoscope sign?

17.* An ESP researcher asks each of five people to choose a number between 1 and 20. After comparing notes, it is determined whether or not any two people chose the same number. After doing this experiment 1000 times, the researcher discovers that 700 times he got two people with the same number. Does this tend to substantiate the claim that ESP exists?

18.* Among a group of n people, compute the probability that at least two persons have their birthday in the same month for

(a)	$n = 2$		(d)	$n = 5$
(b)	$n = 3$		(e)	$n = 6$
(c)	$n = 4$		(f)	$n = 7$

Do your answers surprise you?

8.4 independence

In the previous section we showed how it was possible to obtain an enumeration of the set of outcomes of a multistage experiment arising from a series of experiments or from the repetition of the same experiment several times. It would be convenient if we could find a natural way of assigning probability numbers to the outcomes of a multistage experiment by knowing how probability numbers had been assigned to the component experiments that were performed to obtain the multistage experiment.

Let us consider a concrete simple example. Suppose experiment 1 consists of tossing a nonsymmetric coin and experiment 2 consists of another toss of the same coin. The multistage experiment will consist of tossing a coin twice. Figure 8.11(a), (b) and (c) shows the outcomes of experiments 1 and 2 and the multistage experiment.

Suppose statistics compiled over a long period of time show that the coin results in heads eight-tenths of the cases in which it was tossed. Knowing this, we use the natural assignment of probabilities to the outcomes of experiments 1 and 2 shown in Figure 8.12. We now wish to assign probabilities to the outcomes of the multistage experiment [see Figure 8.11(c)] using the probability assignments in Figure 8.12. Let us consider the simple event *HH* (heads on the first toss and heads on the second toss) which is a typical element in the sample space for the multistage experiment.

It will be convenient to denote by A that a head occurred on the first toss and by B that a head occurred on the second toss. AB will denote that a head occurred on the first toss and a head occurred on the second toss. Note that AB is equivalent to the outcome *HH*.

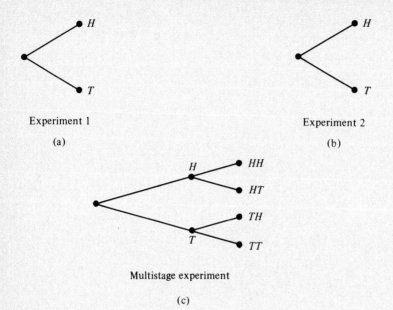

Experiment 1

(a)

Experiment 2

(b)

Multistage experiment

(c)

Figure 8.11

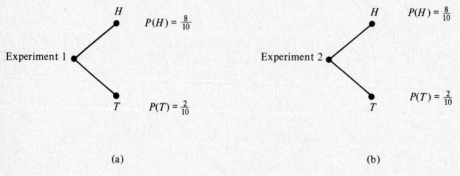

Experiment 1 $P(H) = \frac{8}{10}$

$P(T) = \frac{2}{10}$

(a)

Experiment 2 $P(H) = \frac{8}{10}$

$P(T) = \frac{2}{10}$

(b)

Figure 8.12

Suppose that the multistage experiment is repeated N times, where N is a large number. Let N_{AB} denote the number of times out of the N repetitions of the experiment that AB occurs. Similarly, denote by N_A and N_B the number of times out of N repetitions of the experiment that A and B occur, respectively. We can then start with the algebraic identity,

$$\frac{N_{AB}}{N} = \frac{N_{AB}}{N} \qquad (8.1)$$

294

Now using the fact that

$$\frac{N_A}{N_A} = 1$$

we can rewrite Equation (8.1),

$$\frac{N_{AB}}{N} = \frac{N_A}{N} \cdot \frac{N_{AB}}{N_A} \tag{8.2}$$

Next consider the fraction N_{AB}/N_A. This represents the fraction of the time when the first toss was a head that the second toss was also a head. However, since we are dealing with two different tosses, *there is no reason to believe that the fraction of the time that the second toss is a head when the first toss is a head is different from the fraction of the time that the second toss is head, namely,* N_B/N. Thus, the real-world situation suggests that $N_{AB}/N_A = N_B/N$. Making this substitution into (8.2) we have

$$\frac{N_{AB}}{N} = \frac{N_A}{N} \cdot \frac{N_B}{N}$$

As N gets large, the fractions N_{AB}/N, N_A/N, and N_B/N stabilize at $P(AB)$, $P(A)$, and $P(B)$, respectively, so that

$$P(AB) = P(A) \cdot P(B) \tag{8.3}$$

We have thereby accomplished the task of assigning a probability to the event HH in the multistage experiment on the basis of the assignments in the component experiments; that is, since $P(A) = P$ (heads on first toss) $= \frac{8}{10}$ [see Figure 8.12(a)] and $P(B) = P$ (heads on the second toss) $= \frac{8}{10}$ [see Figure 8.12(b)], we assign as $P(HH) = P(AB)$ the value $\frac{8}{10} \cdot \frac{8}{10}$ as suggested by (8.3). The assignment of probabilities to the remaining outcomes in Figure 8.11(c) is shown in Figure 8.13.

The critical step in deriving the multiplication rule was the observation

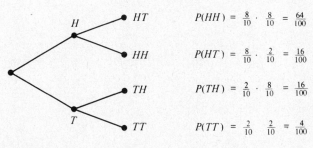

Figure 8.13

that, in the real world, whether there is a head on the first toss has no effect on whether or not there is a head or a tail on the second toss. This notion of the inability of certain occurrences in the real world to affect other occurrences can be formulated as follows:

DEFINITION 2

Two experiments or two repetitions of the same experiment are said to be *independent* if the results of one of the experiments do not influence the results of the other.

Here are some examples of pairs of experiments which seem to be independent:

(1)   An oil company determines whether there is oil far beneath a certain parcel of land. The company also has soil fertility tests made to determine its suitability for agriculture.

(2)   On two successive days you determine whether the temperature reading is an even number or an odd number.

An example of two experiments which are unlikely to be independent is: to measure a man's height and then determine his shoe size. In many cases it is unknown or is a matter of opinion whether two experiments are independent or not. For instance, consider the experiment of determining whether the stock market is "bullish" or "bearish" together with the experiment of determining whether women's hemlines are above the knee, at the knee, or below the knee. Some people claim to see patterns relating stock market fluctuations to hemline fluctuations. If such a pattern exists, it would suggest that these experiments are not independent.

The reasoning used in deriving equation (8.3) in the special case of the coin toss experiment can be extended to justify the following general principle.

PRINCIPLE

If E_1 and E_2 are two independent experiments and A is an event in the sample space of E_1, while B is an event in the sample space of E_2, then

$P\,(A \text{ occurs in } E_1 \text{ and } B \text{ occurs in } E_2)$

$$= P\,(A \text{ occurs in } E_1) \cdot P\,(B \text{ occurs in } E_2)$$

We will call this the *multiplication rule*.

The main importance the multiplication rule has for us is that it allows us to assign probabilities to the outcomes of a multistage experiment on the basis of the probability assignments of the component experiments. Figure 8.13 illustrates this for the coin toss experiment. The rest of this section consists of more examples of the application of the multiplication rule.

Example 1

Statistics suggest that the drug Hexoglom has a probability of $\frac{2}{3}$ of improving a patient suffering from disease X. What is the probability that if the drug is tried on two suffers of disease X, Mr. Gold and Mr. Silver, both will show improvement?

Solution: Figure 8.14 shows a tree for the multistage experiment where the drug is first tried on Mr. Gold and then tried on Mr. Silver. The letter I

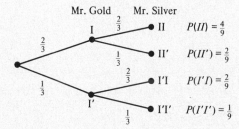

Figure 8.14

denotes improvement and I' denotes the complementary event, nonimprovement. II denotes that both Mr. Gold and Mr. Silver improved. Each branch of the tree is labeled with its appropriate probability. For example, the branch from I to II represents improvement for Mr. Silver and is therefore labeled $\frac{2}{3}$. To find the probability of an outcome by the multiplication rule, multiply the probabilities along the path leading to that outcome from the root. Hence:

$$P(II) = P \text{ (Mr. Gold improved)} \cdot P \text{ (Mr. Silver improved)}$$
$$= \tfrac{2}{3} \cdot \tfrac{2}{3}$$
$$= \tfrac{4}{9}.$$

$$P(II') = \tfrac{2}{3} \cdot \tfrac{1}{3} = \tfrac{2}{9}$$

$$P(I'I) = \tfrac{1}{3} \cdot \tfrac{2}{3} = \tfrac{2}{9}$$

$$P(I'I') = \tfrac{1}{3} \cdot \tfrac{1}{3} = \tfrac{1}{9}.$$

Example 2

In Example 9 of Section 8.2 we considered a problem concerning a married couple planning a family of two children. The second solution developed there had a sample space of four elements. We arrived at an assignment of probabilities by using the equiprobable measure. We now can give an alternative view which yields the same probability assignment. The outcomes can be regarded as belonging to the multistage experiment of one birth after another. (See Figure 8.15.)

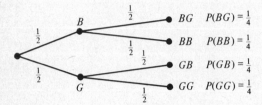

Figure 8.15

On the basis of current genetic theory, we can assume that the sexes of the children are independent. Hence, we can apply the multiplication rule. For example,

$$P(BG) = P \text{ (first child a boy and second child a girl)}$$
$$= P \text{ (first child is a boy)} \cdot P \text{ (second child is a girl)}$$
$$= \tfrac{1}{2} \cdot \tfrac{1}{2} = \tfrac{1}{4}$$

The other three events are similarly computed to be

$$P(BB) = \tfrac{1}{2} \cdot \tfrac{1}{2} = \tfrac{1}{4}$$

$$P(GG) = \tfrac{1}{2} \cdot \tfrac{1}{2} = \tfrac{1}{4}$$

$$P(GB) = \tfrac{1}{2} \cdot \tfrac{1}{2} = \tfrac{1}{4}$$

The concept of independent experiments can be extended to more than two experiments. If we have n experiments, they are said to be independent of one another if none of them is affected by the outcome of any other. One can easily justify a multiplication rule analogous to the one we used for two-stage experiments. The following example illustrates the use of the extended multiplication rule.

Example 3

A manufacturer of airplanes is under government contract to provide the government with a large number of a standard type of aircraft. The manufacturer states that the chance of a plane malfunction is 1 in 1000. The government tests three aircraft and finds two defective. What is the probability of this happening if the manufacturer's claim is correct?

Solution: Letting D denote defective and D' not defective, we obtain as our sample space S,

$$S = \{DDD,\ DDD',\ DD'D,\ DD'D',\ D'DD,\ D'DD',\ D'D'D,\ D'D'D'\}$$

The tree in Figure 8.16 illustrates this multistage experiment.

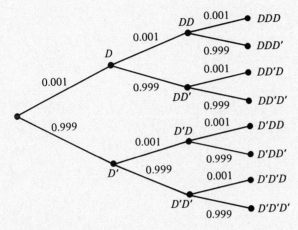

Figure 8.16

The event we are investigating, two of three airplanes defective, has these elements:

$$DDD' \qquad DD'D \qquad D'DD$$

Since we may assume that whether or not a given plane is defective is independent of whether or not the other planes are defective, we can employ the multiplication rule. Thus, for example,

$P(DDD') = P$ (first plane is defective, second plane is defective, and third plane is not defective)

$\quad = P$ (first plane is defective)$\cdot P$ (second plane is defective)$\cdot P$ (third plane is not defective)

$\quad = \frac{1}{1000} \cdot \frac{1}{1000} \cdot \frac{999}{1000}$

$\quad = \frac{999}{1,000.000,000}.$

Similarly,

$$P(DD'D) = \frac{1}{1000} \cdot \frac{999}{1000} \cdot \frac{1}{1000} = \frac{999}{1,000,000,000}$$

and

$$P(D'DD) = \frac{999}{1000} \cdot \frac{1}{1000} \cdot \frac{1}{1000} = \frac{999}{1,000,000,000}$$

Hence,

$$P \text{ (two defective planes and one not defective)} = \frac{3(999)}{1,000,000,000}$$

This number is so small that the manufacturer's claim seems most unlikely.

Our next example emphasizes that it is not necessary to have a repeated experiment to apply independence and the multiplication rule. We may apply the multiplication rule in cases involving two distinct experiments, provided that we are confident they are independent.

Example 4

The Eco-Chem Research and Development Corporation is working on a filtration process to combat water pollution. There is a $\frac{1}{3}$ chance that this process will turn out to be effective. If it is effective, the resulting purchases of the process by industrial plants will keep Eco-Chem from imminent bankruptcy. Alternatively, if Congress passes a pending bill to supply research and development funds to firms like Eco-Chem, then the company will be able to avoid bankruptcy by using these funds alone. The chance of congressional passage for this bill is $\frac{5}{6}$. Assuming that the fate of the bill and the fate of the filtration process are independent, determine the probability that Eco-Chem will avoid bankruptcy.

Solution: The first experiment is to determine whether the filtration system is successful (S) or unsuccessful (U). The second experiment is to determine whether the bill will be passed (P) or not (N). The tree for the multistage experiment is shown in Figure 8.17 and has been labeled with the appropriate

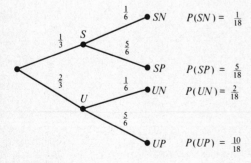

Figure 8.17

probabilities. Since the event "bankruptcy is avoided" is {SN, SP, UP},

$$P(SN, SP, UP) = \tfrac{1}{18} + \tfrac{5}{18} + \tfrac{10}{18} = \tfrac{16}{18} = \tfrac{8}{9}$$

It is worth noting that this problem can also be solved using the tree shown in Figure 8.18. The order in which the experiments are listed has been

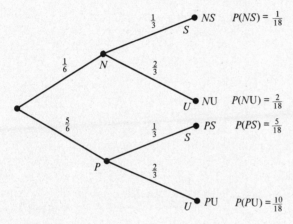

Figure 8.18

reversed. This procedure is valid because the experiments are independent of one another. Using the multiplication rule as usual, we obtain

$$P \text{ (bankruptcy is avoided)} = \tfrac{1}{18} + \tfrac{5}{18} + \tfrac{10}{18} = \tfrac{8}{9}$$

This is the same answer that was obtained using the tree in Figure 8.17.

EXERCISES 8.4

1. A drug is tested on two different sufferers from Parkinson's disease. Extensive testing has shown that the relative frequency of success with the drug is $\tfrac{8}{10}$. Construct an appropriate model and determine the following:
    (a) What is the probability that the drug works on at least one of the two patients?
    (b) What is the probability that the drug fails on both patients?
2. A medium sized city, Middletown, is certain to grow substantially if at least two out of three of the following events occur:
    H: a branch of the interstate highway is constructed nearby.

G: a government installation is built in the area.

U: a university in the vicinity expands.

Assume the probability that H occurs in Middletown is $\frac{1}{2}$ and that each of the other two events has probability $\frac{1}{3}$ of occurring in Middletown. What is the probability that the city will not grow substantially? Assume that independence holds.

3. Jake is taking courses in French and mathematics on a pass-fail basis. If the chance of Jake's failing a course is $\frac{1}{10}$, and if the grades in the two courses are independent, compute:
    (a) The probability of failing both courses.
    (b) The probability of passing both courses.
    (c) The probability of passing at least one course.

4. A manufacturing concern is pursuing two distinct lines of research to try to reduce pollutants in the waste it discharges into local waterways. One line of research involves filtering and treating the discharge. There is a $\frac{1}{3}$ chance that a practical method will result from this work. The other line of research concerns the nature of the production process. There is a $\frac{1}{2}$ chance that this will lead to success. The two lines of research involve totally different scientific principles, so we will assume that the results of the two investigations are independent. What is the probability that the company will not be able to use either method to reduce pollution?

5. Suppose Mary reaches into her dresser and pulls out a blouse at random. She has two blue blouses and one red blouse in the dresser. Now Mary grabs a skirt from a closet that contains three green skirts and one blue skirt.
    (a) Draw a tree diagram for the multistage experiment of choosing a blouse and a skirt.
    (b) Assign probabilities to the component experiments.
    (c) Use the assumption of independence and the multiplication rule to find the probability that:
        (i) The skirt-blouse outfit is all blue.
        (ii) A green skirt and a red blouse were chosen.

6. Mark fires three shots at a target. Past records show that he hits the target four out of five times. If his shots are independent, what is the probability that all three shots are misses?

7. In Act III of Shakespeare's "Romeo and Juliet," Mercutio says: "I am hurt. A plague o' both your houses." Assuming that Mercutio has a $\frac{1}{4}$ chance of calling down a curse on an individual house, what is the chance that both houses will be cursed? Assume that independence holds.

8. A survey of three voters is taken in a town where 40% of the voters are

Independents (I), 25% of the voters are Republicans (R), and the remaining voters are Democrats (D). Construct an appropriate model to determine:

(a) The probability that at least two voters are Independents.

(b) The probability all three voters are Republicans.

(c) The probability all three voters are Democrats.

(d) The probability that at most two voters are Republicans.

9. A hospital has an emergency generator. The chance of the emergency generator working on a given day is $\frac{1}{10}$. The chance of there being a general power failure on a given day is 0.0001. If the failure of the hospital generator and a general power failure are independent, what is the probability that on a given day the hospital will be without power?

10. In any given summer there is a $\frac{1}{4}$ chance that there will be heavy rains and a $\frac{1}{4}$ chance that there will be a plague of locusts. Assuming that the situations are independent, what is the chance that a farmer will avoid both these afflictions?

11. Art Wester, the Prentice-Hall mathematics editor, has a remarkable family. His greatgrandfather, his grandmother, his father, and his son all have their birthdays on the same date. Calculate the probability that four given people have the same birthday by thinking of this as a multi-stage experiment whose component stages are independent.

12.* A new nuclear attack detection system is under development to supplement two existing systems, the satellite based system and the land based radar net. The satellite system has a 0.001 chance of malfunctioning while the radar net malfunctions with probability 0.005. It is desired to refine the new system to the point where there is no more than a 0.000001 chance of two out of three systems malfunctioning at the same time. What is the highest probability of malfunction that is acceptable in the new system, assuming that malfunctions in the three systems will be independent?

8.5 the binomial probability model

We noticed in Chapter 2 that sometimes there were many real-world problems that could be solved simultaneously by using one graph theoretic model. A similar situation occurs in probability theory.

Consider the following problems:

(1) The tubes that a manufacturer makes can be classified as being defective or nondefective. If 10 tubes are tested and the probability that a given tube is defective is $\frac{1}{20}$, find the probability that exactly 6 of the tubes are defective.

(2)   All babies can be classified as males or females. If 20 babies are born on
        a certain day at Algat Hospital and the probability that any given baby
        will be a boy is $\frac{1}{2}$, find the probability that exactly 9 of the babies are
        boys.

(3)   A basketball player is either successful or unsuccessful when he shoots
        a foul shot. If the player attempts 12 foul shots and the probability that
        any given foul shot is successful is $\frac{9}{10}$, what is the probability of his
        making 9 or more shots successfully?

(4)   When Hexaglorifane is tested on a patient, the patient either lives or
        dies. If Hexaglorifane is tried on 15 patients and the probability that
        any given patient lives when treated with the drug is $\frac{1}{20}$, what is the
        probability that at least 11 patients die from the treatment?

The problems above have such strong similarities, that it is possible to
construct one probability model to assist in solving all of them. The model
that we shall construct is called the *binomial probability model*. The problem
under consideration must satisfy the following conditions for the binomial
probability model to be applicable:

(1)   The problem involves N repetitions of an experiment that has exactly
        two outcomes. We shall refer to the outcomes as success (S) and failure
        (F).

(2)   The probability of a success, which we denote by p, and the probability
        of a failure, which we denote by q, ($p + q = 1$) is the same for each
        repetition.

(3)   The outcomes of each repetition of the experiment are independent of
        the outcomes on all the other repetitions.

We shall be interested in computing the following two probabilities in
such cases:

(1)   The probability of exactly r successes in N repetitions of the experi-
        ment.

(2)   The probability of at least r successes in N repetitions of the experi-
        ment.

Example 1

A student has not studied for his history quiz, which consists of three
true-false questions. Suppose the student guesses.

(1)   What is the probability of answering two questions correctly?

(2)   What is the probability of answering at least two questions correctly?

Solution: Let us first investigate why this problem can be treated using the binomial model. We have three ($N = 3$) repetitions of the experiment *guess an answer to a true-false question.* Each repetition of the experiment has exactly two outcomes, a correct answer (C) or an incorrect answer (C'). A correct answer would correspond, in this example, to a success, while an incorrect answer would correspond to a failure. The probability of success equals the probability of a failure equals one-half (i.e., $p = q = \frac{1}{2}$) because the student guesses. Finally, it is reasonable that a correct guess on question 2, say, will not affect whether or not the answers to questions 1 and 3 are correct, i.e., the repetitions are independent. To determine the answers to the questions posed we draw the tree diagram in Figure 8.19. The probability of

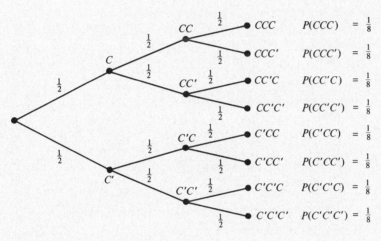

Figure 8.19

the outcomes can be determined by using the multiplication rule. Hence, the probability of exactly two correct responses is given by

$$P \text{ (exactly two correct responses)} = P(CCC') + P(CC'C) + P(C'CC)$$
$$= (\tfrac{1}{2} \cdot \tfrac{1}{2} \cdot \tfrac{1}{2}) + (\tfrac{1}{2} \cdot \tfrac{1}{2} \cdot \tfrac{1}{2}) + (\tfrac{1}{2} \cdot \tfrac{1}{2} \cdot \tfrac{1}{2}) = \tfrac{3}{8}$$

The probability of at least two correct answers is given by

$$P \text{ (exactly two correct answers)} + P \text{ (exactly three correct answers)}$$

The probability of exactly three correct responses equals $P(CCC)$, which from the tree in Figure 8.19 is $\tfrac{1}{2} \cdot \tfrac{1}{2} \cdot \tfrac{1}{2} = \tfrac{1}{8}$. Hence, the probability of at least two correct responses is $\tfrac{3}{8} + \tfrac{1}{8} = \tfrac{1}{2}$. Thus, we see that the student has a 50-50 chance of doing quite well on the quiz merely by guessing.

Example 2

An encyclopedia salesman has found that in one-fifth of the cases where he rings a doorbell, he gets the homeowner to let him into the house. If the salesman calls at four homes in an evening, what is the probability that he enters at least two homes?

Solution: First we note that the binomial model is applicable. Denoting by E (success) the salesman being able to enter a given home, the only other outcome is E', being unable to enter the home. In this problem $N = 4$, $P(E) = \frac{1}{5}$, and $P(E') = \frac{4}{5}$. We may also assume his success in entering any given home is independent of his success or failure in entering any other home. Figure 8.20 shows the outcomes of the experiment and the probability assignment to the edges of the tree. The event the salesman enters at least two

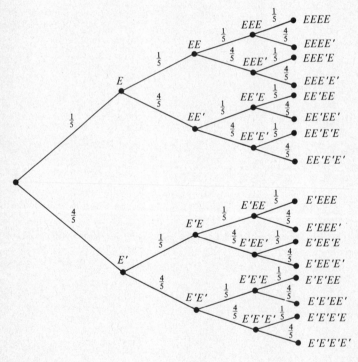

Figure 8.20

homes consists of the outcomes, $EEEE$, $EEEE'$, $EEE'E$, $EEE'E'$, $EE'EE$, $EE'EE'$, $EE'E'E$, $E'EEE$, $E'EEE'$, $E'EE'E$, and $E'E'EE$. Computing the probability of this event using the multiplication rule, we obtain

P (salesman enters at least two homes)

$$= P(EEEE) + P(EEEE') + P(EEE'E) + P(EEE'E') + P(EE'EE)$$
$$+ P(EE'EE') + P(EE'E'E) + P(E'EEE) + P(E'EEE')$$
$$+ P(E'EE'E) + P(E'E'EE)$$

$$= (\tfrac{1}{5} \cdot \tfrac{1}{5} \cdot \tfrac{1}{5} \cdot \tfrac{1}{5}) + (\tfrac{1}{5} \cdot \tfrac{1}{5} \cdot \tfrac{1}{5} \cdot \tfrac{4}{5}) + (\tfrac{1}{5} \cdot \tfrac{1}{5} \cdot \tfrac{4}{5} \cdot \tfrac{1}{5}) + (\tfrac{1}{5} \cdot \tfrac{1}{5} \cdot \tfrac{4}{5} \cdot \tfrac{4}{5}) + (\tfrac{1}{5} \cdot \tfrac{4}{5} \cdot \tfrac{1}{5} \cdot \tfrac{1}{5})$$
$$+ (\tfrac{1}{5} \cdot \tfrac{4}{5} \cdot \tfrac{1}{5} \cdot \tfrac{4}{5}) + (\tfrac{1}{5} \cdot \tfrac{4}{5} \cdot \tfrac{4}{5} \cdot \tfrac{1}{5}) + (\tfrac{4}{5} \cdot \tfrac{1}{5} \cdot \tfrac{1}{5} \cdot \tfrac{1}{5}) + (\tfrac{4}{5} \cdot \tfrac{1}{5} \cdot \tfrac{1}{5} \cdot \tfrac{4}{5}) + (\tfrac{4}{5} \cdot \tfrac{1}{5} \cdot \tfrac{4}{5} \cdot \tfrac{1}{5})$$
$$+ (\tfrac{4}{5} \cdot \tfrac{4}{5} \cdot \tfrac{1}{5} \cdot \tfrac{1}{5})$$

$$= \tfrac{1}{625} + \tfrac{4}{625} + \tfrac{4}{625} + \tfrac{16}{625} + \tfrac{4}{625} + \tfrac{16}{625} + \tfrac{16}{625} + \tfrac{4}{625} + \tfrac{16}{652}$$
$$+ \tfrac{16}{625} + \tfrac{16}{625}$$

$$= \tfrac{113}{625} = 0.2$$

Thus, the salesman has about a 0.2 chance of entering two or more homes.

Since the calculations get increasingly complicated as N increases, it is customary to use tables for the computations in binomial model problems (see Appendix 12.3). For a given value of N and various values of p, the table shows the probability $b(r, N, p)$ that there are exactly r successes.

Example 3

Find the probability that in 10 tosses of a fair coin

(1) We obtain exactly eight heads.
(2) We obtain at least eight heads.
(3) We obtain at least four heads.
(4) We obtain exactly five heads.

Solution: Since N is 10 and $p = q = \tfrac{1}{2}$ (fair coin), we consult the table with $N = 10$ and look in the column with $p = \tfrac{1}{2}$.

(1) In the row labeled 8, we read off the probability

$$b(8, 10, \tfrac{1}{2}) = 0.044$$

(2) To obtain the probability of at least 8 heads, we add up the probability of exactly 8, 9, and 10 heads:

$$P \text{ (at least 8 heads)} = b(8, 10, \tfrac{1}{2}) + b(9, 10, \tfrac{1}{2}) + b(10, 10, \tfrac{1}{2})$$
$$= 0.044 + 0.010 + 0.001$$
$$= 0.055$$

(3)   In order to compute the probability of at least 4 heads:

P (at least 4 heads) $= 1 - P$ (at most 3 heads)

$$= 1 - [P \text{ (0 heads)} + P \text{ (1 head)} + P \text{ (2 heads)}$$
$$+ P \text{ (3 heads)}]$$
$$= 1 - [b(0, 10, \tfrac{1}{2}) + b(1, 10, \tfrac{1}{2}) + b(2, 10, \tfrac{1}{2})$$
$$+ b(3, 10, \tfrac{1}{2})]$$
$$= 1 - [0.001 + 0.010 + 0.044 + 0.117]$$
$$= 1 - 0.172$$
$$= 0.828$$

(4)   In the row labeled 5, we find

$$b(5, 10, \tfrac{1}{2}) = 0.246$$

The fact that the probability of obtaining 5 heads and 5 tails in 10 tosses of a fair coin is *not* $\tfrac{1}{2}$ but 0.246 sometimes surprises people.

Let us illustrate some further examples that can be treated by the binomial probability model with the aid of tables (see Appendix 12.3).

Example 4

The ABC Corporation, a manufacturer of transistors, knows that on the average 1 out of 10 of its transistors is defective. ABC has a government contract to provide the government with a shipment of transistors. The government will take a sample of 12 transistors and reject the shipment if there are 2 or more defectives. What is the probability the government will *accept* the delivery?

Solution: The situation can be analyzed using the binomial model. The delivery will be accepted, provided there is 0 or 1 defective. Hence,

$$P \text{ (delivery accepted)} = P \text{ (0 defective)} + P \text{ (1 defective)}$$
$$= b(0, 12, \tfrac{1}{10}) + b(1, 12, \tfrac{1}{10})$$
$$= 0.282 + 0.377$$
$$= 0.659$$

Note that the probability the delivery will be rejected is 0.341 or approximately one-third.

Example 5

A multiple-choice exam has 10 questions and five choices for each question. Each question is worth 10 points, and 50 points is passing. If a student merely guesses, what is the probability of passing the exam?

Solution: The problem can be analyzed in the framework of the binomial model. $N = 10$ and since there is one chance in five of guessing the correct answer to a question, $p = 0.2$. To pass the exam would require getting *five or more* correct answers. Hence,

$$P \text{ (passing)} = P \text{ (5 or more correct)}$$
$$= P \text{ (5 correct)} + P \text{ (6 correct)} + P \text{ (7 correct)} + P \text{ (8 correct)}$$
$$+ P \text{ (9 correct)} + P \text{ (10 correct)}$$
$$= b(5, 10, 0.2) + b(6, 10, 0.2) + b(7, 10, 0.2) + b(8, 10, 0.2)$$
$$+ b(9, 10, 0.2) + b(10, 10, 0.2)$$
$$= 0.026 + 0.006 + 0.001 + 0.000 + 0.000 + 0.000$$
$$= 0.033$$

You may have been surprised that the table gave P (8 correct), P (9 correct), and P (10 correct), as 0.000. These probabilities are not actually zero but when their true values are rounded to three decimal places, their values are 0.000.

Note that the chance of passing by guessing is very small.

EXERCISES 8.5

1. Solve the first problem on p. 303.
2. Solve the second problem on p. 304.
3. Solve the third problem on p. 304.
4. Solve the fourth problem on p. 304.
5. A psychology professor is recruiting a random sample of volunteers for an experiment and discovers that the first five of his volunteers are science majors. He wonders whether this coincidence means that his sample isn't random. What is the probability of this happening by chance in a random sample, assuming that one-fifth of the students at this school are science majors?
6. A key legislative proposal is being voted on. There are only seven senators who have not announced how they will vote. Suppose that if at least five of these senators vote for the proposal, that will be enough for

its approval. If their votes are independent of one another and each key senator announces that it is a "toss-up" how he will vote, what is the probability the proposal will pass? (Assume that we have ruled out abstentions in the voting.)

7. Use a tree diagram to calculate $b(0, 4, \frac{1}{3})$, $b(1, 4, \frac{1}{3})$, $b(2, 4, \frac{1}{3})$, $b(3, 4, \frac{1}{3})$, and $b(4, 4, \frac{1}{3})$.

8. A noted gambler, the Chevalier de Mere, was accustomed to winning money by betting that a six would occur at least once in 4 throws of a die. Then he changed his offer and volunteered to bet that one or more double sixes would occur in 24 rolls of a pair of dice. He thought the odds would be in his favor to the same degree as in his first bet and he was greatly surprised when he began losing money. Can you justify his experiences on these two bets through probability theory? (Hint: $(\frac{35}{36})^{24} \approx 0.509$.)

9.* Why is it true, that for any r, N and p

$$b(r, N, p) = b(N - r, N, 1 - p)$$

10.* Why do the numbers calculated in Exercise 7 add up to 1? Is it an accident or must it be so?

11.* Phil and Kevin are going to settle a dispute in a basketball game by taking a series of foul shots to see who makes more of them. Phil's foul shooting percentage is 0.500, while Kevin's is 0.600. They are undecided about whether to shoot three shots apiece or just two shots. Does it make any difference and, if so, which setup gives Phil the better chance of winning?

8.6 expected value

Would you participate in a coin toss experiment in which you win each time heads comes up but lose when tails result? Assuming the coin is fair, it seems you would win as often as you would lose so there is nothing to be gained or lost. Now suppose that we attach different payoffs to the outcomes. Imagine that heads wins you $3 while tails loses you $2. Using v to symbolize value (payoff) we write $v(H) = 3$, $v(T) = 2$. Now you would undoubtedly play since you stand to gain in the long run. For a more problematical case, suppose we still have $v(H) = 3$ and $v(T) = 2$ but now the coin is unfair and the probability of heads is one-third. Is it worth playing? For cases such as this, where the elements of a sample space have payoffs as well as probabilities attached to them, it is convenient to have a way of calculating some sort of average payoff that you can expect to win.

To work out the last example, let's suppose you play the game N times. Then heads comes up about $\frac{1}{3}N$ times and each of these times you win \$3 for a total of $\frac{1}{3}N \times 3$ dollars. Similarly, there will be a \$2 loss about $\frac{2}{3}N$ times for a net loss of $\frac{2}{3}N \times 2$. Combining these, we get

$$\text{Total payoff} = (\tfrac{1}{3}N \times 3) - (\tfrac{2}{3}N \times 2) = -\tfrac{1}{3}N$$

Thus you lose a total of $\frac{1}{3}N$. On a per game basis the payoff is $-\frac{1}{3}$. We call this per game payoff the *expected payoff* and we note, as motivation for the definition we shall give shortly, that it can be calculated without using N:

$$\text{Expected payoff} = (\tfrac{1}{3} \times 3) - (\tfrac{2}{3} \times 2) = -\tfrac{1}{3}$$

Before proceeding to our formal definition of expected value, let's look at an example that shows that it need not be money whose expected value might interest us.

Example 1

The telephone inquiries desk of a large library handles four basic types of questions, which we shall symbolize by q_1, q_2, q_3, q_4. The records of relative frequency suggest the following probabilities for these questions:

$$P(q_1) = \tfrac{1}{3} \qquad P(q_2) = \tfrac{1}{4} \qquad P(q_3) = \tfrac{1}{4} \qquad P(q_4) = \tfrac{1}{6}$$

(e.g., whenever the phone rings, the probability the question is of type q_1 is one-third, the probability it is of type q_2 is one-fourth, etc.). The times required to handle these types of inquiries are, respectively, 3, 2, 4, and 10 minutes; that is, $v(q_1) = 3$, $v(q_2) = 2$, $v(q_3) = 4$, and $v(q_4) = 10$. What is the average (expected) time for an inquiry?

Solution: Out of a large number of calls, say N, we would have about

$$\tfrac{1}{3}N \text{ calls of type } q_1$$
$$\tfrac{1}{4}N \text{ calls of type } q_2$$
$$\tfrac{1}{4}N \text{ calls of type } q_3$$
$$\tfrac{1}{6}N \text{ calls of type } q_4$$

$$\text{Total time} = (\tfrac{1}{3}N \times 3) + (\tfrac{1}{4}N \times 2) + (\tfrac{1}{4}N \times 4) + (\tfrac{1}{6}N \times 10)$$
$$= \tfrac{25}{6}N$$

$$\text{Average (expected) time per call} = 4\tfrac{1}{6} \text{ minutes}$$

Again we note that the last calculations could be performed without using N:

$$\text{Average (expected) time} = (\tfrac{1}{3} \times 3) + (\tfrac{1}{4} \times 2) + (\tfrac{1}{4} \times 4) + (\tfrac{1}{6} \times 10)$$
$$= 4\tfrac{1}{6}$$

These examples suggest the following definition.

DEFINITION 3

Suppose $S = \{o_1, o_2, o_3, \ldots, o_n\}$ is a sample space, with probabilities $p(o_i)$ assigned to the outcome o_i. Suppose also that there are other numbers, representing some important quantity, denoted $v(o_i)$, assigned to the o_i. Then the *expected value E* of the quantity represented by the numbers $v(o_i)$ is defined

$$E = p(o_1)v(o_1) + p(o_2)v(o_2) + \cdots + p(o_n)v(o_n)$$

Example 2

In 1972 New York doubled the toll on the Whitestone Bridge from 25¢ to 50¢. In addition to raising the monetary cost to commuters, this action also may have imposed a hidden cost in time. The following analysis, using hypothetical figures, shows how this might come about.

A motorist will either have exact change (E) or need to get change at a manned toll booth (C). These two outcomes make a sample space $\{E, C\}$ that will have different probability distributions depending on what the toll is. It is much more likely that a motorist will have exact change for 25¢ than for 50¢. Let P_q and P_f denote the probability assignments for the 25¢ and 50¢ tolls, respectively. We shall assume that

$$P_q(E) = \tfrac{3}{4} \qquad P_q(C) = \tfrac{1}{4}$$
and
$$P_f(E) = \tfrac{1}{4} \qquad P_f(C) = \tfrac{3}{4}$$

Now we shall assume that for the outcome E the time required to go through the toll gate is 5 seconds while the time to go through a toll gate when change needs to be made (C) is 7 seconds. Now for each of the two tolls we calculate the expected value of the time required to go through the toll gate.

25¢ toll: Expected time $= [P_q(E) \times 5] + [P_q(C) \times 7]$
$$= (\tfrac{3}{4} \times 5) + (\tfrac{1}{4} \times 7) = \tfrac{22}{4} = 5.5 \text{ seconds}$$

50¢ toll: Expected time $= [P_f(E) \times 5] + [P_f(C) \times 7]$
$$= (\tfrac{1}{4} \times 5) + (\tfrac{3}{4} \times 7) = \tfrac{26}{4} = 6.5 \text{ seconds}$$

Thus, we see that the expected time through the toll gate increases by 1 second per motorist as we increase the toll to 50¢. How significant is this? There are a number of points of view that can be taken on this question:

(1) The increase of 1 second over the previous time of 5.5 seconds is a significant increase in time that a car spends *at the toll gate*.

(2) One second is an insignificant addition to the *total trip time* for the occupants of one automobile.

(3) Assuming that approximately 200,000 persons cross this bridge in automobiles in a day, there will be a significant total of about $55\frac{1}{2}$ man-hours lost.

Our remaining examples will involve binomial models and will launch us toward some useful theoretical results involving expected values in binomial models.

Example 3

Suppose a fair coin is tossed three times. Find the expected number of heads.

Solution: We take as our sample space {*HHH, HHT, HTH, HTT THH, TTH, THT, TTT*}. Each probability is one-eighth. Since we are interested in the expected number of heads, we associate with each outcome the number of heads:

$$\text{Expected number of heads} = (\tfrac{1}{8} \times 3) + (\tfrac{1}{8} \times 2) + (\tfrac{1}{8} \times 2) + (\tfrac{1}{8} \times 1)$$
$$+ (\tfrac{1}{8} \times 2) + (\tfrac{1}{8} \times 1) + (\tfrac{1}{8} \times 1) + (\tfrac{1}{8} \times 0)$$
$$= \tfrac{12}{8}$$
$$= 1.5$$

This result appeals to our intuition since there are three tosses and $1.5 = P(\text{heads}) \times 3$. Indeed it can be shown that:

THEOREM 2

Given a binomial probability model involving N repetitions of an experiment where the probability for a success in one trial is p, the expected value E of the number of successes is given by

$$E = pN$$

Example 4

A drug is used on 1000 patients. If the probability of improvement is $\frac{1}{20}$, what is the expected number of patients who improve?

Solution: Using Theorem 2, the value of $N = 1000$, and the value of $p = \frac{1}{20}$, the expected number of patients who show improvement is

$$\frac{1}{20} \times 1000 = 50$$

When we discussed the concept of a mean for data in Chapter 7, we saw that without some measurement of the dispersion of the data we could not recover adequate information from the data. A similar problem concerns the concept of expectation. We have no idea how much dispersion the $v(o_i)$ can have about the expected value. This problem can be treated by defining the standard deviation σ analogously to the way it was defined in Chapter 7. Rather than give formal definitions, we shall restrict ourselves to the binomial model and state the probability analogue of Chebyshev's theorem.

THEOREM 3

Given a binomial probability model involving N repetitions of an experiment where the probability for a success is p, then the standard deviation σ (for the number of successes) is given by

$$\sigma = \sqrt{(p)(1 - p)(N)}$$

We shall not be much concerned with the definition or conceptual nature of this standard deviation. We are interested in it mainly so that we can use it in the calculations discussed in the next theorem, which is a variant of Chebyshev's theorem. It is quite useful and you should master its meaning and application.

THEOREM 4

Given a binomial model involving N repetitions of an experiment, where the expected value and standard deviation for the number of successes are given by E and σ, then the probability that the number of successes differs from E by more than k standard deviations is less than $1/k^2$.

The following three examples suggest the inferential possibilities of these theorems.

Example 5

A student claims to know 90% of the material in his economics course. When a true-false examination consisting of 100 questions is given to the students, he gets 77 questions correct. The student claims that the result is a "fluke." Does probability theory support his claim?

Solution: We shall assume that if the student knows 90% of the material in the course, then his chance of answering any question correctly is nine-tenths. Hence, if asked 100 questions, his expected number of correct responses would be, by Theorem 2, 3 (100)(0.9) = 90.
By Theorem 3,

$$\sigma = \sqrt{(\tfrac{9}{10})(\tfrac{1}{10})(100)}$$
$$= \sqrt{9}$$
$$= 3$$

Now we can apply Chebyshev's theorem (Theorem 4). The probability that the student does not get between 78 and 102 questions correct (i.e., an interval of 4 standard deviations around $E = 90$) is less than $\tfrac{1}{16} = 0.063$. The student's claim that his performance was a fluke seems unlikely.

Example 6

A symmetrical coin is tossed 100 times.

(1)   What is the expected number of heads?
(2)   What is the *probability* that the expected number of heads will occur?

Solution: Since the coin is symmetrical, we may assume that

$$P(\text{heads}) = \tfrac{1}{2} = P(\text{tails})$$

Hence, in 100 tosses, the expected number of heads is

$$\tfrac{1}{2}(100) = 50 \qquad (\text{by Theorem 2})$$

The probability that in 100 tosses of a coin one gets exactly 50 heads can be computed on theoretical grounds; it turns out to be 0.08, not a very large number. Since the expected value is 50, it may confuse you that P (exactly 50 heads) is so small. Perhaps you may get more insight if we apply Chebyshev's theorem. Calculating, $\sigma = \sqrt{(\tfrac{1}{2})(\tfrac{1}{2})(100)} = 5$. Hence, for example, the probability of getting more than 60 or less than 40 heads (we are using 2 standard deviations from $E = 50$) in 100 tosses is *less* than $1/(2)^2 = \tfrac{1}{4} = 0.25$. Thus, we have little chance of getting exactly 50 heads but a good chance of getting close to 50 heads.

EXERCISES 8.6

1. In the experiment of rolling a fair die once, what is the expected payoff if the payoff for each outcome equals the number showing uppermost on the die for that outcome?

2. In the experiment of rolling a pair of fair dice once, what is the expected payoff if the payoff for each outcome equals the difference between the larger number showing and the smaller (or 0 if the numbers showing on each die are the same)?

3. Jack flips a coin three times or until he gets a head, whichever happens first. Each time a tail is obtained, Jack gives Jill a kiss; while if a head is obtained, he rolls down the hill and bumps his crown. Assuming the coin is rigged so that the probability of a head is one-quarter, what is the expected number or kisses given by Jack? What is the expected number of bumps Jack's crown sustains?

4. Do Exercise 3 under the assumption that the coin is flipped three times regardless of whether a head appears.

5. A die is tossed 500 times and the number on the face that appears is recorded. Compute the expected number of times that four dots appear if the die is loaded so that the probability of four dots appearing is
    (a) $\frac{1}{2}$
    (b) $\frac{1}{3}$
    (c) $\frac{1}{10}$

6. A fair coin is tossed 10 times. What is the expected number of heads? What is the probability of the event that the expected number of heads occurs?

7. Consider the experiment of determining the sexes of 100 randomly selected newborn children. Assuming that the probability that a randomly selected newborn child is a boy is one-half, calculate
    (a) the expected number of boys,
    (b) the standard deviation.
    Use Chebyshev's theorem to give an upper estimate on the probability that more than 60 or less than 40 of the 100 newborn children are boys.

8. Ten thousand mousetraps are tested for proper functioning. Assuming that there is a one-tenth probability that any given mousetrap is defective, determine the expected number of defectives, the standard deviation, and an upper estimate on the probability that the number of defectives differs from 1,000 by more than 20.

9. A political candidate assumes the population is evenly divided pro and con on a certain legislative proposal. When he takes a poll of 300 people to make sure, however, he discovers that 176 of the 300 are opposed. Is this result serious grounds for questioning his assumption

that the electorate is evenly divided? Suppose 157 of the 300 were opposed?

8.7 Markov analysis

If a coin is flipped twice, we have been assuming that the outcome of the second experiment is uninfluenced by the first; in technical language, the two experiments are independent. In addition to coin tossing, many other repeated experiments have been handled with the modeling assumption of independence—notably when we studied the binomial model. There are examples, however, where it is more reasonable to assume the opposite, that the outcome of one experiment affects the outcome of the repetition of that experiment.

Consider, for example, an electric power company that checks its main generator once each quarter year to forestall blackouts due to equipment failure. For simplicity we shall assume that there are two outcomes for each quarterly inspection: W, the generator is in good working order and needs no repair; D, the generator is defective and needs repair. If the outcome in one quarter is D, repairs will be made and it is extremely likely that the next quarter's outcome will be W. On the other hand, if the outcome in one quarter is W, no repairs are made and there is a fair chance that the next quarter's outcome will be D. *Thus the outcome in the second quarter is not independent of the outcome in the first quarter.*

Let us continue the discussion with some specific numbers: If a given quarter's outcome is W, we assume:

the probability that the next quarter's outcome is W is 0.6

the probability that the next quarter's outcome is D is 0.4

If a given quarter's outcome is D, we assume:

the probability that the next quarter's outcome is W is 0.9

the probability that the next quarter's outcome is D is 0.1

With this data we shall try to answer this question: If this quarter's outcome is D, what is the probability of having a D quarter two quarters from now— or, to be more general, n quarters from now?

Our goal in this section is to find a mechanical method for answering this question but as preparation we shall attack the question using a tree diagram. Figure 8.21 shows a tree displaying three repetitions of this experiment where the outcome in the first quarter is D. The sample space for this

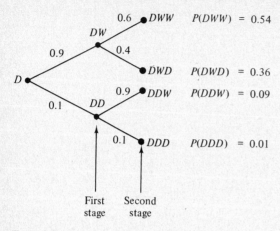

$P(DWW) = 0.54$

$P(DWD) = 0.36$

$P(DDW) = 0.09$

$P(DDD) = 0.01$

First
stage

Second
stage

Figure 8.21

multistage experiment is $\{DDD, DDW, DWD, DWW\}$ and we would now like to determine a suitable probability assignment for it. It is tempting to try to find the probabilities for the end points of the tree by the multiplication rule as we did in Section 8.4 while studying independence. This rule would give, for example,

$$P(DWD) = (0.9)(0.4) = 0.36$$

Since our events are not independent, however, we had better check that it makes as much sense to use the multiplication rule here as it did in the case where the outcomes in the various stages were independent of one another.

We shall give an example of the kind of reasoning that justifies the multiplication rule under the present circumstances. We shall regard the tree of Figure 8.21 as representing the two-stage experiment of checking the generator twice after an initial finding of defective (D). We shall imagine that this two-stage experiment is repeated N_D times. If N_{DWD} is the number of times DWD is the result, then the ratio N_{DWD}/N_D will be regarded as a good approximation to what $P(DWD)$, the probability of DWD, should be. Let us further denote by N_{DW} the number of times the outcome in the first stage (after the initial D) is W. Clearly,

$$P(DWD) = \frac{N_{DWD}}{N_D} = \frac{N_{DW}}{N_D} \times \frac{N_{DWD}}{N_{DW}} \tag{8.4}$$

However, N_{DW}/N_D is approximately the probability that W occurs on the first stage of our experiment, namely nine-tenths. The ratio N_{DWD}/N_{DW} is approximately the probability that D occurs on the second stage, assuming that W has occurred on the first stage, namely four-tenths. Substituting these num-

318

bers in (8.4) gives the multiplication rule we seek:

$$P(DWD) = (0.9)(0.4) = 0.36$$

In a similar way we can calculate that $P(DWW) = 0.54$, $P(DDW) = 0.09$, and $P(DDD) = 0.01$. The event D *at the second stage* is $\{DWD, DDD\}$ and we can now calculate the probability of this event:

$$P(DWD, DDD) = P(DWD) + P(DDD) = 0.36 + 0.01 = 0.37$$

The multiplication rule can be used to calculate the probabilities of the end points of a tree for any of the problems in this section (the so-called Markov processes defined below). Suppose, for example, we wanted the probability that three quarters after the initial D we obtain a D quarter again. Now we need a tree with three stages and the probability of each end point is the product of the three probabilities on the three edges leading to that end point. As we calculate further into the future, our tree becomes larger and the calculations more difficult to manage. This difficulty will be overcome by the mechanical method we shall develop. Before reaching this, it will be useful to abstract the main features of the foregoing problem insofar as they typify the problems we deal with in this section.

(1) We have an experiment with sample space $\{o_1, \ldots, o_r\}$ that is repeated a number of times. We often call the o_i *states* instead of outcomes and we think of the succession of states as a *process*, often called a *Markov process*. In our example, the states were W and D.

(2) In any repetition, the probability of an outcome o_i is dependent on which outcome came about on the previous experiment. We denote by p_{ij} the probability that outcome j will come about assuming that outcome i arose in the previous experiment. These numbers, which are generally part of the given data of the problem, are often arranged in a *transition* matrix like this:

$$
\begin{array}{c}
\\
o_1 \\
o_2 \\
\cdot \\
\cdot \\
\cdot \\
o_r
\end{array}
\begin{array}{cccc}
o_1 & o_2 & \cdots & o_r \\
\left[\begin{array}{cccc}
p_{11} & p_{12} & \cdots & p_{1r} \\
p_{21} & p_{22} & \cdots & p_{2r} \\
& & & \\
& & & \\
& & & \\
p_{r1} & p_{r2} & \cdots & p_{rr}
\end{array}\right]
\end{array}
$$

In our electric generator example, since the states were W and D rather than subscripted o's, we would use the notations $p_{WW}, p_{WD}, p_{DW}, p_{DD}$. They would be arranged in a matrix like this:

$$\begin{array}{cc} & W \qquad D \\ \begin{array}{c} W \\ D \end{array} & \begin{bmatrix} p_{WW} & p_{WD} \\ p_{DW} & p_{DD} \end{bmatrix} \end{array}$$

The particular data given were

$$\begin{array}{cc} & W \quad D \\ \begin{array}{c} W \\ D \end{array} & \begin{bmatrix} 0.6 & 0.4 \\ 0.9 & 0.1 \end{bmatrix} \end{array}$$

It is also useful to formalize some notation for our question. We denote by $p_{ij}^{(n)}$ the probability that after n steps in time (quarter years in our example) a process starting in state o_i winds up in state o_j. In our electric generator example, $p_{DD}^{(2)} = 0.37$. Note that $p_{ij}^{(1)} = p_{ij}$. We can think of these $p_{ij}^{(n)}$ as being arranged, according to their subscripts, into a matrix that we denote $M^{(n)}$:

$$M^{(n)} = \begin{bmatrix} p_{11}^{(n)} & p_{12}^{(n)} & \cdots & p_{1r}^{(n)} \\ p_{21}^{(n)} & p_{22}^{(n)} & \cdots & p_{2r}^{(n)} \\ \cdot & \cdot & & \cdot \\ \cdot & \cdot & & \cdot \\ \cdot & \cdot & & \cdot \\ p_{r1}^{(n)} & p_{r2}^{(n)} & \cdots & p_{rr}^{(n)} \end{bmatrix}$$

It should be noted that the (n) that appears in the exponent's place in $p_{ij}^{(n)}$ and $M^{(n)}$ is not really an exponent but just another index; it appears in this location because there is no other convenient place for it. In the case of $M^{(n)}$, however, we have this remarkable coincidence:

THEOREM 5

If M is a transition matrix for a Markov process, then $M^{(n)} = M^n$.

Proof: We give the proof for the case where $n = 2$ and M is the 2×2 matrix:

$$M = \begin{bmatrix} p_{11} & p_{12} \\ p_{21} & p_{22} \end{bmatrix}$$

By direct calculation we find that

$$M^2 = \begin{bmatrix} (p_{11} \cdot p_{11}) + (p_{12} \cdot p_{21}) & (p_{11} \cdot p_{12}) + (p_{12} \cdot p_{22}) \\ (p_{21} \cdot p_{11}) + (p_{22} \cdot p_{21}) & (p_{21} \cdot p_{12}) + (p_{22} \cdot p_{22}) \end{bmatrix}$$

Now we need to show that

$$(p_{11} \cdot p_{11}) + (p_{12} \cdot p_{21}) = p_{11}^{(2)}$$

$$(p_{11} \cdot p_{12}) + (p_{12} \cdot p_{22}) = p_{12}^{(2)}$$

$$(p_{21} \cdot p_{11}) + (p_{22} \cdot p_{21}) = p_{21}^{(2)}$$

$$(p_{21} \cdot p_{12}) + (p_{22} \cdot p_{22}) = p_{22}^{(2)}$$

(8.5)

To verify the first equality, note that $p_{11} \cdot p_{11}$ is the probability of the outcome $o_1 o_1 o_1$, while $p_{12} \cdot p_{21}$ is the probability of $o_1 o_2 o_1$, so $(p_{11} \cdot p_{11}) + (p_{12} \cdot p_{21}) = P(o_1 o_1 o_1, o_1 o_2 o_1)$. But $\{o_1 o_1 o_1, o_1 o_2 o_1\}$ is precisely the event that the process starting in state o_1 ends in o_1 two steps later. Therefore,

$$(p_{11} \cdot p_{11}) + (p_{12} \cdot p_{21}) = P(o_1 o_1 o_1, o_1 o_2 o_1)$$
$$= p_{11}^{(2)}$$

A similar analysis establishes the other three equations in (8.5).

Example 1

In our electric generator example, the transition matrix is

$$M = \begin{array}{c} \\ W \\ D \end{array} \begin{array}{cc} W & D \\ \left[\begin{array}{cc} 0.6 & 0.4 \\ 0.9 & 0.1 \end{array} \right] \end{array}$$

Therefore,

$$M^{(2)} = M^2 = \left[\begin{array}{cc} 0.72 & 0.28 \\ 0.63 & 0.37 \end{array} \right]$$

Thus, $p_{DD}^{(2)} = 0.37$, which we already knew from our tree analysis. But we now also know that $p_{WD}^{(2)} = 0.28$, $p_{DW}^{(2)} = 0.63$, and $p_{WW}^{(2)} = 0.72$, facts that we did not work out using the tree method.

Example 2

A sociologist wishes to examine the relationship between education level of parents and their children. He classifies the adults into three groups: those with no more than a grade school education (G), those with no more than a high school education (H), and those who have been to college (C). If a person is in one group, then there is, for any of his children, a certain probability that the child will belong to group G, a certain probability that the child will belong to group H, and a certain probability that the child will belong to group C. These probabilities, we shall assume, depend only on the group of the parent, as in the transition matrix:

$$
\begin{array}{cc}
 & \begin{array}{ccc} G & H & C \end{array} \\
\begin{array}{c} G \\ H \\ C \end{array} &
\begin{bmatrix}
\frac{2}{3} & \frac{1}{3} & 0 \\
\frac{1}{3} & \frac{1}{3} & \frac{1}{3} \\
0 & \frac{1}{3} & \frac{2}{3}
\end{bmatrix}
\end{array}
$$

What is the probability that the great grandson of a high school graduate will receive a college education?

Solution: Since we are looking for $p_{HC}^{(3)}$, we compute $M^{(3)} = M^3$, which is

$$
M^3 =
\begin{bmatrix}
\frac{13}{27} & \frac{9}{27} & \frac{5}{27} \\
\frac{9}{27} & \frac{9}{27} & \frac{9}{27} \\
\frac{5}{27} & \frac{9}{27} & \frac{13}{27}
\end{bmatrix}
$$

$$
p_{HC}^{(3)} = \frac{9}{27} = \frac{1}{3}
$$

Example 3

Suppose that each day the air quality or your hometown is checked to see if it is clear (C) or dirty (D). Assume that if a given day is C, the probability that the next day is C is two-fifths, while if a given day is D, the probability that the next day is C is one-fifth. What is the probability that the fourth day is D, assuming that the first day is D?

Solution: Only two probabilities are given in the problem and four are needed for the transition matrix. For example, the probability that D will follow C is not given. D following C is the complementary event to C following C, however, so $p_{CD} = 1 - p_{CC} = 1 - \frac{2}{5} = \frac{3}{5}$. Similarly, $p_{DD} = 1 - p_{DC} = 1 - \frac{1}{5} = \frac{4}{5}$. Thus, the transition matrix is

$$
\begin{array}{cc}
 & \begin{array}{cc} C & D \end{array} \\
\begin{array}{c} C \\ D \end{array} &
\begin{bmatrix}
\frac{2}{5} & \frac{3}{5} \\
\frac{1}{5} & \frac{4}{5}
\end{bmatrix}
\end{array}
$$

Now we calculate M^2.

$$
M^2 =
\begin{bmatrix}
\frac{7}{25} & \frac{18}{25} \\
\frac{6}{25} & \frac{19}{25}
\end{bmatrix}
=
\begin{bmatrix}
0.28 & 0.72 \\
0.24 & 0.76
\end{bmatrix}
$$

Therefore,

$$M^4 = (M^2)^2 = \begin{bmatrix} 0.2512 & 0.7488 \\ 0.2496 & 0.7504 \end{bmatrix}$$

From this matrix we see that $p_{DD}^4 = 0.7504$.

EXERCISES 8.7

1. Calculate, using matrix multiplication, $p_{22}^{(2)}$, $p_{12}^{(3)}$, and $p_{11}^{(3)}$, assuming that

$$M = \begin{bmatrix} \frac{1}{3} & \frac{2}{3} \\ \frac{2}{3} & \frac{1}{3} \end{bmatrix}$$

2. Do Exercise 1 using the method of tree diagrams instead of matrix multiplication.

3. Jack's spending pattern for entertainment is such that he plans two kinds of weekends, extravagant and stingy. After an extravagant weekend, there is a two-thirds probability of next having a stingy weekend, while after a stingy weekend the chances are even that the next weekend will also be stingy. Describe this Markov process with a transition matrix. Assuming that he starts the year with an extravagant weekend, what is the probability that four weekends later will also be extravagant?

4. A certain computer has a 0.0001 probability of recording an input 0 as a 1 and a 0.0002 probability of recording a 1 fed into it as a 0. A 0 in the machine has a 0.0001 probability of being printed out as a 1, while a 1 in the machine has a 0.0002 probability of coming out as a 0. If a 1 is fed in, what is the chance that it will be fed out properly?

5. A traveling salesman serves three cities, A, B, C. To avoid boredom, he doesn't always cover them in the same order but instead constructs a Markov process of transitions from one city to another. His transition matrix is

$$M = \begin{matrix} & \begin{matrix} A & B & C \end{matrix} \\ \begin{matrix} A \\ B \\ C \end{matrix} & \begin{bmatrix} 0 & \frac{7}{12} & \frac{5}{12} \\ \frac{1}{2} & 0 & \frac{1}{2} \\ \frac{1}{3} & \frac{2}{3} & 0 \end{bmatrix} \end{matrix}$$

Assuming he starts in A, what are the probabilities of being in A, B, or C, respectively, after three transitions?

6. At the beginning of President X's term in office, which commences with a recession year, he is faced with choosing a set of economic

advisers. There are two competing economic philosophies he can choose from: the Reds, whose policies ensure that the probability of a boom year following a recession year is two-thirds, while the probability of a recession year following a boom year is three-fifths; or the Blues, whose policies ensure that the probability of a boom year following a recession year is one-half and the probability of a recession year following a boom year is one-quarter. Which philosophy, if followed consistently, gives the highest chance of a boom after 3 years, during the President's reelection campaign?

7.   Commuter trains arriving at Grand Central Station on a certain line of the railroad are always either early or late. The probability that the train following a late train is also late is three-fourths, while the probability that the train following an early train is early is one-half. Suppose the first train of the morning is early and you are meeting someone on the fourth train. What is the probability that your friend will be late?

8.*   Do you think the salesman in Exercise 5 will visit each city equally often in the long run? Justify your answer.

9.*   In Exercise 5, why would it make no sense to have

$$M = \begin{bmatrix} \frac{1}{3} & \frac{1}{3} & \frac{1}{3} \\ \frac{1}{2} & 0 & \frac{1}{2} \\ \frac{1}{3} & \frac{2}{3} & 0 \end{bmatrix}$$

10.* Express $p_{11}^{(3)}$ as a formula involving some or all of p_{11}, p_{12}, p_{21}, and p_{22} (assuming a Markov process with just two states).

11.* Whether or not a child grows up emotionally stable often depends on the emotional stability of his or her parents. If one or both parents are emotionally unstable, the chance that the child will be emotionally unstable is four-tenths. If both parents are stable, the probability that their child will be too is nine-tenths. Compute the probability that a child will be unstable if the chance that his parents are both emotionally stable is seven-tenths.

8.8   long-term behavior
of Markov processes

Suppose we are training a rat to run the maze shown in Figure 8.22. Each run begins at A and ends at either B or C, depending on the choice the rat makes at D. At C there is a piece of cheese as a reward and at B there is a jolt of elec-

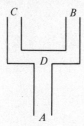

Figure 8.22

tricity as a punishment. The experiment is to determine how soon and how reliably the rat learns to make a left turn at D.

We can build a theoretical model involving a Markov process to try to predict and explain the rat's behavior. We assume that at any try there is a definite probability that the rat will choose a left turn (L) and that this probability depends upon only one thing: whether the last trial was successful or not. The idea is that we assume that the rat has an imperfect but functional memory that reaches back one trial into the past. Specifically, let's assume a transition matrix

$$T = \begin{matrix} & \begin{matrix} L & \ R \end{matrix} \\ \begin{matrix} L \\ R \end{matrix} & \begin{bmatrix} \frac{4}{5} & \frac{1}{5} \\ \frac{3}{5} & \frac{2}{5} \end{bmatrix} \end{matrix}$$

Assuming that the rat starts his career with R, what is the probability that after four trials the rat's choice is L? Suppose instead he starts his career with L. What is the probability that after four trials the rat chooses L? These questions require us to calculate T^4.

$$T^4 = (T^2)^2 = \begin{bmatrix} 0.76 & 0.24 \\ 0.72 & 0.28 \end{bmatrix}^2 = \begin{bmatrix} 0.7504 & 0.2496 \\ 0.7488 & 0.2512 \end{bmatrix}$$

The matrix T^4 has the remarkable property that the top and bottom rows are practically identical. They are so close that we might as well round these numbers off and imagine that we are dealing with

$$T^4 = \begin{bmatrix} 0.75 & 0.25 \\ 0.75 & 0.25 \end{bmatrix}$$

This is interesting because it means that on the fourth trial the chances of choosing L or R are 0.75 and 0.25, respectively, *regardless of what was done on the first trial*. Thus, despite our assumption that the rat's behavior depends on the immediate past, it does not, at the fourth trial, depend on the distant past. It is reasonable to ask whether this forgetfulness of the distant past would be manifested in T^5 and the other higher powers of T.

325

At this point, instead of restricting ourselves to this special example, let us be more general. Let us suppose Q is any transition matrix for a Markov process and that some power of Q, say, Q^n, is a matrix whose rows are identical. We can show that $Q^{n+1} = Q \times Q^n$ also has identical rows. Let

$$Q = \begin{bmatrix} a & b \\ c & d \end{bmatrix} \quad \text{and} \quad Q^n = \begin{bmatrix} e & f \\ e & f \end{bmatrix}$$

By calculation we find

$$Q^{n+1} = Q \times Q^n = \begin{bmatrix} a & b \\ c & d \end{bmatrix} \begin{bmatrix} e & f \\ e & f \end{bmatrix}$$

$$\begin{bmatrix} ae + be & af + bf \\ ce + de & cf + df \end{bmatrix} = \begin{bmatrix} e(a + b) & f(a + b) \\ e(c + d) & f(c + d) \end{bmatrix}$$

Since Q is a transition matrix, however, $a + b = 1$ and $c + d = 1$ so

$$Q^{n+1} = \begin{bmatrix} e & f \\ e & f \end{bmatrix} = Q^n$$

In the same way, we can conclude that $Q^{n+2} = Q^{n+1}$ and that, in general, all higher powers of Q will be the same. This means that in all trials beginning with the nth trial the rat has the same probability of completing the maze successfully. He has reached the peak of his learning. In the specific example we started with, after about the fourth trial the rat always has a three-fourths probability of completing the maze successfully. We can summarize what we have demonstrated by saying *as we take powers of a transition matrix, at the point when the rows become identical, successive powers give the same matrix from then on.* Taking this statement from the model and interpreting it in the rat's learning situation, we have this amusing state of affairs: When the rat gets to the point where it has "forgotten" its initial choice, it has reached the peak of its learning.

Strictly speaking, the foregoing analysis is not correct. The discrepancy from the truth is minor, however, and can be ignored for practical purposes. The source of the theoretical difficulty is that the rows of T^4 were not exactly equal—they were rendered equal by rounding off. If we refrain from rounding off and take higher and higher powers of T, we may never reach a power T^n which has rows which are exactly equal. However, the successive powers would have rows which were more and more identical and which would approach a certain row vector w, never exactly reaching it but coming closer and closer all the time. The mathematical jargon for this is that the powers T^n "converge" to a certain "limit" matrix W, whose rows are all equal to a certain row vector w. A general summary of the situation is contained in the following theorem, stated for a general n-state Markov process.

THEOREM 6

If Q is a matrix with the property that either Q or some power of Q has no zero entries, then

(1) The powers Q^m converge to a limit matrix W, whose rows are all the same vector $w = (w_1, w_2, \ldots, w_n)$.

(2) w satisfies the following two properties:

$$w_1 + w_2 + \cdots + w_n = 1$$

$$wQ = w$$

We shall not give a proof of the theorem. We have already seen an illustration of part (1) of the theorem. We shall now see how part (2) is reflected in the trials of our rat. The vector w to which the rows of T^n appeared to be converging was $(\frac{3}{4}, \frac{1}{4})$. Part (2) states that

$$\tfrac{3}{4} + \tfrac{1}{4} = 1$$

and

$$[\tfrac{3}{4}, \tfrac{1}{4}] \begin{bmatrix} \frac{4}{5} & \frac{1}{5} \\ \frac{3}{5} & \frac{2}{5} \end{bmatrix} = [\tfrac{3}{4}, \tfrac{1}{4}]$$

which are both true statements. Therefore, part (2) holds for our example.

It is handy to have some terminology to describe the phenomena in this theorem.

DEFINITION 4

Q is called *regular* matrix if either Q or some power of Q has no zero entries.

Example 1

Determine whether the following matrix is regular:

$$Q = \begin{bmatrix} 0 & 1 \\ \frac{2}{5} & \frac{3}{5} \end{bmatrix}$$

Solution: Q itself has zero entries but

$$Q^2 = \begin{bmatrix} 0 & 1 \\ \frac{2}{5} & \frac{3}{5} \end{bmatrix} \cdot \begin{bmatrix} 0 & 1 \\ \frac{2}{5} & \frac{3}{5} \end{bmatrix} = \begin{bmatrix} \frac{2}{5} & \frac{3}{5} \\ \frac{6}{25} & \frac{19}{25} \end{bmatrix}$$

Since Q^2 has no zero entries, Q is regular.

Example 2

Determine whether the following matrix is regular:

$$Q = \begin{bmatrix} 0 & 1 \\ 1 & 0 \end{bmatrix}$$

Solution: Q has zero entries so we calculate Q^2:

$$Q^2 = \begin{bmatrix} 0 & 1 \\ 1 & 0 \end{bmatrix} \cdot \begin{bmatrix} 0 & 1 \\ 1 & 0 \end{bmatrix} = \begin{bmatrix} 1 & 0 \\ 0 & 1 \end{bmatrix} = I$$

Q^2, which turns out to be the identity matrix, also has zero entries. Higher powers of Q now need to be checked but fortunately we don't actually have to do the matrix multiplications since we have determined that $Q^2 = I$. For example, $Q^3 = (Q^2)Q = IQ = Q$, which has zero entries; $Q^4 = (Q^3)Q = QQ = I$, which has zero entries; $Q^5 = (Q^4)Q = IQ = Q$, which has zero entries; $Q^6 = (Q^5)Q = QQ = I$, which all have zero entries.

As you can see, the pattern perpetuates itself and the powers are alternately equal to I and Q. Thus, all powers have zero entries and Q is not regular.

DEFINITION 5

A vector w with the property that $wQ = w$ for a certain transition matrix Q is called a *fixed vector for Q*.

The significance of a fixed vector is that its entries give the long-term probabilities of the various states of the Markov process. These probabilities are independent of the state in which the process began.

Part (2) of the theorem provides a handy way of calculating the fixed vector for a given transition matrix Q. The equation $wQ = w$ is actually a set of n equations in the unknowns $w_1, w_2, \ldots, w_n$. These equations turn out to be linear; when we include in our system the equation $w_1 + w_2 + \cdots + w_n = 1$, we can solve for the variables w_i. Once the vector w is known, we can also determine the limiting matrix toward which the powers Q^n converge since this matrix has for its rows the vector w.

Example 3

What is the limiting matrix toward which Q^n tends if

$$Q = \begin{bmatrix} \frac{2}{3} & \frac{1}{3} \\ \frac{1}{5} & \frac{4}{5} \end{bmatrix}$$

Solution: Each row vector $w = [w_1, w_2]$ of the limit matrix must satisfy

$$[w_1, w_2]\begin{bmatrix} \frac{2}{3} & \frac{1}{3} \\ \frac{1}{5} & \frac{4}{5} \end{bmatrix} = [w_1, w_2] \tag{8.6}$$

and $w_1 + w_2 = 1$. However,

$$[w_1, w_2]\begin{bmatrix} \frac{2}{3} & \frac{1}{3} \\ \frac{1}{5} & \frac{4}{5} \end{bmatrix} = \left[\frac{2w_1}{3} + \frac{w_2}{5}, \ \frac{w_1}{3} + \frac{4w_2}{5}\right]$$

According to (8.6) we must have

$$[w_1, w_2] = \left[\frac{2w_1}{3} + \frac{w_2}{5}, \ \frac{w_1}{3} + \frac{4w_2}{5}\right]$$

which means

$$w_1 = \frac{2w_1}{3} + \frac{w_2}{5}$$

$$w_2 = \frac{w_1}{3} + \frac{4w_2}{5}$$

Simplifying these equations, we discover they are really the same equation:

$$\frac{w_1}{3} - \frac{w_2}{5} = 0$$

Solving this simultaneously with $w_1 + w_2 = 1$ gives the solution

$$w_1 = \tfrac{3}{8}$$

$$w_2 = \tfrac{5}{8}$$

Thus, the limit matrix toward which the powers Q^n converge is

$$\begin{bmatrix} \frac{3}{8} & \frac{5}{8} \\ \frac{3}{8} & \frac{5}{8} \end{bmatrix}$$

If this problem had arisen in connection with our rat trials, we would conclude that, in the long run, the rat would choose the left turn three-eighths of the time.

Example 4

Suppose we have an intelligent rat with a two-trial memory in the maze of Figure 8.22. He runs the first two trials at random but after that his behavior depends on the last *two* trials. Now let us regard a state as being not the outcome of one trial but rather the outcomes of a consecutive pair of trials. Thus, we have three states: two successes on the last two trials (*LL*); one success out of the last two trials (*LR* or *RL*); and no successes on the last two trials (*RR*). We symbolize these by S_2, S_1, and S_0, respectively. For example, let's suppose the series of choices actually made by the rat is *LRRLRLLLR*. After the first two choices we are in state S_1. When the rat makes his third choice *R*, he is in state S_0 since his last two choices were both *R*. The succession of states corresponding to his series of choices is $S_1 S_0 S_1 S_1 S_1 S_2 S_2 S_1$. Let us assume the transition matrix

$$
\begin{array}{c}
 \\
S_0 \\
S_1 \\
S_2
\end{array}
\begin{array}{ccc}
S_0 & S_1 & S_2
\end{array}
\begin{bmatrix}
\frac{1}{2} & \frac{1}{2} & 0 \\
\frac{1}{3} & \frac{1}{3} & \frac{1}{3} \\
0 & \frac{1}{4} & \frac{3}{4}
\end{bmatrix}
$$

Determine whether or not the matrix is regular. If it is, find the fixed vector of long-term probabilities for the rat to be in states S_0, S_1, or S_2.

Solution: Denoting the transition matrix by T, we calculate

$$
T^2 = \begin{bmatrix}
\frac{1}{2} & \frac{1}{2} & 0 \\
\frac{1}{3} & \frac{1}{3} & \frac{1}{3} \\
0 & \frac{1}{4} & \frac{3}{4}
\end{bmatrix} \cdot \begin{bmatrix}
\frac{1}{2} & \frac{1}{2} & 0 \\
\frac{1}{3} & \frac{1}{3} & \frac{1}{3} \\
0 & \frac{1}{4} & \frac{3}{4}
\end{bmatrix}
$$

$$
= \begin{bmatrix}
\frac{1}{4} + \frac{1}{6} & \frac{1}{4} + \frac{1}{6} & \frac{1}{6} \\
\frac{1}{6} + \frac{1}{9} & \frac{1}{6} + \frac{1}{9} + \frac{1}{12} & \frac{1}{9} + \frac{1}{4} \\
\frac{1}{12} & \frac{1}{12} + \frac{3}{16} & \frac{1}{12} + \frac{9}{16}
\end{bmatrix}
$$

We have left some arithmetic undone because it is enough to observe that all entries are nonzero, and therefore T is regular,

To calculate the fixed vector,

$$
[w_1, w_2, w_3] \begin{bmatrix}
\frac{1}{2} & \frac{1}{2} & 0 \\
\frac{1}{3} & \frac{1}{3} & \frac{1}{3} \\
0 & \frac{1}{4} & \frac{3}{4}
\end{bmatrix} = [w_1, w_2, w_3]
$$

This yields the first three equations in the system below.

$$
\frac{w_1}{2} + \frac{w_2}{3} = w_1
$$

$$
\frac{w_1}{2} + \frac{w_2}{3} + \frac{w_3}{4} = w_2
$$

$$\frac{w_2}{3} + \frac{3w_3}{4} = w_3$$

$$w_1 + w_2 + w_3 = 1$$

The solution of the system is $(w_1, w_2, w_3) = (\frac{2}{9}, \frac{1}{3}, \frac{4}{9})$. Therefore, in the long run, the rat is in state S_2 (two consecutive successes) four-ninths of the time, in S_1 (one success in last two tries) one-third of the time, and in S_0 (no successes in last two tries) two-ninths of the time.

EXERCISES 8.8

1. State why each of the following matrices is regular.

   (a) $\begin{bmatrix} \frac{1}{2} & \frac{1}{4} & \frac{1}{4} \\ \frac{1}{3} & \frac{1}{3} & \frac{1}{3} \\ \frac{2}{5} & \frac{2}{5} & \frac{1}{5} \end{bmatrix}$
   (b) $\begin{bmatrix} 0 & 1 \\ \frac{1}{5} & \frac{4}{5} \end{bmatrix}$

   (c) $\begin{bmatrix} 0 & 0 & 1 \\ \frac{1}{2} & 0 & \frac{1}{2} \\ \frac{1}{2} & \frac{1}{2} & 0 \end{bmatrix}$
   (d) $\begin{bmatrix} 0 & 1 & 0 \\ 0 & 0 & 1 \\ \frac{1}{4} & \frac{3}{4} & 0 \end{bmatrix}$

   (e) $\begin{bmatrix} 0 & \frac{1}{2} & \frac{1}{2} \\ \frac{1}{4} & \frac{1}{2} & \frac{1}{4} \\ \frac{1}{3} & \frac{1}{3} & \frac{1}{3} \end{bmatrix}$

2. For each of the following nonregular transition matrices, calculate the first few powers and see if you can find a pattern that shows that all powers have zero entries.

   (a) $\begin{bmatrix} \frac{1}{5} & \frac{4}{5} \\ 0 & 1 \end{bmatrix}$
   (b) $\begin{bmatrix} 1 & 0 & 0 \\ 0 & \frac{1}{2} & \frac{1}{2} \\ 0 & \frac{1}{2} & \frac{1}{2} \end{bmatrix}$

   (c) $\begin{bmatrix} \frac{1}{3} & 0 & \frac{2}{3} \\ 0 & 1 & 0 \\ \frac{1}{2} & \frac{1}{4} & \frac{1}{4} \end{bmatrix}$

3. For each of the following matrices, determine whether or not it is regular. If it is, find the fixed vector.

(a) $\begin{bmatrix} 1 & 0 \\ \frac{1}{2} & \frac{1}{2} \end{bmatrix}$ (b) $\begin{bmatrix} 1 & 0 & 0 \\ 0 & 1 & 0 \\ \frac{1}{3} & \frac{1}{3} & \frac{1}{3} \end{bmatrix}$

(c) $\begin{bmatrix} 0 & \frac{1}{2} & \frac{1}{2} \\ 0 & 0 & 1 \\ \frac{1}{2} & \frac{1}{2} & 0 \end{bmatrix}$ (d) $\begin{bmatrix} 0.6 & 0.4 \\ 0.3 & 0.7 \end{bmatrix}$

4.(a)–(d) For each of the matrices of Exercise 1, determine the fixed vector.

5. Would you regard the intelligent rat of Example 4 as more intelligent or less intelligent if his transition matrix, instead of being as given in the example, has the $\frac{1}{4}$ and the $\frac{3}{4}$ entries in the last row of the matrix interchanged?

6–10. For Exercises 5 through 9 on pp. 323–324, determine whether or not the transition matrix of those exercises is regular. If it is, determine, through the fixed vector, the long-term probabilities of the various states. Is it reasonable to suppose that the Markov process goes on for a long time?

11. Describe in simple nonmathematical language what is going on in the case of a rat whose transition matrix is

$$\begin{bmatrix} 0 & 1 \\ 1 & 0 \end{bmatrix}$$

12.* Show, in the case of a 3×3 matrix Q, that if Q^n has equal rows, then $Q^n = Q^{n+1}$.

13.* For the intelligent rat of Example 4, find the long-run probability of taking a single left turn.

8.9* probability and genetics

Genetics is the branch of biology that deals with the mechanisms by which offspring inherit traits such as hair color or eye color from their parents. No area within biology depends as heavily on probability models as does genetics. After a brief discussion of some basic principles, we shall pose some simple genetic questions which can be answered using probabilistic methods.

The inheritance of a trait (e.g. albinism, some kinds of dwarfism, and short fingeredness) generally is governed by a pair of genes. An individual gets one gene of the pair from his (or her) father and the other from his (or her) mother. We shall discuss the simplest situation, where each gene for a certain trait X comes in two varieties, A and a. Using the genes A and a, there are three possible gene pairs, *AA*, *aa*, and *Aa* (= *aA*). The various possible gene pairs with *A* and *a* are known as *genotypes*.

The only traits we shall discuss are those where the individual either has the trait or does not have it. For example, one is either a dwarf or is not. These two physical manifestations, dwarfism and ordinary stature, are called the two *phenotypes* for this trait. Many traits do not satisfy our assumptions that there are just two phenotypes. For instance, eye color comes in various phenotypes: blue, brown, green, hazel, and shades in between.

When persons with genotype *AA* or *Aa* have the same phenotype but have a different phenotype from persons with genotype *aa*, then the gene *A* is called a *dominant gene* and the gene *a* is called a *recessive gene*. Individuals with genotype *Aa* are called *hybrids* or *carriers* for the *a* gene.

Example 1

Albinism is a trait characterized by milky skin color, very light hair color, and pinkish eye color. The relative frequency of albinos in the United States is 1 person in 20,000. Whether a person is an albino or a non-albino depends on a single pair of genes which we will denote by *A* and *a*. Persons who are non-albino can have genotype *AA* or *Aa*. Albinos always have genotype *aa*. Using the dominant-recessive terminology, *A* is a dominant gene and *a* is a recessive gene. For convenience, we shall refer to *A* as the non-albino gene and to *a* as the albino gene in future discussion.

Here are some of the questions we shall answer in this section. Try to guess or deduce some of the answers yourself.

QUESTION 1

Can albino children be born to non-albino parents?

QUESTION 2

In view of the dominance of the non-albino gene, *A*, will the proportion of albinos in the population become smaller as time goes on?

QUESTION 3

If 1 person out of 20,000 is an albino (*aa*), what is the relative frequency of carriers (*Aa*)?

Let's begin with Question 1, which is comparatively easy to answer. Consider a mating between two hybrids (*Aa*). Determining the genotypes of the offspring is a multistage experiment in which the first stage is to determine the gene contributed by the father and the second stage is to determine the gene contributed by the mother. In each experiment we assume the equiprobable model and we also assume that the two experiments are independent. Figure 8.23 shows the possibilities for the various outcomes of this multi-stage experiment. We see, for example, that there is a $\frac{1}{4}$ chance that the child will be an albino. Thus, Question 1 is answered affirmatively.

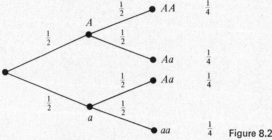

Figure 8.23

In pursuit of the answer to Question 2, for convenience we shall do calculations for other types of matings analogous to those done in Figure 8.23. The results are shown in the following table.

Father	Mother	(Conditional) Probability of Child's Being		
		AA	$Aa (= aA)$	aa
AA	AA	1	0	0
AA	Aa	$\frac{1}{2}$	$\frac{1}{2}$	0
AA	aa	0	1	0
Aa	AA	$\frac{1}{2}$	$\frac{1}{2}$	0
Aa	Aa	$\frac{1}{4}$	$\frac{1}{2}$	$\frac{1}{4}$
Aa	aa	0	$\frac{1}{2}$	$\frac{1}{2}$
aa	AA	0	1	0
aa	Aa	0	$\frac{1}{2}$	$\frac{1}{2}$
aa	aa	0	0	1

Before using this table to answer Question 2, we need to make some further assumptions. We are going to imagine that we have a population with the following characteristics:

HW1.   The only matings possible are within the group (i.e. no migration).
HW2.   Mating partners are chosen at random.
HW3.   No mutations occur. (A *mutation* is the appearance of a new gene other than A or a which affects the trait in question.)
HW4.   No "selection" occurs, that is, persons of all genotypes have equal chances to survive, to mate, and to have fertile offspring.

HW5.   On the average there are as many males as females in each generation and each sex has the same distribution of genotypes.

Now we can answer Question 2 on the basis of the following theorem.

THEOREM 7

If the relative frequency of AA, Aa, and aa individuals in the population is equal to u, v, and w, respectively, then in the next generation the relative frequency of AA, Aa, and aa individuals is $(u + v/2)^2$, $2(u + v/2)(w + v/2)$, and $(w + v/2)^2$, respectively (see the table below).

	Original Generation	Next Generation
AA	u	$(u + v/2)^2$
Aa	v	$2(u + v/2)(w + v/2)$
aa	w	$(w + v/2)^2$

Proof: We imagine a random mating of man and a woman and we wish to determine the probabilities of various genotypes in the offspring. We conceive of this as the multistage experiment shown in Figure 8.24. The first stage of the experiment is to select a man at random; the second stage is to select a woman at random. The probabilities that the man (or the woman) is AA, Aa, or aa can be regarded as equal to the respective relative frequencies of these genotypes, namely u, v, and w. This accounts for the labelings on the branches in the first two stages of the tree. The branches and labelings of the third stage are determined by reference to the table on p. 324.

The probabilities of the various outcomes $O_1, O_2, \ldots, O_{15}$ in the tree are calculated by the multiplication rule. Note that each genotype can occur in a number of different ways and so is represented by several outcomes in the multistage experiment. We calculate the probabilities of the various genotypes by adding the probabilities of the appropriate outcomes (see Figure 8.24):

$$P(AA) = P(O_1) + P(O_2) + P(O_5) + P(O_7)$$

$$= u^2 + \frac{1}{2}uv + \frac{1}{2}uv + \frac{1}{4}v^2$$

$$= u^2 + uv + \frac{1}{4}v^2$$

$$= \left(u + \frac{1}{2}v\right)^2$$

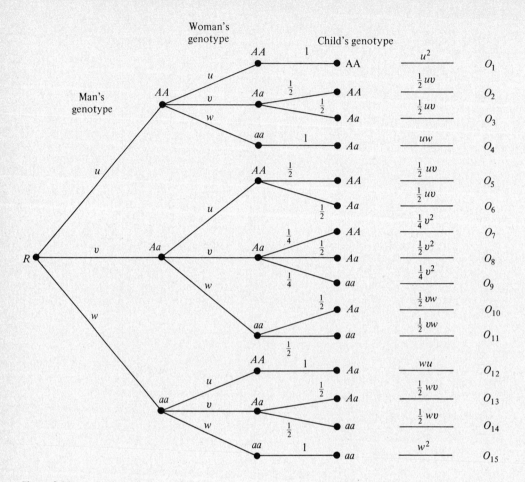

Figure 8.24

$$P(Aa) = P(O_3) + P(O_4) + P(O_6) + P(O_8) + P(O_{10}) + P(O_{12}) + P(O_{13})$$

$$= \frac{1}{2}uv + uw + \frac{1}{2}uv + \frac{1}{2}v^2 + \frac{1}{2}vw + wu + \frac{1}{2}wv$$

$$= 2\left(uw + \frac{uv}{2} + \frac{wv}{2} + \frac{1}{4}v^2\right)$$

$$= 2\left(u + \frac{v}{2}\right)\left(w + \frac{v}{2}\right)$$

$$P(aa) = P(O_9) + P(O_{11}) + P(O_{14}) + P(O_{15})$$

$$= \frac{1}{4}v^2 + \frac{1}{2}vw + \frac{1}{2}wv + w^2$$

$$= w^2 + vw + \frac{1}{4}v^2$$

$$= \left(w + \frac{v}{2}\right)^2$$

Example 2

Suppose that 0.5 of a population is *AA*, 0.2 is *Aa*, and the remaining 0.3 of the population is *aa*. What will be the proportions of the genotypes in the next generation?

Solution: We have $u = 0.5$, $v = 0.2$, and $w = 0.3$. Therefore, $(u + v/2)^2 = 0.36$; $2(u + v/2)(w + v/2) = 2(0.6)(0.4) = 0.48$; and $(w + v/2)^2 = 0.16$. Notice that the distribution of genotypes is quite different in the two generations.

Example 3

Let us carry Example 2 a generation further and calculate the distribution of genotypes in the third generation. Thus, we take $u = 0.36$, $v = 0.48$, and $w = 0.16$. Hence:

$$\left(u + \frac{v}{2}\right)^2 = (0.36 + 0.24)^2 = (0.6)^2 = 0.36$$

$$2\left(u + \frac{v}{2}\right)\left(w + \frac{v}{2}\right) = 2(0.36 + 0.24)(0.16 + 0.24) = 0.48$$

$$\left(w + \frac{v}{2}\right)^2 = (0.16 + 0.24)^2 = 0.16$$

The distribution of genotypes in the second and third generations is identical. Furthermore, it is clear that there will be no further change in the distribution in subsequent generations.

Example 3 gives a partial answer to Question 2 by showing a case where the recessive albino genotype does not die out or decrease. The next theorem will give a more complete answer to Question 2.

THEOREM 8

Suppose that genotypes *AA*, *Aa*, and *aa* constitute the fractions *u*, *v*, and *w* of the population respectively.

(1) If $v^2 = 4uw$, then the distribution of genotypes in all subsequent generations will be the same.
(2) Conversely, if the distribution of genotypes is the same in all subsequent generations, then $v^2 = 4uw$.

Proof: The population in the next generation will have the same distribution of genotypes, provided that the following equations hold.

$$u = \left(u + \frac{v}{2}\right)^2 \tag{8.7}$$

$$v = 2\left(u + \frac{v}{2}\right)\left(w + \frac{v}{2}\right) \tag{8.8}$$

$$w = \left(w + \frac{v}{2}\right)^2 \tag{8.9}$$

It can be shown by straightforward algebra that if $v^2 = 4uw$, then all these equations hold. Conversely, if any of these equations holds, then $v^2 = 4uw$. Here is one of the verifications required:

If $v^2 = 4uw$, then

$$\left(u + \frac{v}{2}\right)^2 = u^2 + uv + \frac{v^2}{4}$$

$$= u^2 + uv + uw$$

$$= u(u + v + w) = u(1) = u$$

Conversely, if

$$\left(u + \frac{v}{2}\right)^2 = u$$

then

$$u^2 + uv + \frac{v^2}{4} = u$$

and

$$\frac{v^2}{4} = u(1 - u - v)$$

$$= u(w)$$

Example 4

If $u = 0.5$, $v = 0.2$, and $w = 0.3$, then $v^2 = 0.04$ and $4uw = 0.6$. Since $0.04 \neq 0.6$, our theorem tells us that the distribution in the next generation will be different. We have already found this to be true in Example 2.

Example 5

If $u = 0.36$, $v = 0.48$, and $w = 0.16$, then $v^2 = 0.2304$ and $4uw = 0.2304$ also. Theorem 8 tells us that subsequent generations will have the same distribution of genotypes. This was verified in Example 3.

What happens to a population for which $4uw \neq v^2$ and whose distribution consequently is going to change in the next generation? Will it continue to change in the third, fourth, and fifth generations? Will it ever reach an equilibrium? The next theorem provides the startling answer.

THEOREM 9

Suppose a population consists of AA's, Aa's, and aa's in proportions u, v, and w, respectively. Regardless of the values of u, v, and w, the population will reach an equilibrium distribution in one generation (after one mating).

Proof: After one generation the proportions are $u' = (u + v/2)^2$, $v' = 2(u + v/2)(w + v/2)$, and $w' = (w + v/2)^2$. We only need to calculate $4u'w'$ and v'^2 to see that they are equal and then to apply Theorem 8.

The three theorems we have proved, based on assumptions HW1-HW5, are referred to as the Hardy-Weinberg Law in honor of the mathematician G. H. Hardy and the geneticist W. Weinberg, who both discovered it independently in 1908.

We now show how Theorem 9 can be used to answer Question 3. Since 1 in every 20,000 Americans is an albino, we can take $w = \frac{1}{20,000}$. Theorem 9 implies that for all practical purposes we can assume that the distribution of genotypes is in equilibrium. Thus:

$$4uw = v^2$$

or

$$\frac{4u}{20,000} = v^2$$

or

$$u = 5{,}000 \, v^2 \tag{8.10}$$

Also:

$$u + v + w = 1$$

whence

$$u + v + \frac{1}{20,000} = 1$$

and

$$u = 1 - \frac{1}{20,000} - v$$

Hence,

$$u \approx \text{(is approximately equal to) } 1 - v \tag{8.11}$$

Comparing (8.10) and (8.11):

$$5,000 \, v^2 = 1 - v$$

$$5,000 \, v^2 + v - 1 = 0$$

Applying the quadratic formula:

$$v = \frac{-1 + \sqrt{1 + 4(5,000)}}{2(5,000)}$$

$$\approx \frac{\sqrt{20,000}}{10,000} = \frac{100\sqrt{2}}{10,000} \approx \frac{1.414}{100}$$

$$= .01414 \approx \frac{1}{71}$$

Thus, there is a 1 in 71 chance that you are a carrier for albinism. To put it another way, in each randomly selected group of 71 people in the population, on the average one is a carrier for the albinism gene. This seems surprisingly high; so high that one might think that the number of albinos should be larger than 1 in 20,000. However, the multiplication rule implies that the chances of two carriers' mating would be

$$\left(\frac{1}{71}\right)\left(\frac{1}{71}\right) \approx \frac{1}{5,000}$$

Furthermore, only $\frac{1}{4}$ of these matings between Aa's would produce an albino (aa). Since $\frac{1}{4} \cdot \frac{1}{5,000} = \frac{1}{20,000}$, and since $\frac{1}{20,000}$ is the actual incidence of albinism, the $\frac{1}{71}$ incidence of carriers doesn't really contradict the 1 in 20,000 incidence of albinos.

Example 6

Suppose a physical trait X occurs in individuals with genotype rr but not in individuals with genotypes RR or Rr. Suppose that the fraction of the population with this trait is $\frac{1}{4}$. Assuming that the conditions for the Hardy-Weinberg Law hold, find the number of carriers (Rr).

Solution: Assuming equilibrium and using the notation u = fraction of RR, v = fraction of Rr, and w = fraction of rr:

$$w = \frac{1}{4} = 0.25$$

$$v^2 = 4uw = \frac{4u}{4} = u. \tag{8.12}$$

Also

$$1 = u + v + w$$
$$= u + v + 0.25$$

and so

$$0.75 - v = u. \tag{8.13}$$

Comparing (8.12) and (8.13):

$$v^2 = \frac{3}{4} - v$$

$$4v^2 + 4v - 3 = 0$$

$$v = \frac{-4 + \sqrt{16 + 4 \cdot 4 \cdot 3}}{8}$$

$$= \frac{-4 + \sqrt{64}}{8}$$

$$= \frac{4}{8}$$

$$= \frac{1}{2}.$$

One-half of the population consists of carriers.

EXERCISES 8.9

1. Using a tree as in Figure 8.23, verify that each line in the table on p. 334 is correct.
2. State which of the following populations would be in equilibrium if the assumptions of the Hardy-Weinberg Law are valid:
   (a) 100% of the population is AA.
   (b) 100% of the population is Aa.
   (c) 100% of the population is aa.
   (d) 1% of the population is AA, 98% of the population is Aa, and 1% of the population is aa.

(e) 64% of the population is *AA*, 32% of the population is *Aa*, and 4% of the population is *aa*.

(f) 1% of the population is *AA*, 16% of the population is *Aa*, and 83% of the population is *aa*.

3. For what values of x, if any, is a population with x% of the population *AA*, $(100 - 2x)$% of the population *Aa*, and x% of the population *aa*, in an equilibrium, assuming the conditions HW1 through HW5 hold?

4. For a group of 300 people, it is determined that 150 are *AA*, 120 are *Aa*, and 30 are *aa*.

    (a) What proportion of genotypes will the first generation of children have?

    (b) Is the original population in equilibrium?

5. Is a population of 500 individuals in which 40 are *AA*, 150 are *Aa*, and 310 are *aa* in equilibrium?

6. A trait X is governed by a single pair of genes, each of which can be R or r. Suppose that R is dominant and r is recessive and that only rr individuals show trait X. State which of the following situations are possible:

    (a) The great grandchild of two parents who do not have trait X has this trait.

    (b) The child of two parents, each of whom has trait X, does not have this trait.

    (c) All the children of parents who do not have trait X have the trait.

    (d) All the children of a man who has trait X and a woman who does not have the trait have trait X.

    (e) All the children of a man who has trait X and a woman who does not have trait X do not have this trait.

7.* Given the relative frequencies of the genotypes *AA*, *Aa*, and *aa* as u, v, and w, respectively, one can compute the relative frequency of gene A and of gene a as follows:

$$p = \text{relative frequency of gene } A = u + \frac{v}{2}$$

$$q = \text{relative frequency of gene } a = \frac{v}{2} + w$$

(a) If there are 30 individuals of genotype *AA*, 50 individuals of genotype *Aa*, and 20 individuals of genotype *aa*, what are the relative frequencies of A and a?

(b) Show by constructing an example that the same gene frequencies could arise from two populations having different relative frequencies for the genotypes.

(c) Using the numbers in part (a) of this exercise, find the proportions of the genotypes for the first generation. What are the gene frequencies for this generation? Have the gene frequencies changed from one generation to another? Does the answer surprise you?

SUGGESTED READING

BARTOS, O. J., *Simple Models of Group Behavior*, Columbia University Press, New York, New York, 1967. Discusses the use of Markov models in sociology.

GOLDBERG, S., *Probability: An Introduction*, Prentice-Hall, Inc., Englewood Cliffs, N.J., 1960. A solid and readable introduction to the theory of probability.

KLINE, M. (ED.), *Mathematics in the Modern World*, W. H. Freeman and Company, San Francisco, Calif., 1968. Available in paperback. This book contains many provocative articles about probability.

MESSICK, D. (ED), *Mathematical Thinking in Behavioral Sciences*, W. H. Freeman and Company, San Francisco, Calif., 1968. Available in paperback. This book also contains many provocative expository articles about probability.

NIVIN, I., *Mathematics of Choice*, Random House, New York, N.Y., 1965. Elaborate discussion of counting principles and their use in probabilistic arguments.

9
games and decisions

9.1   decisions

Throughout history philosophers have argued vigorously about whether or not man has free will—the power to influence the course of events by conscious decisions. Regardless of the answer, we generally behave as if we do have this power. Recently, a mathematical theory of decisions has been developed to help people make important decisions.

As we shall use the word, a decision situation is one where a person has two or more courses of action (called *strategies*) available and where the consequences of his action depend not only on the act itself but on contingencies in the outside world. We think of these contingencies as future states of nature. We assume that we cannot predict the future state of nature with any confidence.

Example 1

A recently created duplicating and printing service is thinking of leasing an additional photocopy machine for the next year as a means of increasing its income. The value of leasing an additional machine will depend on the general economic climate, that is, whether the next year is a prosperous year or a recession year. Thus, the company's prospects depend on a situation beyond its control. We indicate this by stating that nature's future state is either prosperity or recession. The firm has two choices (strategies)—to lease a new machine or not to lease a new machine.

Table 9.1 shows a matrix that gives the firm's estimates of the payoffs, i.e., business gains (or losses) for the various combinations of choices made by the decision maker (firm) and of the future states of nature.

Table 9.1

		Nature's Future States	
		Prosperity	Recession
Decision Maker's Choice	Lease machine	$20,000	$16,000
	Don't lease it	$17,000	$15,000

What should the firm do? It is not hard to see that it should lease the additional machine because, regardless of the state of the economy, the firm will be better off with the extra machine than without it. In technical language we would say that the strategy of leasing the extra machine *dominates* the strategy of not leasing it.

Example 1 is typical of the decision problems we shall deal with. The decision maker has various choices available, each of which yields a certain payoff depending on the future state of nature. The payoff, positive when the decision maker makes gains and negative when he loses, is entered in a matrix, the rows of which correspond to the decision maker's choices and the columns of which are the future states of nature.

DEFINITION 1

In a matrix for a decision problem, one row *dominates* another if each entry in the dominat*ing* row is greater than or equal to the corresponding (same column) entry of the dominat*ed* row.

In Example 1, the first row dominated the second because 20,000 > 17,000 and 16,000 > 15,000. The significance of this concept of domination is that a dominated row should never be considered as a possible course of action. This narrows the choices, sometimes to one row, as for the duplicating company, or sometimes to more than one, as in Example 2, which follows shortly.

Before going on to Example 2 it is worth observing that our analysis involves an assumption that the future state of nature is independent of the choice made by the decision maker. Suppose independence does not hold and, for example, leasing the machine would cause nature's future state to be recession while not leasing would have the inevitable effect of bringing about prosperity (never mind that this is absurd). Now the only outcomes are the lower left and upper right corners of the matrix and the decision maker's best choice is not to lease. This conclusion is exactly opposite to the one we reached with our dominant strategy analysis because our dominant strategy analysis is actually based on the assumption of independence. In Section 9.3 we shall discuss decision problems where independence does not apply but in Sections 9.1 and 9.2 we assume that the choice of the decision maker does *not* influence the future state of nature.

Example 2

Simplify the matrix for the decision problem in Table 9.2.

Table 9.2

Future States of Nature

	A	B	C
	$1	$2	$0
	$0	$2	$−1
	$3	$1	$1
	$0	$0	$1

Decision Maker's Choices

Solution: The first row dominates the second ($1 > 0, 2 \geq 2, 0 > -1$), while the third dominates the fourth ($3 > 0, 1 > 0, 1 \geq 1$). Therefore, we strike the second and fourth rows from the matrix. The remaining two rows (see the matrix below) have no dominance relationships so we have simplified the problem as much as possible. In the next section we shall study a method for choosing between the remaining rows.

Table 9.3

Future States of Nature

	A	B	C
	$1	$2	$0
	$3	$1	$1

Decision Maker's Choices

Notice that there is no need for us to consider the notion of dominance for columns. After all, nature does not actually make decisions and certainly has no motivation to frustrate the ambitions of particular decision makers.

One characteristic of both of our examples thus far is that the payoffs have been in money. One could equally well imagine decision situations in which the payoffs come in more subjective and less measurable forms, such as emotional reactions.

Example 3

Jack wants to marry Jill but is afraid to ask for fear that she would not say yes. Here is a payoff matrix modeling Jack's predicament. (Note that in this example, since Jack does not know how Jill will react to his proposal, Jill's role is that of "nature.")

Table 9.4

	Jill Would Say Yes	Jill Would Say No
Jack Asks	Joy	Hurt
Jack Doesn't Ask	Neutral	Neutral

Mathematicians have suggested ways in which the feelings in the matrix could be gauged and replaced by numbers called *utilities*, as in the matrix in Table 9.5. The manner of doing this is complicated and controversial and we won't go into further detail. However, we shall sometimes present examples where the payoffs are utilities. In interpreting these numbers, one should keep in mind that the higher the utility, the better the decision maker likes that payoff.

Table 9.5

	Jill Would Say Yes	Jill Would Say No
Jack Asks	1	-1
Jack Doesn't Ask	0	0

Example 4   (Pascal's Wager)

The philosopher and mathematician Blaise Pascal gives the following argument in the *Pensées* for believing in God. God exists or does not exist. Reason cannot tell us the answer so we are playing a game against nature, as shown in the matrix in Table 9.6. Pascal makes some attempt to evaluate the payoffs numerically. One interpretation of his writings suggests that he conceives of the payoffs as shown in Table 9.7. In this case, believing in God is the dominant strategy. Some people would argue that the payoffs "religious life" and "poisonous pleasures" don't have equal value if God does not exist. They believe the latter is more valuable than the former; therefore, a better matrix would be Table 9.8. Pascal seems to feel that the difference between these payoffs is an illusion but he concedes that some people may perceive the values of the payoffs to be as they are in this last matrix. There

Blaise Pascal (1623–1662) was a child prodigy who made important discoveries about geometry while still a teenager. At the age of eighteen he developed the plans for a calculating machine, which he built and marketed over the next few years. In 1654 between 10:30 and 12:30 at night, Pascal underwent a religious ecstasy which made him abandon mathematics for philosophy and theology. He returned to mathematics briefly once before his death at age 39. Apparently he was able to ease the pain of a toothache by thinking about mathematics, and he took this as a sign from God that He approved of mathematics.

(Courtesy The Wolff-Leavenworth Collection, George Arents Research Library at Syracuse University.)

Table 9.6

	God Exists	God Doesn't Exist
Believe	A religious life and an eternity of happiness	A religious life
Don't Believe	A life of "poisonous pleasures" of the flesh and an eternity of suffering	A life of "poisonous pleasures" of the flesh

Table 9.7

	God Exists	God Doesn't Exist
Believe	∞	1
Don't Believe	$-\infty$	1

Table 9.8

	God Exists	God Doesn't Exist
Believe	∞	1
Don't Believe	$-\infty$	2

The symbol ∞ stands for infinity.

is no dominant strategy for this matrix. Pascal has a separate argument for this case, which we shall touch on in the next section.

Find an atheist and ask him what he thinks of Pascal's wager.

EXERCISES 9.1

1. For each of the following decision situations, eliminate as many rows as possible using dominant strategy analysis.

(a)

1	−3
0	−4

(b)

2	1
−1	3

(c)

1	−1
2	−1
3	1

(d)

1	0	−3
2	1	4
3	1	2

(e)

0	−4	2
1	−2	3
2	1	1

(f)

1	−1	3
0	−2	1
−1	−1	0
4	2	2

2. Suppose that in matrix (a) of Exercise 1 the column choices were being made not by an impersonal outside world but by a human opponent who was motivated to keep your payoff to a minimum and who would make his column choice according to this motivation. Would you change your decision? Answer the same question for matrix (b).

3. In which of the examples in this section is it reasonable to challenge the modeling assumption that the decision maker's choice has no effect on the future state of nature?

9.2 *decisions and probabilities*

In this section we shall study a method of making decisions when dominant strategy analysis is inapplicable.

Example 1

A farmer has the choice of growing wheat or corn on his land. The profits for these crops depend on rainfall levels during the coming year, as shown by the matrix in Table 9.9 (payoffs in thousands of dollars).

Table 9.9

	Level I	Level II
Wheat	10	12
Corn	13	11

Since neither row dominates the other, it is not obvious which choice to make. To analyze the situation, let us add two new assumptions to our model:

(1) Repeatability—we assume that the farmer is confronted with exactly this decision situation year after year.

(2) Probabilities for nature's future states—we assume that we can estimate the probability that nature will be in a given state.

In our example, suppose level I occurs three-fourths of the time. First we shall determine how much profit there will be if the farmer always chooses wheat, for example, over a 20-year period. The probability three fourths gives the relative frequency of level I's occurrence. Hence, over a period of 20 years, level I can be expected to occur 15 times ($\frac{3}{4} \times 20$), while level II can be expected to occur 5 times. The total profits expected from level I years come to 15×10, while the total profits expected from level II years will be 5×12, giving a total profit (in all years) of 210. Schematically,

	Level I $\frac{3}{4} \times 20$	Level II $\frac{1}{4} \times 20$
Wheat	10	20

Wheat profit $= (\frac{3}{4} \times 20 \times 10) + (\frac{1}{4} \times 20 \times 12) = 210$

Doing the same calculation under the assumption the farmer always plants corn,

	Level I $\frac{3}{4} \times 20$	Level II $\frac{1}{4} \times 20$
Corn	13	11

Corn profit $= (\frac{3}{4} \times 20 \times 13) + (\frac{1}{4} \times 20 \times 11) = 250$

The best strategy is, therefore, always to plant corn.

If we wish to find the average (mean) profit per year, instead of the total profit over 20 years, we could divide by 20 and find

$$\text{Mean wheat profit} = \tfrac{210}{20} = 10.5$$

$$\text{Mean corn profit} = \tfrac{250}{20} = 12.5$$

A glance at our earlier calculations shows that the 20 could have been divided out earlier or, even more conveniently, never multiplied in at all. For example, we could simplify the calculations by doing them this way:

	Level I $\tfrac{3}{4}$	Level II $\tfrac{1}{4}$
Wheat	10	12

Expected wheat profit $= (\tfrac{3}{4} \times 10) + (\tfrac{1}{4} \times 12) = 10.5$

	Level I $\tfrac{3}{4}$	Level II $\tfrac{1}{4}$
Corn	13	11

Expected corn profit $= (\tfrac{3}{4} \times 13) + (\tfrac{1}{4} \times 11) = 12.5$

One question we have not considered in connection with decision problems without dominant strategies is this: Might it not be better to mix our strategies rather than to play one strategy all the time? For example, could the farmer in Example 1 do better by sometimes planting corn and sometimes wheat rather than always planting corn? The answer is no because we calculated that however many times we plant wheat, the expected payoff in each wheat year is 10.5 and however many times we plant corn, the expected payoff is 12.5 in each corn year. Clearly it is better to expect 12.5 each year by always planting corn than to expect 12.5 some of the time (during corn years) and 10.5 the rest of the time (during wheat years). This argument can be formalized and applied to any decision-making problem of the sort we are discussing here.

Experiment

See how an experimental subject unfamiliar with decision theory makes his decisions in the situation of Example 1 (in order to simulate the states of nature with the proper probabilities, use the table of random numbers as described in Appendix 12.4). Does your subject mix his strategies or does he stick to one and play it all the time? Does he behave differently for other matrices? Try a matrix, for example, where the numerical entries are more dramatically different.

Example 2

Workers in an industrial firm are faced with the choice of renewing the existing wage contract, aiming for one with a modest increase or aiming for one with a large increase. The firm's economists claim that a wage increase would lead to a price increase for the product, which would, in turn, lead to a decrease in demand. The consequent decrease in production would necessitate curtailing overtime and firing some workers. The union has an economist whose figures tell a different story. The workers are engaged in a decision against nature where nature's choices are either that the company economist is right or that the union economist is right. The payoffs in this problem, shown in the matrix in Table 9.10 are in total dollars paid out to workers. Suppose this situation happens repeatedly and the union's economist is right two-thirds of the time. What should the union do to maximize total dollars paid to workers?

Table 9.10

	Company Economist Correct	Union Economist Correct
No Increase	1	1
Modest Increase	0.75	0.9
Large Increase	0.8	1.2

(Payoffs in millions of dollars)

Solution: The second row can be immediately disregarded because it is dominated by the third. The expected payoffs for each of the remaining two rows are shown in Table 9.11.

Table 9.11

	Company Economist Correct $\frac{1}{3}$	Union Economist Correct $\frac{2}{3}$
No Increase	1	1

Expected payoff $= (\frac{1}{3} \times 1) + (\frac{2}{3} \times 1) = 1$

	Company Economist Correct $\frac{1}{3}$	Union Economist Correct $\frac{2}{3}$
Large increase	0.8	1.2

Expected payoff $= (\frac{1}{3} \times 0.8) + (\frac{2}{3} \times 1.2) = 1.067$

Thus, our calculations show that the best strategy is to aim for the large increase.

Although we have focused in our examples so far on decision situations that are repeatable, many theorists believe that this type of calculation is also applicable for situations occurring only once. The next example illustrates this.

Example 3

An automobile company is thinking about making syncopated steering an optional feature on next year's models. Sales prospects will depend not only on the outcome of this decision but also on what other manufacturers decide to do with respect to syncopated steering. The decisions of other manufacturers are not known at the time the decision must be made. The payoffs in the matrix in Table 9.12 represent percentage change from current

Table 9.12

	Few	About Half	Many
Syncopated Steering	−5	4	2
No Syncopated Steering	0	−3	−10

sales. Nature's future states are abbreviations with this significance: *few* means few other manufacturers adopt syncopated steering; *about half* means that about half adopt it; and *many* means that many do. The company realizes that this situation will not occur again and again so the probabilities for *few*, *about half*, and *many* cannot be estimated from relative frequencies. On subjective grounds (hunches), however, the company decides to assign probabilities of 1/4, 3/8, and 3/8 to these choices, respectively. What should the company do?

Using the same techniques as before, we calculate that the expected payoff for syncopated steering is

$$(-5 \times \tfrac{1}{4}) + (4 \times \tfrac{3}{8}) + (2 \times \tfrac{3}{8}) = 1$$

while the expected payoff in the other case is

$$(0 \times \tfrac{1}{4}) + (-3 \times \tfrac{3}{8}) + (-10 \times \tfrac{3}{8}) = -\tfrac{39}{8}$$

Thus, the company should develop syncopated steering.

Since the situation described will never occur again, a "subjective" interpretation of expected value must be used.

Example 4   (Pascal's Wager Again)

Suppose we conceive of the payoffs in Pascal's wager as in the matrix that follows (see the discussion in Example 4 of Section 9.1). Since we have no dominant strategy, a probability approach is warranted. Note once again that our decision situation is not a repeatable one: We do not over and over again decide on whether or not to believe and then have the answer revealed to us. In fact, some might assert that although we may decide, the answer never becomes clear. Nevertheless, Pascal proceeds with a verbal argument very close to the kind of numerical analysis we have been doing in this section. He assumes "there is an equal risk of gain and loss," which suggests that we take the probability of the existence of God to be 1/2. We now carry out our probability analysis, assuming some plausible rules for manipulating infinite quantities.

	God Exists	God Doesn't Exist
Believe	∞	1
Don't Believe	$-\infty$	2

$$\text{Expected value of believing} = (\infty \times \tfrac{1}{2}) + (1 \times \tfrac{1}{2}) = \infty$$

$$\text{Expected value of not believing} = (-\infty \times \tfrac{1}{2}) + (2 \times \tfrac{1}{2}) = -\infty$$

EXERCISES 9.2

1. For each of the matrices below, determine the best choice of row for the decision maker. The probabilities of the various columns are indicated above the columns.

(a)

	$\tfrac{1}{2}$	$\tfrac{1}{2}$
	1	2
	2	-3

(b)

	$\tfrac{2}{3}$	$\tfrac{1}{3}$
	-1	1
	2	-3

(c)

	$\tfrac{1}{4}$	$\tfrac{3}{4}$
	-1	1
	2	-3

2. For the matrix of part (a) of Exercise 1, determine what probabilities on the columns (replacing the ones given) would make either of the last two rows equally advantageous for the decision maker. Do the same for matrices (b) and (c).

3. If a tornado is sighted 20 mi from town, there is a probability of 0.05 that it will touch down in town and do damage. The problem is to determine whether to sound the alarm each time a tornado is sighted at the 20-mi limit or whether to wait until it is sighted closer to town. Suppose the utilities are as in the matrix below:

	Touch Down 0.05	Won't Touch Down 0.95
Alarm	−20	−1
No Alarm	−50	0

(Payoffs in utility numbers)

What should the policy be?

4. In which of the examples and exercises in this section is it reasonable to challenge the modeling assumption that the decision maker's choice has no effect on nature's future state?

5. You are a widget manufacturer and need to decide whether or not to invest in new equipment. If the next year is a booming one, a decision to invest will be worth $100,000 but postponing investment costs you $10,000. If it is a recession year and you don't invest, you come out even but if you do invest in the equipment, you lose $20,000. Model this in a matrix for a decision against nature. If the probability of a boom is 3/4, what should you do?

6. You are entrusted with the problem of making a suitable investment with a large sum of money. You decide to buy either common stocks or bonds. Which of these will turn out to be more sensible depends on which of two possible economic situations will prevail in the next year. Let us call these two economic situations *A* and *B*. Suppose that under *A* a commitment to bonds will yield 5%, while stock purchases will lose 1%. Under situation *B* the bond and stock gain percentages are 4% and 8%, respectively. Model this as a decision problem against nature by drawing a matrix. If each economic situation is equally likely, what should you do?

7. You are appealing a $10 traffic ticket in traffic court. Since you do not know which judge you will receive, you are unsure of whether the judge will be a stern or lenient one. You figure you have two strategies: You can be apologetic and plead guilty with an excuse or you can claim innocence aggressively. If you are apologetic, either judge will cut the fine in half. But if you are aggressive, a stern judge will leave the fine at $10, while a lenient one will cancel it all. Draw the decision matrix, using

the fines as payoffs. If the probability of getting a lenient judge is 1/3, what should you do?

8.* Suppose, in reference to Exercise 3, that town officials want to determine the exact mileage limit so that tornados sighted at that distance from town would sound the alarm but nothing sighted farther away would sound the alarm. Suppose it has been determined that a tornado sighted r ($r \geq 5$) miles from the center of town has a probability of $20/r^2$ of touching down in town. What mileage limit should be set?

9.3   decision trees

> *Two roads diverged in a yellow wood*
> *And sorry I could not travel both . . . .*
>
> —*ROBERT FROST**

There are two respects in which real decision problems are more complicated than the ones we discussed in the last two sections. First, the decision maker's decision may influence the future state of nature. Second, the decision maker may be faced with a sequence of related decisions. The following examples illustrate these phenomena and show how we can use graphs called *decision trees* to deal with them.

The first example is adapted directly from a 1972 study of hurricane control ("The Decision to Seed Hurricanes," by Howard, Matheson, and North, in *Science*, June 16, 1972).

Example 1

The government is faced with the decision of whether or not to seed a severe hurricane threatening a coastal area. *Seeding* refers to the use of silver iodide crystals dropped from planes to break up a hurricane. Seeding often reduces the wind speed of the hurricane and may consequently reduce property damage. However, sometimes hurricanes react to seeding by increasing rather than diminishing their wind speed. Of course, wind speed may either increase, decrease, or stay the same even if there is no seeding. The situation is illustrated in Figure 9.1. The square vertex at the left represents the point at which the government needs to make its decision to seed or not to seed, each of these choices being represented by an edge branching to the right. At

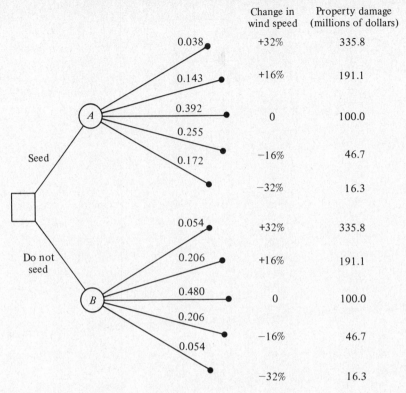

	Change in wind speed	Property damage (millions of dollars)
0.038	+32%	335.8
0.143	+16%	191.1
0.392	0	100.0
0.255		
0.172	−16%	46.7
	−32%	16.3
0.054	+32%	335.8
0.206	+16%	191.1
0.480	0	100.0
0.206		
0.054	−16%	46.7
	−32%	16.3

Figure 9.1

the ends of these edges there are circular vertices representing places where there will be a choice regarding nature's future state. (In general, we always use circular vertices to indicate decision points for nature's state and reserve square vertices for decision points of the decision maker.) These circular vertices branch out into five edges representing a sample of possible changes in wind speed. The numerical labels on these edges are the probabilities for the various changes in wind speed. For example, following the top path in Figure 9.1 would mean the hurricane was seeded and the wind speed of the hurricane increased by 32%, thereby causing 335.8 million dollars in property damage. The fact that the distribution of probabilities on the edges branching from A is different from the distribution on the edges branching from B is a reflection of the fact that the decision about seeding has an effect on nature's future state.

To decide whether or not to seed we need to compare the expected property damage at A with the expected property damage at B. For each of A and B we need to do an expectation calculation of the same sort as we did when we dealt with decision matrices:

Expected property damage with seeding

$$= 0.038(335.8) + 0.143(191.1) + 0.392(100.0)$$
$$+ 0.255(46.7) + 0.172(16.3)$$
$$= \$94.00$$

Expected property damage without seeding

$$= 0.054(335.8) + 0.206(191.1) + 0.480(100.0)$$
$$+ 0.206(46.7) + 0.054(16.3)$$
$$= \$116.00$$

At each of A and B we "pinch off" the branches, leaving the vertex and labeling the vertex with the expected property damage. In this way we reduce our
tree to that of Figure 9.2. Now the decision is clear—we choose to seed
because the expected property damage is 22.00 million dollars less.

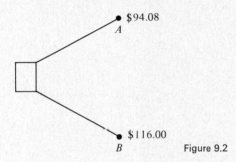

Figure 9.2

Example 2

Figure 9.3 shows a map of part of the state college campus where the
mathematician Linus Meander teaches. Each morning he enters campus at
A and is immediately faced with the choice of driving to X where there is
a 1/5 chance of getting a parking space right in front of the math building
or of driving to parking lot Siberia where there is always space available.
If he goes to Siberia, he takes the bus back along the diagonal road and his
total trip is 13 min, as opposed to the 4 min needed to go directly to X.
Another alternative is to park illegally in the loading zone behind the mathematics building where there is always space available. The fine for parking
there is $10 and there is a 3/10 chance of incurring this fine any given time
one parks there. Meander estimates that his time is worth $10/hr to him.
What is his best strategy, assuming that being at L or X is equally desirable to
him?

Figure 9.4 shows how we build the tree for this problem in a step-by-

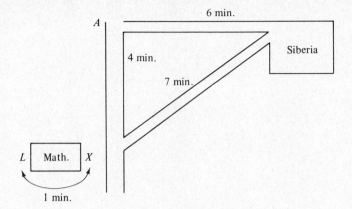

Figure 9.3

step way. When Meander is at A, he has to decide between proceeding to X or to Siberia. If he chooses Siberia, there are no further decisions to be made by him or by nature so we can label that endpoint with the time cost of 13 min. If Meander chooses to go to X, we have a decision by nature [see Figure 9.4(b)]: There is a space or there is not. If there is, we label that end point with 4 min. Otherwise, Meander must now decide between driving to Siberia (which makes the time cost a sure thing of 18 min) and parking at L. This decision is shown in Figure 9.4(c). In the latter case there is a final decision of nature, regarding whether or not Meander gets a ticket. This brings us to Figure 9.4(d) as our final tree. Unfortunately, the outcomes are not directly comparable since one involves money as well as time. We can convert the \$10 to a time equivalent of 60 min, however, and get a total time cost of 65 min for this outcome.

We begin the analysis by pinching off at D. If Meander finds himself at D, his expected time for the trip is $(\frac{3}{10})(65) + (\frac{7}{10})(5) = 23.0$ min. Thus, we produce the tree of Figure 9.5.

Clearly if Meander finds himself at C, he will choose the lower branch, to go to Siberia, so we get the further simplifications shown in Figure 9.6. To pinch off at B we calculate the expected time for the trip at B as $(\frac{1}{5})(4) + (\frac{4}{5})(18) = \frac{76}{5} = 15.2$. Thus, we produce the final tree in Figure 9.7 and this shows that Meander should choose Siberia immediately.

The preceding example illustrates two important principles in decision problems: the sequencing of decisions and the principle of conversion between different measures of cost (or benefit). The sequencing of decisions in the tree is not necessarily in the order in which they occur in reality. For example, whether or not Meander gets a parking ticket if he parks at L may have

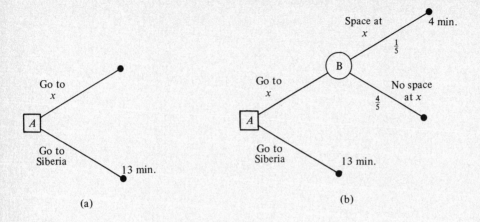

(a)

(b)

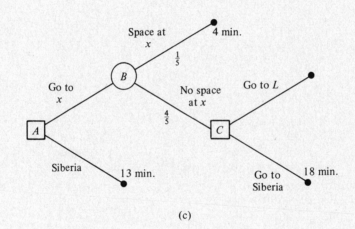

(c)

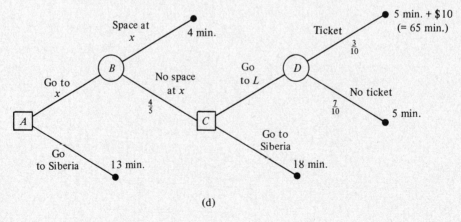

(d)

Figure 9.4

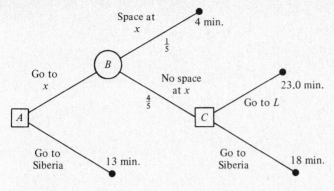

Figure 9.5

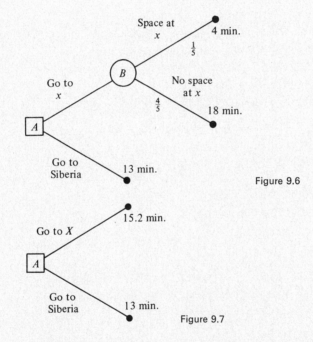

Figure 9.6

Figure 9.7

been effectively determined the day before Meander's trip when the campus parking patrol fixed its route and schedule for the next day. This is of no consequence to Meander because he doesn't find out whether or not he will be ticketed unless he first parks illegally. *In general, decisions in the decision tree are sequenced in the order in which they would be encountered by the decision maker.*

In this problem it was necessary to reduce time and money to a common denominator in order to compare outcomes and do expectation calculations. This necessity for conversion of costs or benefits to common units is frequent in decision problems. Unless such conversion is possible, the pinching off process based on calculations of expected values is inapplicable.

Both of the previous examples involved minimizing losses and contained no possibilities for gain. The following short example shows the more common situation where losses and gains occur in the same problem and need to be combined in the calculations.

Example 3

A business man is contemplating a trip to try to arrange a business deal. If he doesn't make the trip, there is no chance for the deal; while if he does, there is a 1/4 chance of concluding the deal. The deal is worth $2000 and the cost of the trip is $400. What should he do?

Solution: There are two ways we can solve this decision problem, as shown in Figure 9.8(a) and (b). The trees are the same but the labelings are different. In Figure 9.8(a) we combine the cost of the trip with the profits of the outcomes involving the trip to find the total values of these outcomes. In Figure 9.8(b) we place the cost of the trip (as a negative number) on the edge where that cost is incurred but we do not combine that cost with the profits from the deal at the outset. Then we pinch off at *B* and only then do we amalgamate the cost of the trip with the expected profit. It is not accidental that the results turn out the same, namely, that the trip offers an expected profit of $100. We can use either method for combining costs and benefits.

EXERCISES 9.3

1. The Kobalevsky Tool Company has a contract to produce 50,000 ding-a-ling pins in 4 months time for a payment of 1 million dollars. If the pins are delivered in 3 months, there is a bonus of $100,000. The Kobalevsky Company doesn't have a crucial machine that is needed in the manufacture of the pins. It has the option of buying one or two of them at $75,000 a piece or remodeling a related machine in its possession at essentially no cost. There is a one-quarter chance that the remodeled machine might not work at all and there is a $\frac{3}{4}$ chance that the remodeled machine will get the job done in 4 months. If it does not work, either the company can withdraw from the contract by paying a penalty clause of $100,000 or it can pay $85,000 a piece for quick delivery of two machines that will just complete the work by the end of the 4-month period. If the company buys one new machine immediately, there is a three-fourths probability that it will get the job done in 3 months. Two machines will certainly get the job done in 3 months. What should the company do? If remodeling is the answer, what should be done if the machine doesn't work?

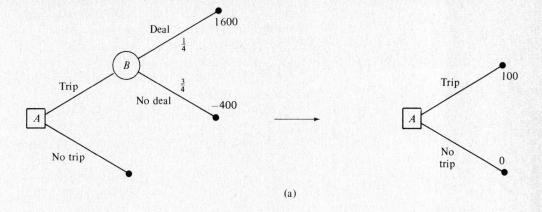

(a)

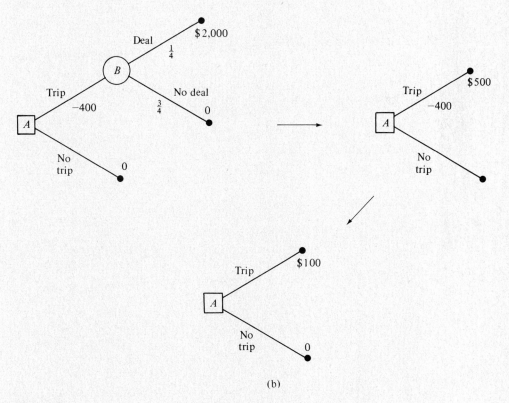

(b)

Figure 9.8

2. The head of the Hibernian Import Company is considering a business trip to try to sell his stock of plastic shamrocks to various novelty chain stores. His trip would cost only $400 and there is a probability of one-half of unloading his shamrocks at a profit of $3000. Unfortunately, if he goes on his trip, there is a probability of one-fourth he will miss the long awaited novelty item auction, which is presently unscheduled. Being at this auction is worth $2000 to him. If he does not sell his shamrocks on a trip, he can sell them at a $1000 profit to a pop art manufacturer who will melt them down to form figurines. Should he go on the trip or not?

3. A regional planning commission must decide whether or not to spend a year and $100,000 for a study planning an expanded water supply system for the county. At the end of 1 year there will be county elections in which there is a three-fifths chance that DeSicca will be elected county executive. DeSicca is strongly opposed to expanding the water supply system and, if elected, he will block expansion for the full 5 years of his term. If DeSicca is elected, by the end of his term the population may have increased (with probability of one-third), necessitating expansion that would cost 1.7 million dollars plus $120,000 for a rush study, if a study hasn't already been done. If the population does not increase during his term, no expansion will be needed. If DeSicca loses, his opponent will order a rush study costing $120,000 if a study has not already been done, followed by immediate construction costing 1.5 million dollars, or his opponent will simply order immediate construction costing 1.5 million dollars if the study has been done. What should the planning commission do to minimize total expected cost to the county?

4. The gimcrack industry is trying to decide whether or not to lobby on behalf of protective tariffs aimed at reducing the influx of cheaper foreign gimcracks. A lobbying effort sufficient to produce a one-tenth chance of ensuring protective legislation would entail a one-shot lobbying cost of $500,000. Protective legislation would ensure annual profits of 10 million dollars. The industry feels that present estimates and calculations are reliable only for about 4 years into the future so total profit calculations should be done on this basis. Assume that without lobbying there is no chance that protective legislation will be passed. Even without protective legislation, however, there is only a one-quarter chance that foreign gimcracks can crack the fiercely loyal United States market. If they do crack the market, however, foreign gimcracks will cut United States gimcrack profits in half. What should the gimcrack industry do to maximize its net profits?

5. You are considering whether or not to look for a new salesman's job. You are presently unhappy about the nonfinancial aspects of your

position and you decide that being happy in your job is worth $2000 a year to you. You feel that there is a three-fourths chance of getting a new position but you will have to spend $100 for resumés and travel and put in 80 hr of your time to ensure this probability. You estimate that your time is worth $5/hr. If you receive an offer, you will definitely take it, but you estimate that there is only a one-third chance you will be happy in the new job. Assume that you would expect to hold the new job for 3 years, no matter how things work out. Draw a decision tree and decide what your decision should be.

6. The ABC Advertising Agency is interested in the potentially profitable Milky Chocolate Company account. It realizes that it has no chance to land the account unless it prepares a substantial sample of advertising copy for the perusal of Milky Chocolate executives. Such an effort would cost $20,000 but would ensure a one-fourth chance of obtaining the account. If the agency gets the account, it can then develop the rest of the ad campaign immediately or proceed first with an in-depth market survey of customer responses to the sample material. This survey would cost $10,000. There is a one-fourth chance that the sample material has shortcomings. A survey would reveal these shortcomings, but if no survey is taken they will not be discovered. If there are shortcomings, eliminating them would cost another $10,000. The company must decide whether or not this is worth doing. The way ABC will be paid for its efforts is as follows: There will be a flat fee of $30,000 if ABC gets the account and this fee will be tripled if Milky Chocolate sales double. The chances of sales doubling in response to the advertising are as follows: a three-fifths chance if the advertising campaign has no shortcomings and a two-fifths chance if there are shortcomings. Should ABC prepare the sample copy to try for the Milky Chocolate account? If it gets the account, should it do the survey and, if so, should it act upon the changes suggested by the survey?

7. The Ferro-Hippus Auto Company is considering whether or not to make a syncopated steering option in the next model change. An engineering study would cost one million dollars. By the time the engineering study is completed the company can determine how many other companies are fully committed to produce the syncopated steering option: few, some, or many. These outcomes have probabilities $\frac{1}{4}$, $\frac{3}{8}$, and $\frac{3}{8}$ respectively. The anticipated profits for various combinations of decisions by Ferro-Hippus and its competitiors are listed in Table 9.13. These profits assume that the engineering study is performed.

It is possible for Ferro-Hippus to make its decision without first doing the engineering study, and without knowing what the competitors will do. In this case, the profit figures are the same as in the table except that those in the right-hand column need to be increased by one million dollars each.

Table 9.13

| | | Ferro-Hippus | |
		Syncopated Steering Option	No Syncopated Steering Option
	Few	$10 million*	$8 million
Competitors	Some	$12 million	$6 million
	Many	$5 million	$3 million

There is a final complication. Congress is currently discussing making currently optional safety equipment mandatory in the next model change. If this is done, and there is a $\frac{3}{4}$ probability that it will be done, there will be little demand for syncopated steering and the starred profit should be cut in half.

Should Ferro-Hippus make the engineering study or not? If not, should it decide to produce the syncopated steering option or not? If it makes the engineering study, how should it react to the three possible actions of the competitors?

8. Suppose in Example 1 that seeding a hurricane costs 25 million dollars. Incorporate this cost into the calculations and redo the example so as to minimize total cost rather than just property damage cost.

9. Do Example 2 of the text by converting all time estimates to monetary equivalents.

10. Many of the uncertainties in the problems of this section concern situations that seem unrepeatable. Therefore, the probabilities must be thought of as subjective estimates. If this is the case, does it seem reasonable to reason and calculate so precisely on the basis of these numbers? Would an intuitive approach be better? Can you conceive of any experiments that would shed light on these questions?

9.4 games

We now turn our attention to decision problems where the decision maker has another human being, or a group of them, to contend with rather than blind fate.

Example 1

Pat and Mike play a game in which Pat chooses a row in the matrix shown in Table 9.14 and Mike chooses a column without knowing Pat's choice. Mike then pays Pat the sum listed in the intersection of the chosen row and column if this sum is positive and receives a payment of the indicated magnitude if the sign is negative. For example, if the first row and the first column are chosen, Mike pays Pat $2; while if the first row and second column are chosen, Pat pays Mike $100.

Table 9.14

	Mike's	Choices
Pat's Choices	2	−100
	3	50

We call such situations *matrix games*. This game undoubtedly seems simple to you in comparison to more complex games such as checkers, chess, or poker, which are more familiar to you. Unfortunately, the mathematical theory of games is not yet powerful enough to give much insight into chess, checkers, or poker so we shall have to content ourselves with simpler situations, such as that of Example 1. Even though our examples will be simpler, however, they share aspects with more complicated games: conflict between two opponents with conflicting goals, intellectual challenge, and payoffs involving something of value.

Occasionally matrix games occur in a slightly disguised form, as in the following example.

Example 2

The offensive side in a crucial play of a football game has the option of planning a passing play or a running play, while the defense can choose a "pass rush" or a normal defense. If the pass rush is chosen and the play is a passing play, the offensive side loses 7 yards, but if the play is a running play, the pass rush gains 5 yards for the offense. A normal defense against a running play results in a 1-yard gain for the offense, while a normal defense against a pass play yields a gain of 15 yards for the offense.

We can think of the teams as two players engaged in the matrix game shown in Table 9.15, where the offensive team chooses a row and the defense

Table 9.15

		Defense	
		Pass Rush	Normal
Offense	Run	5	1
	Pass	−7	15

chooses a column. The payoffs are in yards gained by the offense, negative payoffs signifying losses.

Both of these examples illustrate the general pattern of games we shall study in this section and the next:

(1)   There are two players, each intelligently seeking his own self-interest.
(2)   Each player has two or more courses of action to choose from. The choices are made secretly and revealed simultaneously.
(3)   For any pair of choices by the players, there is a numerical payoff that is gained by one player and lost by the other. The payoffs listed in the matrix are, by convention, always the row player's payoffs. Thus, a positive payoff means a gain for the row player but a loss for the column player. If we wish to list the payoffs for both players, we might use a matrix like the one in Table 9.16, which describes Example 2, where the first entry in each box refers to the row player, while the second entry gives the payoff for the column player. Since we are at first only considering games where one player's gain is the other's loss, the payoffs in each box would always be negatives of one another and, therefore, add up to zero. Consequently, for these so-called *zero-sum games*, it is customary to enter just the row player's payoff.

Table 9.16

		Defense	
		Pass Rush	Normal
Offense	Run	(5, −5)	(1, −1)
	Pass	(−7, 7)	(15, −15)

In the analysis of matrix games, we begin with some very simple considerations which you may already have discovered if you thought about Example 1. In that game, it makes no sense for Pat to play the first row because, no matter what column Mike picks, Pat is better off with the second row (by contrast, this situation does not exist in Example 2). We say that the second row dominates the first. This is really the same concept of domination we used in decision problems against nature. The difference here is that we also must perform a similar analysis for the other player. From Mike's point of view, neither column is uniformly better than the other. Column 2 is better than column 1, provided Pat chooses row 1, but if Pat chooses the second row, then Mike would be better off with the first column. We are going to assume that Mike is clever enough not to restrict his analysis to these observations, however, but that he will realize that the row player will never play the first row. Consequently, we are effectively playing a game in which there is only one choice for the row player, while the column player has his choice between losing 3 and losing 50. Clearly the column player would rather lose only 3 so he should play column 1. This type of analysis leads us to make the following definitions.

DEFINITION 2

(1) A row of a matrix game is said to *dominate* another row if each entry of the domina*ting* row is greater than or equal to the corresponding (i.e., same column) entry of the domina*ted* row.

(2) A column of a matrix game is said to *dominate* another column if each entry of the domina*ting* column is smaller than or equal to the corresponding (i.e., same row) entry of the domina*ted* column.

The value of dominating rows or columns in analyzing matrix games lies in the following principle: If a row (or column) dominates another row (column), then a rational player will always prefer the dominating row (column) to the dominated row (column).

Consequently the dominated row (column) can be stricken from the matrix, thus simplifying the analysis.

Example 3

In this example we show the simplification of a matrix game using dominant strategy analysis. Consider the game matrix on the left below. Under each transition we list the inequalities that justify the striking out of the row or column in question. In each stage, the two rows or columns involved in a dominance relationship are indicated with an asterisk.

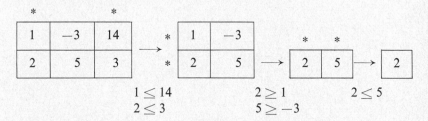

At the outset there is no dominance among the rows but we can strike column 3 out because it is dominated by column 1. Once we have done this, we have checked once for row dominance and once for column dominance; one might think we had gone as far as one could with dominance analysis. That, however, is not so in this example. Even though we have checked both kinds of dominance, we should check again. *As a result of striking out column 3, a row dominance has appeared that did not exist earlier.* Thus, we strike the first row, producing the third matrix. In this third matrix it is clear that the first column dominates the second. Our conclusion is that with intelligent play on both sides, each time this game is played the row player will choose the second row and the column player will choose the first column, producing a payoff of 2 for the row player and, thus, a loss of 2 for the column player. Since we have narrowed the possible payoffs down to a single one, we call this payoff of 2 the *value of the game*.

The algorithm for simplifying a matrix game by using dominant strategy analysis is shown in the flow chart in Figure 9.9.

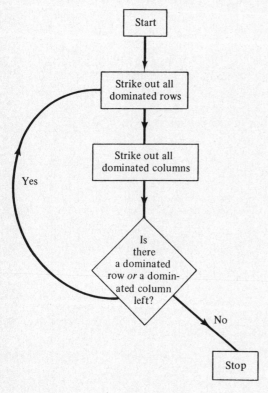

Figure 9.9

Example 4

Smith and Jones each intends to buy a new car. If they were guided by practical consideration, each would buy the well-engineered, inexpensive Cheapmobile. However, the American folk custom of keeping up with the Joneses (Smiths) brings the more expensive Sleekmobile and the outrageously expensive Delux into consideration for each family. Each family makes its decision independently, without knowledge of the other family's deliberations. The payoffs to Jones, the row player, for the various combinations of choices that can occur are given in the matrix below. The matrix in Table 9.16 can be simplified as in Table 9.17.

Table 9.16

		Smith's Choice		
		Cheap	Sleek	Delux
Jones' Choice	Cheap	0	-1	$+2$
	Sleek	$+1$	0	-2
	Delux	$+2$	$+2$	0

Table 9.17

0	-1	2
* 1	0	-2
* 2	2	0

$\rightarrow$

*	*	
0	-1	2
2	2	0

$\rightarrow$

-1	2
2	0

In this example, dominant strategy analysis has not narrowed things down to a single value. Nevertheless, we can say that with intelligent play the only reasonable outcomes are 2, -1, or 0. In effect, the players are really playing the game represented by the last matrix above.

It is worthwhile to examine the modeling aspects of this matrix game to see whether the game satisfies the three criteria we set up as assumptions about the kinds of situations we are going to deal with in this section. As for the first criterion, since it is possible that Smith and Jones are only dimly aware that they are playing this game, it is questionable whether each is intelligently seeking to obtain the best payoff in the matrix that he possibly can. The second criterion, about secret decisions simultaneously announced, may also be debatable. Smith and Jones may discuss their plans or one may wait for the other to make a purchase and then make his own decision with this information in mind. A final important question, relevant to the third criterion, concerns the payoffs in this game. In what units are the payoffs measured? Clearly no money changes hands in this game. Status points perhaps? This is a suggestive name, but merely giving something a name doesn't tell us much about it. Moreover, if we do not know what a status point is, we can hardly hope to measure it to fill in our table. This is, of course, the same problem of utilities that we touched upon in discussing decision problems against nature.

All in all, there are many reasons to regard Example 4 as more fantasy than practical modeling. Do you think the same is true of Example 2?

EXERCISES 9.4

1. If possible, simplify each of the following matrix games by using the algorithm for eliminating dominated strategies. Indicate matrices for which no simplification is possible.

(a)

3	2	5
0	−1	1
−2	0	6

(b)

−1	0	2
−2	−3	4
−1	0	−2

(c)

1	−1	2
−1	1	3
−1	−1	0

(d)

2	−4	3	2
−7	3	3	0
0	−4	2	1
1	−5	3	0

(e)

1	2	−3
−1	0	4
2	1	−3

(f)

−3	2	7
1	−2	0
0	5	−4

(g)

−1	0	3
2	−1	5
0	6	3

2. In a crucial baseball game, team I has pitchers I_1 and I_2 available, while team II has pitchers II_1 and II_2 available. The team managers are hard pressed to announce pitcher choices for the following reason:

I_1 generally beats II_1 but loses to II_2

I_2 generally beats II_2 but loses to II_1

If either manager announces a choice, the other can counter with a pitcher who will beat that choice. Therefore both managers hit upon the following device: No previous announcement will be made and the pitcher choice will be apparent only when the lineup card is handed to the umpire, at which point the lineup is official. Since the lineups are handed in simultaneously, neither manager can react to the other's choice. Represent this as a matrix game between the two managers. Use a pay-off of 1 to indicate a win and -1 to indicate a loss. If there is any simplification possible through dominant strategy analysis, do it.

3. Discuss the effect on Exercise 2 of taking into account baseball's substitution rule: At any time in the game any player may be replaced by another but with the provision that once a player is removed from the game, he may not return to that game.

4. Tony and Dominic play the following game. Simultaneously each shows either one or two fingers. If the sum of fingers shown is even, Tony pays that amount to Dominic whereas if the sum is odd, Dominic pays that amount to Tony. Represent this situation as a matrix game. (Remember that the payoffs in your matrix should be to the row player, whomever you choose him to be.) If there is any simplification possible through dominant strategy analysis, do it.

5. A defending army has the choice of two important locations to defend, locations I and II. The attacking army must decide whether to attack I or to attack II without knowing which is being defended. If I is defended and is attacked, the pay-off is $+2$ for defenders (D); if I is defended but not attacked, the payoff is $+3$ for the attackers (A). If, instead, II is defended and attacked, A wins $+3$, while if II is defended but I is attacked, D wins 4. Devise a matrix to represent this game. (Remember, payoffs must all be from the row player's point of view.) If dominant strategy analysis applies, use it to simplify the game matrix.

6.* Find a cooperative student not taking this course who is willing to play some of the games in Exercise 1 with you. See whether he uses dominant strategy reasoning to secure the best result for himself. Does it help to play one game over and over again? Compare results for a number of subjects. Discuss your results with someone in the Psychology Department.

7.* Prove that if row I_1 dominates row I_2, and row I_2 dominates row I_3, then row I_1 dominates row I_3. Is there an analogous result for columns?

8.* Is the following situation a zero-sum game? Explain your answer. Two competing widget stores, *A* and *B*, have sales at the same times of year. Each of *A* and *B* has the option of using either newspaper (*N*) or television (*TV*) for advertising, but not both. We suppose that:

(a) *A* and *B* are the only widget stores in town.

(b) The total demand for widgets in this town is constant from week to week and is unaffected by sales or advertising.

9.* Explain why dominant strategy analysis might not be relevant in a situation where criteria 1 and 2 (listed after Example 2) are not satisfied.

9.5 mixed strategies

Here is a game where analysis seems hopeless:

	II_1	II_2
I_1	1	-1
I_2	-1	1

It seems apparent that neither row is better than the other for the row player and that neither column has any claim to preference over the other. Games such as this, in which there are no dominant strategies for either player, are quite common. This very game arose in the context of choosing pitchers for a baseball game (Exercise 2, p. 372). It also arises in a time-honored method of choosing sides and settling disputes in street games: An *evens* player and *odds* player each simultaneously displays either one or two fingers. If the total of fingers displayed is even, the *evens* player wins; if the total is odd, then the *odds* player wins. If we think of the first row and the first column (I_1 and II_1) as both representing one finger, while the second row and second column (I_2 and II_2) represent displaying two fingers, then the matrix above is the payoff matrix for the *evens* player as the row player.

Let us return to the question of how one ought to play this game. You may feel that one may as well make the choices randomly and forget about further analysis. We shall soon see that not only is this approach tempting but there is also a certain wisdom to it.

Let us make things a bit more interesting by supposing that you will play this game against the same opponent 15 times. Is it then reasonable to pick one row and stick with it all 15 times? Clearly this is not sensible unless you assume your opponent is a fool who cannot detect patterns in your play. (For example, if the row player always plays row 1, the column player can

John von Neumann (1903–1957) was born in Budapest, but after 1930 was associated with Princeton University and The Institute for Advanced Study. He made mathematical contributions in logic set theory, algebra, quantum theory, mathematical economics, game theory, and computer science. Game theory was essentially created by von Neumann at age 25. He was also largely responsible for the development of the digital computer as we know it today by his invention of the concept of the internally stored program computer. Von Neumann had a remarkable combination of almost superhuman intellectual powers. He could do incredible mental calculations at lightning speed and apparently possessed total recall, yet his engaging personality endeared him to his colleagues. The story was told of him at Princeton that he was indeed a demi-god but had made a detailed study of human beings and could imitate them perfectly.
(*Courtesy The Bettmann Archive.*)

always win by playing column 2.) Is it reasonable to alternate your rows like this: $I_1, I_2, I_1, I_2, I_1, I_2, \ldots$? Once again, we must assume your opponent will catch on and anticipate your moves. He can then select his column choices so as to win all the time. Thus, we seem forced to the conclusion that you should make your choices with *no pattern*. However, this is not all there is to it. Here are two patternless series of choices for a series of 15 repetitions of the game. Do you think they would be equally effective?

$$I_1, \quad I_1, \quad I_2, \quad I_2, \quad I_2, \quad I_1, \quad I_2, \quad I_2, \quad I_1, \quad I_1, \quad I_1, \quad I_2, \quad I_1, \quad I_2, \quad I_2$$

$$I_2, \quad I_1, \quad I_2, \quad I_1, \quad I_2, \quad I_2, \quad I_2, \quad I_2, \quad I_1, \quad I_2, \quad I_2, \quad I_1, \quad I_2, \quad I_2, \quad I_1$$

If you want to try an experiment, use each of these series of choices in turn against a column player who plays II_1 every third play. The results are not identical. The difference between the two series is that in one series about half of the choices were I_1, while in the second series only one-third of the choices were I_1. The question that we have been trying to motivate is this one:

QUESTION 1

Granting the necessity of avoiding any pattern to your row choices, is it also necessary to maintain a certain numerical ratio between the two possible row choices?

Here is an algebraic argument that suggests that the answer is *yes* (which you discovered if you tried the experiment). In the first matrix below we have labeled each row and column with a fraction that represents the proportion of times out of 15 in which that row or column is chosen (we are actually doing a "theoretical" study of part of the experiment earlier suggested). In the next matrix, each box is divided into two parts by a diagonal line. The upper part of the box contains the payoff, while the lower part contains the portion of times out of 15 that that particular box is actually the outcome of the row and column choices. We illustrate how these numbers are calculated by doing it for the upper left box. This box is determined as the outcome when you, the row player, choose I_1 and the column player chooses II_1. Now you choose I_1 one-third of the 15 plays of the game. Since your opponent plays II_1 one-third of the time overall, we shall also assume that approximately one-third of the ($\frac{1}{3} \times 15$) times you played I_1 he will play II_1. Thus, the upper left box is the outcome a total of approximately $(\frac{1}{3})(\frac{1}{3})(15)$ times. The other boxes are filled in similarly.

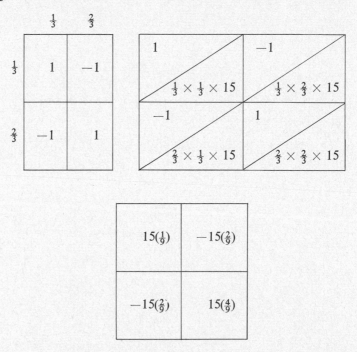

To determine the winnings for the row player from any particular box, multiply the number of times that outcome comes about (determined as previously described) by the payoff for that box. The results of this calculation are shown in the third matrix, in which each box is filled in with the total winnings that can be expected from that box over 15 plays. To determine the total winnings (overall boxes) over 15 plays for the row player, simply add the expected winnings from each box to obtain

$$15(\tfrac{1}{9}) - 15(\tfrac{2}{9}) - 15(\tfrac{2}{9}) + 15(\tfrac{4}{9})$$
$$= 15(\tfrac{1}{9} - \tfrac{2}{9} - \tfrac{2}{9} + \tfrac{4}{9})$$
$$= 15(\tfrac{1}{9})$$
$$= \tfrac{5}{3}$$

To compute the expected winning for the row player in a single play, we must divide by 15. Taking $\tfrac{5}{3}$ and dividing by 15 we obtain $\tfrac{1}{9}$, as the expected payoff.

Let us do the same type of calculation for the second series of choices on p. 375, keeping the column player's play exactly as it was. We make one modification of the basic method. We are really interested in the mean payoff of the game. In the last example this caused us to divide the final figure of $\tfrac{5}{3}$ by 15. We could have accomplished the same objective by never multiplying by the 15 to start with. For example, the entry in the upper left box in the second matrix on p. 376 would be $(\tfrac{1}{2})(\tfrac{1}{3})$ instead of $(\tfrac{1}{2})(\tfrac{1}{3})(15)$. We do the example involving the second series this way.

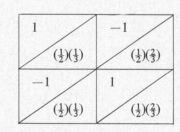 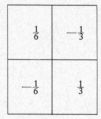

Computing the expected payoff, we obtain

$$\text{Expected payoff} = \tfrac{1}{6} - \tfrac{1}{3} - \tfrac{1}{6} + \tfrac{1}{3}$$
$$= 0$$

Thus, we see that the expected payoff of the game for the row player is less with the second series than the expected payoff for the first series. Presumably you are now convinced that not only is it necessary to avoid patterns in your play but also that in any given matrix game each player has an optimal ratio that should be maintained between the two choices (strategies). It is important to remember that if a player—say, the column player—chooses any fixed ratio of times that he plays his strategies, then the expected payoff to the row player will vary according to what ratio he chooses for playing his strategies. How does one find the exact optimal ratio? The following example shows how. We have done it in detail so that you should be able to follow the pattern for other matrices as well.

Example 1

For the matrix in Table 9.18 determine the best ratio p of I_1 choices to total choices for the row player. Determine the best ratio q of II_1 choices to total choices for the column player. Notice that the ratio of I_2 choices to the total must be $1 - p$ and the ratio of II_2 choices to the total must be $1 - q$.

Table 9.18

	II_1	II_2
I_1	1	-2
I_2	-1	3

Solution: Using the same procedure as in the previous example,

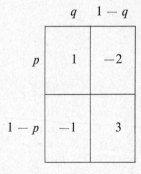
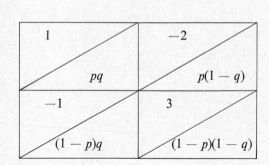

pq	$-2p(1-q)$
$-q(1-p)$	$3(1-p)(1-q)$

Expected payoff $= pq - 2p(1-q) - q(1-p) + 3(1-p)(1-q)$
$= 7pq - 5p - 4q + 3$

This expression doesn't seem very enlightening, involving as it does two unknowns, p and q. To make sense out of it we need to transform the form of the algebraic expression. This is done in the left column below. The right column contains a step-by-step description of the method to follow when doing this transformation.

$7pq - 5p - 4q + 3$

$= p(7q - 5) - 4q + 3$

$= 7p(q - \frac{5}{7}) - 4(q) + 3$

$= 7p(q - \frac{5}{7}) - 4(q - \frac{5}{7}) + 3$
$\quad + (-4)(\frac{5}{7})$

$= 7p(q - \frac{5}{7}) - 4(q - \frac{5}{7}) + \frac{1}{7}$

$= \underbrace{7(p - \frac{4}{7})(q - \frac{5}{7})}_{R} + \frac{1}{7}$

1. Factor out p from the terms that include it.
2. By factoring constants, arrange the q terms to have coefficient 1.
3. By adding or subtracting a constant, arrange second q term to be the same as first q term.
4. Factor the q term to get the expression into the form $k(p - a)(q - b) + c$.

Notice that if the row player, who controls the choice of p, chooses $p = \frac{4}{7}$, then the expression $7(p - \frac{4}{7})(q - \frac{5}{7})$, which we denote by R, is zero, and the expected payoff is one-seventh. The column player can ensure the same expected payoff of one-seventh by picking $q = \frac{5}{7}$.

We claim that both players have good reason to make exactly these choices, for if the row player makes $p \neq \frac{4}{7}$, then $p - \frac{4}{7}$ is not equal to zero and hence either positive or negative. Suppose, for example, $p - \frac{4}{7} < 0$. What if the row player is a pessimist? He might fear that the column player will choose q so as to make $q - \frac{5}{7} > 0$ (picking $q = \frac{11}{12}$, for example). R would then be a product of a positive and negative number and hence negative. Combining this negative value of R with one-seventh, the average payoff to the row player would be worse than the one-seventh that could have been guaranteed by picking $p = \frac{4}{7}$. If, on the other hand, $p - \frac{4}{7} > 0$, a pessimistic row player would worry that perhaps q was chosen so that $q - \frac{5}{7}$ is negative whence R would once again be negative, leading once more to an outcome for the row player that is worse than one-seventh. Thus, the row player, if he wants to be safe, should pick $p = \frac{4}{7}$ and settle for the small average gain of one-seventh. The column player can reason the same way and conclude that he should pick $q = \frac{5}{7}$ in order to ensure an average loss of one-seventh.

The general rule is: *in a matrix game without dominant strategies, once the expected payoff is determined from the matrix and transformed using the four steps listed above, each player's strategy is to pick his fraction so as to make $R = 0$. The expected payoff that comes about by making $R = 0$ is called the value of the game and is denoted V. We use the term* mixed strategy *to describe* *the strategy of alternating one's choices in a patternless way but with a certain fraction (determined as above) devoted to each choice.*

Example 2

Here is an example, done concisely in the matrix below with the reason for the steps omitted. This example illustrates the fact that before doing a mixed strategy analysis we should eliminate as many rows and columns as possible using dominant strategy analysis. Indeed, in this example we could not immediately attempt a *mixed strategy* analysis because we have not learned to do this for a 3×3 matrix game (it would be theoretically possible, however.)

-3	1	2
4	-1	0
3	-2	0

$\longrightarrow$

-3	1	2
4	-1	0

$\longrightarrow$

-3	1
4	-1

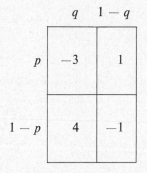

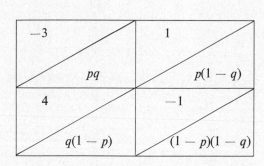

$-3pq$	$p(1-q)$
$4q(1-p)$	$-(1-p)(1-q)$

Average payoff $= -3pq + p(1-q) + 4q(1-p) - (1-p)(1-q)$

$= -9pq + 2p + 5q - 1$

$= p(-9q + 2) + 5q - 1$

$= -9p(q - \tfrac{2}{9}) + 5(q) - 1$

$$= -9p(q - \tfrac{2}{9}) + 5(q - \tfrac{2}{9}) + \tfrac{1}{9}$$
$$= (-9p + 5)(q - \tfrac{2}{9}) + \tfrac{1}{9}$$
$$= \underbrace{-9(p - \tfrac{5}{9})(q - \tfrac{2}{9})}_{R} + \tfrac{1}{9}$$

The value of the game is one-ninth. The optimal mixed strategy for the row player is to play the first row five-ninths of the time. The optimal mixed strategy for the column player is to play the first column two-ninths of the time.

EXERCISES 9.5

1. Determine the optimal mixed strategies for the players and the values of the games in each of the following cases.

(a)

2	1
0	3

(b)

−4	2
2	−1

(c)

1	3	−2
0	2	−3
−4	0	−1

(d)

0	−1	2
1	0	−2
2	2	0

2. Determine the optimal mixed strategies and the value of the game of Exercise 4 or p. 373.
3. Determine the optimal mixed strategies and the value of the game of Exercise 5 on p. 373.
4.* Discuss what happens when mixed strategy analysis is applied to a game in which there is a dominant row or column. Can one determine the value of the game from this mixed strategy analysis?
5.* Determine the formula, in terms of a, b, c, and d, for the value of the following game, assuming that there are no dominant rows or columns:

a	b
c	d

6.* Use Appendix 12.4 (random number table) to play the first game in the text of this section twenty times using the following mixed strategies for the players:

(a) I_1 two-thirds of the time; II_2 one-third of the time.

(b) I_1 one-half the time; II_1 one-half the time.

(c) I_1 one-fifth of the time; II_1 one-sixth of the time.

How do your experimental results compare with the theory we have developed?

9.6   nonzero sum games

Until now we have considered games where the amount won by one player is lost by the other. There are games, however, where both players may gain simultaneously or both players may lose simultaneously.

Example 1

Consider an arms race involving two countries (players), each of which has the option of developing a new missile system in the next year. Furthermore, suppose it is possible for each side to keep its decision secret from the other for a long period of time. Consequently each side feels it must decide immediately whether to develop the system (the hawk position) or not to develop it (the dove position) rather than waiting for intelligence estimates of the intentions of the other side. Each side can be either a hawk or a dove so the four possible situations are (hawk, hawk), (hawk, dove), (dove, hawk), and (dove, dove). In the first and last cases the military situation is a standoff. It is certainly not true, however, that the situations are identical, for in (dove, dove) both sides have more money available for nonmilitary domestic uses. Now it is reasonable to suppose that if each country is left to itself, it will choose the hawk position just to be "safe" and so (hawk, hawk) will come about. Clearly both countries will profit if they can cooperate and negotiate an agreement whereby both will take the dove position. Of course cooperation takes communication, which may not always be possible. Although it is difficult to measure with numbers the advantages and disadvantages of these situations, we can attempt to illustrate the principles in the payoff matrix in Table 9.19. Instead of having one number in each box as we did with zero-sum games, we need to enter both players' payoffs in the cells of the matrix. We do so according to the convention that the first number in a cell is the payoff to the row player.

Notice that this game is not a zero-sum game. Although in three of the outcomes (cells) the sum of the two payoffs is zero, in the fourth cell

Table 9.19

	Hawk	Dove
Hawk	0, 0	10, −10
Dove	−10, 10	5, 5

both players simultaneously gain five because they have money for peaceful domestic use.

Example 2

Two manufacturers compete to sell the most toasters. Each manufacturer can set his price at $10 or $15. The profits obtained by a manufacturer depend on the prices he charges and on the price of the competitor's toaster, as indicated in the matrix shown in Table 9.20. The first entry in a cell is the payoff to the row player.

Table 9.20

		Price for Company II	
		$15	$10
Price for Company I	$15	100, 100	20, 120
	$10	120, 20	50, 50

(Profits in thousands of dollars per year)

If there is no communication for the purpose of cooperation, presumably both companies will inevitably wind up setting the price at $10. Suppose that both start by setting the price at $15. At this point both get $100 as payoff. However, one of the companies is likely to notice that it can get 120, which is a greater profit, if it lowers the price. Hence, this company is likely to lower its price. Now the other company, which is still selling at $15, is getting a diminished profit of 20. To improve its position, the second company must also lower its price to $10, at which time both companies will make a profit of 50. This 50, 50 payoff is worse for both companies than the 100, 100 with which they started. Consequently, the companies might find it in their best interests to agree between themselves to fix the price at $15 for each toaster; that is, they will agree that neither will lower the price to $10. This kind of cooperative behavior, referred to as price-fixing, has usually been frowned upon and regarded as illegal. The reason is that the benefit to the companies is achieved at the expense of the public interest, a factor that does not appear in the original model. This illustrates that if a model is to be useful, it must reflect all the important factors in a situation. From the companies' points of view the model may be adequate but this position may not be shared by the man in the street. If we try to expand the present model to include a third "player," the public interest, we will be dealing with a three-person game. Games with more than two players will not be considered.

The previous examples suggest that cooperation between players is a crucial factor in nonzero-sum game theory. We can get a better understanding of this by first examining these games under the assumption that there is *no cooperation* between the players.

For example, if we examine the matrix in Example 1 under the hypothesis of total competition, we can show that if both players play rationally, there is good reason to believe that the outcome will be (hawk, hawk), which is clearly not the most desirable condition for the world. The row player will reason like this: If my opponent, the column player, plays hawk, I am better off with my hawk row than with my dove row because zero is better than −10; but even if my opponent plays dove, I am better off with hawk than with dove since 10 is better than 5. Consequently, I will play hawk since this is the best course of action for me, regardless of how my opponent plays. Using the same reasoning, the column player discovers that hawk is a better position for him, no matter what the row player does. Thus, both sides will develop the new weapon if they reason *rationally* and are *unable to cooperate.*

You may have noticed that our analysis of the matrix from Example 1 in the noncooperative case was essentially a dominant strategy analysis of the sort that we studied when we investigated zero-sum games. Here are definitions of dominance, adapted to the case in hand, nonzero-sum games.

DEFINITION 3

One row in the matrix of a nonzero-sum game *dominates* another if each *first entry* of the domina*ting* row is greater than or equal to the corresponding (i.e., same column of the matrix) entry of the domina*ted* row.

One column *dominates* another if each *second entry* of the dominating column is greater than or equal to the corresponding (i.e., same row of the matrix) entry of the dominated column.

As an illustration of the definition, consider the following example:

Example 3

In the matrix shown below, row 1 dominates row 3 because $1 \geq -1$ and $2 \geq 1$. There is no row that dominates the second row nor does the second row dominate another row. Column 1 dominates column 2 because $3 \geq 1$, $5 \geq 2$, and $3 \geq -3$.

1, 3	2, 1
4, 5	0, 2
-1, 3	1, -3

The significance of dominance is this: If a row (or column) is dominated by another row (or column), that dominated row (or column) is not likely to be played if the players do not communicate and cooperate. Consequently, we can simplify the analysis of a nonzero-sum, noncooperative game by the following procedure:

(1) Strike out all dominated rows from the game matrix and go to the next step.
(2) In the new matrix, strike out all dominated columns and return to step 1.
(3) If at some stage there are neither dominated rows nor dominated columns, then stop. The flow chart for this algorithm is the same as that shown in Figure 9.9.

If this technique is applied to the matrix in Example 3, we have the sequence of simplifications shown below.

1, 3	2, 1	$\longrightarrow$	1, 3	2, 1	
4, 5	0, 2		4, 5	0, 2	
-1, 3	1, -3				

1, 3	$\longrightarrow$	4, 5
4, 5		

Hence, in Example 3, we may expect the row player to play row 2 and the column player to play column 1. The outcome would be a gain of 5 for the column player and a gain of 4 for the row player.

Sometimes the simplification obtained by this procedure will not reduce the game to a single best outcome but will only narrow the choices somewhat.

386 games and decisions

This is illustrated in Example 4 which shows the simplification of such a noncooperative nonzero-sum game.

Example 4

1, 1	−4, 2	3, 2
0, 1	3, 0	−2, −1

→

1, 1	−4, 2
0, 1	3, 0

We reiterate that dominant strategy analysis is reasonable only where there is no cooperation between the players. It is possible that after eliminating some choices by this method, we may have ruled out an outcome that would be no worse than the remaining alternatives *for both players* and actually better for at least one player. Such an outcome is called a *cooperative improvement* over the remaining outcomes. It is called a cooperative improvement because it can be achieved only if the players cooperate instead of each separately applying the self-centered dominant strategy analysis.

To illustrate the idea of cooperative improvement, we shall reconsider the arms race problem of Example 1. Dominant strategy analysis leads both players to choose the hawk position, which leads to the outcome 0, 0. However, 5, 5, the outcome from both players' selection of the dove strategy, is a cooperative improvement over 0, 0 since both players would have a greater payoff. As another example, consider the matrix of Example 4. The outcome 3, 2 is a cooperative improvement over all the outcomes that remain after applying dominant strategy analysis, namely, 1, 1; −4, 2; 0, 1; and 3, 0. The outcome 3, 2 is better for both players when compared with 1, 1 or 0, 1. Compared to −1, 2 and 3, 0, the outcome 3, 2 is better for one player and no worse for the other.

There are a number of other contexts in which cooperation in a game-like situation is crucial. Since these examples involve more than two players, they do not fit conveniently into our discussion but some of them are quite striking and deserve mention.

(1) Suppose you decide not to have a polio vaccination and thereby avoid the trouble, minor pain, and risk of allergic reaction. You reason that since almost everyone else is vaccinated, there will be nobody from whom you could catch the disease. Suppose, though, that everyone were to come to this conclusion. Since there are still occasional cases of polio reported, there would be some risk of a polio epidemic in the event that a substantial part of the population fails to acquire immunity

through vaccination. Interestingly, the situation with smallpox is somewhat different. There has not been a case of smallpox reported in the United States since 1949 but ironically there have been deaths due to smallpox vaccination. Consequently, starting in 1972 it has not been routine procedure to give smallpox vaccinations.

(2)   It is highly unlikely that any individual vote in a national election will be the deciding vote. An individual may, therefore, decide to save himself the trouble of voting. If everyone reasoned this way, however, our political institutions would be very different.

(3)   An individual may feel it is wise to save 20% of his income because this would put him ahead of the typical United States family, which saves about 7% of its after-tax income. (See *Understanding Macroeconomics* by Robert Heilbroner, Prentice-Hall, Inc., Englewood Cliffs, N.J.) It is an economic fact, however, that if everyone suddenly switched to saving 20 percent of his income, there would be a quick decline in demand for consumer goods, leading to unemployment and economic chaos.

The common thread of these examples is that a large number of people must and do act cooperatively to secure certain benefits (the suppression of polio, the maintenance of democratic institutions, the maintenance of economic stability). What mechanism induces such large-scale cooperative behavior? Are sociological principles at work? Emil Durkheim, the French sociologist, thought that division of labor brought about a type of cooperation he referred to as "mechanical solidarity." Are biological principles at work? Have we inherited a "herd instinct" as part of our biological makeup? Are ethical principles at work? The Golden Rule ("Do unto others as you would have others do unto you") certainly suggests cooperation. An interesting example to consider is the very different reactions to police strikes in two major North American cities. During a police strike one might hypothesize that many persons would be tempted to further their self-interest by looting stores and indulging in other criminal activities. To an extent this is the behavior that occurred during the 1969 police strike in Montreal, Canada. New Yorkers, on the other hand, seem to have opted for cooperative behavior during a police strike in 1971. Which of the considerations raised above might have been operative is difficult to assess.

EXERCISES 9.6

1.   For each of the following matrices, determine whether or not there is a cooperative improvement over the outcomes that remain after a dominant strategy analysis.

(a)

−1, 2	2, 4
1, 0	3, 1

(b)

4, 3	2, 4
5, 0	3, 2

(c)

1, 2	1, 0	7, −1
0, 0	0, 4	4, 3

(d)

2, 3	−6, 6	6, 5
0, 4	7, 0	−3, −2

2.* Give an algorithm for deciding when there is cooperative improvement over outcomes that remain after a dominant strategy analysis. Phrase your algorithm to deal with a matrix whose entries are as below.

a, a'	b, b'
c, c'	d, d'

3.* Give examples, in addition to those in the text, where an individual might gain in the short run by refusing to cooperate with a group of other people but benefit himself as well as others in the long run by cooperating.

4.* Discuss the structure of the tax laws of the United States in connection with group cooperation.

9.7 the credibility of cooperation

In the last section we discovered that in some nonzero-sum games the best results were often obtained if the players cooperatively agreed to play certain choices that they might otherwise avoid. Under some circumstances, however, this cooperation might be hard to bring about, as the following example illustrates.

Example 1 (Prisoner's Dilemma)

The district attorney suspects two persons of having committed a crime together but he does not have conclusive proof. In the presence of both, he announces the following offer. In the morning he will separately visit each suspect in his cell and give him an opportunity to confess or remain silent. In the interview with the second suspect, the district attorney will not reveal the results of the first interview. If both suspects confess, they each get 8-year jail terms. If neither confesses, each will get 1 year on a minor charge for which evidence exists. If only one confesses, he will go free for turning state's evidence, while the one who doesn't confess gets 10 years. The matrix for this game is

	C	D
C	$-8, -8$	$0, -10$
D	$-10, 0$	$-1, -1$

C = confess
D = don't confess

If cooperation is not allowed, both players will probably confess, in accordance with dominant strategy analysis. Suppose, however, that the suspects are allowed a brief conference before being led to their separate cells. It seems likely that they will agree to cooperate and both not confess, producing the outcome $-1, -1$.

The new aspect of the analysis is this: Neither will know whether the other will abide by the agreement. Would you abide by it? A glance at the matrix shows that a player will gain if he does not abide by the agreement while the other player does. Of course, if both are treacherous, we are back at $-8, -8$, the worst set of payoffs.

Unfortunately, game theory can shed no light on what would or should happen. It is entirely a matter of the ethics of the players. However, we can isolate and define the problem.

DEFINITION 4

A pair of strategies is called an *equilibrium pair* if the payoff p_r to the row player and the payoff p_c to the column player corresponding to these strategies have these properties:

(1) If the column player changes his strategy but the row player *does not*, then the new payoff for the column player is less than or equal to p_c.

(2) If the row player changes his strategy but the column player *does not*, then the new payoff for the row player is less than or equal to p_r.

In other words, a pair of strategies is an *equilibrium pair* if neither player gains if he alone changes his strategy. It should be noted that this concept of an equilibrium pair applies to both zero-sum and nonzero-sum games.

Example 1 (continued)

In the Prisoner's Dilemma game the pair of strategies (C, C), leading to the outcome $-8, -8$, is an equilibrium pair. There is no gain for either player if he alone switches strategies. We have seen, however, that if both cooperate and simultaneously switch strategies to (D, D), this is a cooperative improvement. If we examine (D, D), we see that it is not an equilibrium pair.

Example 2

Kenneth Clark Asks New Drugs to Curb Hostility of Leaders

By BRUCE RENSBERGER
special to the New York Times

Washington, Sept. 4—The President of the American Psychological Association proposed today the creation of new drugs that could routinely be given to people, especially leaders holding great power, to subdue hostility and aggression and, thereby, allow more humane and intelligent behavior to emerge.

This example is motivated by a suggestion by a noted psychologist that scientists develop a drug that could be taken by world leaders to reduce their aggressive tendencies and to stimulate cooperative ones instead. Suppose two leaders have arranged a summit conference between them to negotiate a way out of a potentially explosive international situation. That morning each leader has two strategies: to take the pill (P) or not to take the pill (D). It is reasonable to suppose that if one leader chooses P while the other selects D, then the one who chose D and skipped the pill has an advantage. The matrix below could conceivably be a payoff matrix for this situation.

	P	D
P	0, 0	$-4, 4$
D	4, -4	$-2, -2$

Unfortunately, (P, P) is not an equilibrium pair and each player may be tempted to play D. If, however, both play D, we obtain an unpleasant outcome, even though (D, D) is an equilibrium pair.

Does this example demolish the psychologist's suggestion?

The question of whether to cooperate with or double-cross one's opponent in a game with a matrix such as in Example 1 or Example 2 is influenced by how many times the game will be repeated. If there are many plays of the game, it is conceivable that the players may communicate or interact through the medium of the pattern of choices made. For example, we might decide to play only the cooperative strategy (C in Example 1 or P in Example 2) every single time in the hope that our opponent will read this as a signal that says "I want to cooperate." This approach was called the *Ghandi approach* by Lester Lave ("Factors Affecting Cooperation in Prisoner's Dilemma," Behavioral Science, 1965. p. 26–38.) who discovered that when this is used against an opponent, he will initially cooperate but then pick the double-cross strategy (D) all the time, thus making Ghandi a perpetual sucker. Lave also studied the opposite approach, called *Stalin*, in which one always plays D. Oddly enough, when the Stalin approach is used against an opponent, most opponents occasionally play the cooperative strategy (thus becoming a sucker) in the hope of inducing cooperation. Somewhat more effective in tricking opponents was the *Krushchev* strategy, in which one plays the uncooperative D strategy most of the time but occasionally chooses the cooperative strategy to rouse the hopes of the opponent and to induce him to become a sucker.

Try to verify these results by experimentation.

EXERCISES 9.7

1. In each matrix below, identify all equilibrium strategy pairs. Are there cooperative improvements over any of these equilibrium pairs?

(a)

	II$_1$	II$_2$
I$_1$	10, 9	−100, 100
I$_2$	120, −120	0, 0

(b)

	II$_1$	II$_2$
I$_1$	1, 3	−2, 4
I$_2$	4, −1	1, −3

(c)

	II_1	II_2
I_1	$1, -1$	$-3, 3$
I_2	$2, -2$	$0, 0$

2.* In the zero-sum game below, suppose that every one of the four strategy pairs is an equilibrium pair. Show that $a = b = c = d$.

	II_1	II_2
I_1	a	b
I_2	c	d

3.* In the zero-sum game of Exercise 2, suppose (I_1, II_1) and (I_2, II_2) are equilibrium pairs and $a = d$. Show that $a = b = c = d$.

4.* Answer the question at the end of Example 2 in Section 9.7.

5.* Get two friends to play the Prisoner's Dilemma game 25 times with no spoken or written communication allowed. See whether cooperation occurs. Does the tendency to double-cross die out?

6.* Suppose in a 2×2 game, row 1 dominates row 2 and column 1 dominates column 2.
   (a) Show that (row 1, column 1) is an equilibrium pair.
   (b) Suppose, in addition, that (row 2, column 2) is a cooperative improvement over (row 1, column 1). Is it possible that (row 2, column 2) is an equilibrium pair? If so, give an example. If your answer is no, back it up with reasons.

SUGGESTED READING

Davis, M., *Game Theory*, Basic Books, New York, N.Y., 1970.

Gardner, M., "Mathematical Games," *Scientific American*, July 1973, p. 104. This article discusses a recently discovered paradox involving game theory, free will, and an omniscient supreme being.

Kline, M. (ed.), *Mathematics in the Modern World*, W. H. Freeman and Company, San Francisco, Calif., 1968. Available in paperback. This book contains many provocative expository articles about games.

Luce, R. D. and H. Raiffa, *Games and Decisions*, John Wiley & Sons, Inc., New York, N.Y., 1957. Well written and touches all bases.

Messick, D. (ed.), *Mathematical Thinking in Behavioral Sciences*, W. H. Freeman and Company, San Francisco, Calif., 1968. Available in paperback. This book also contains many provocative expository articles about games.

Raiffa, H., *Decision Analysis*, Addison Wesley Pub. Co., Reading, Mass., 1968. An intensive but readable discussion of the material in our Section 9.3.

Rapoport, A., *Strategy and Conscience*, Schocken Books, New York, N.Y., 1969. The author discusses how game theory does and does not apply to political and military conflict.

Persons contributing to the theory of elections demonstrate the wide range of individuals giving input to some mathematical problems: Kenneth Arrow (1921–   ) is an economist; Jean-Charles de Borda (1733–1799) was, among other things, a cavalry officer and a naval captain; Marie Jean Antoine Nicolas Caritat, Marquis de Condorcet (1743–1794) was a philosopher and social scientist who died in prison; and the Reverend Charles Dodgson (1832–1898), although he held a teaching position in mathematics at Oxford, is best known under the *nom de plume* Lewis Carroll for having written *Alice in Wonderland*. (Photo of Arrow courtesy of Photography Department of United Press International. Photos of Borda, Condorcet, and Dodgeson courtesy of The Bettmann Archive.)

10
the theory
of elections

10.1 the board of elections problem

With a clocklike regularity, each November the United States voting public goes to the polls to elect its numerous public officials at the many levels of government. It is characteristic of the voters that though (hopefully) they give much thought to which candidate they will vote for, they give little attention to the mechanics of how the election officials will decide who the winner of the election is on the basis of the votes cast. The purpose of this chapter is to discuss the problems faced by a hypothetical board of elections.

Suppose Mr. A_1 and Mr. A_2 are running for the office of district attorney in Springfield. Also up for grabs is the post of mayor of Springfield. Mr. B_1, Mr. B_2, and Mr. B_3 are seeking this post. Furthermore, there are 972 voters in Springfield. Election day arrives, 451 voters go to the polls, and the Board of Elections of Springfield duly count the votes. The results of the count for district attorney is shown in Table 10.1 and for mayor in Table 10.2. On the basis of the election results, the Board of Elections must declare a winner in the elections. Being believers in democracy, the members of the board know that one of the important principles of democratic elections is

Table 10.1 Vote for District Attorney

Mr. A_1	Mr. A_2
225 votes	226 votes

Table 10.2 Vote for Mayor

Mr. B_1	Mr. B_2	Mr. B_3
152 votes	151 votes	148 votes

the majority rules. The board is aware that *majority rule* may not guarantee that the "best man" will be elected. After all, what it takes to be "best man" is unclear but surely the winner of the election must represent the "choice of the people." After consulting the numbers in Table 10.1, the Board of Elections unanimously decide to declare Mr. A_2 the winner in the race for district attorney. Admittedly, he received only one more vote than his opponent but he did have a majority of the votes cast; that is, he had at least one more than half the total votes cast. Majority rule requires that he be declared the winner. When it came to decide the winner of the mayoralty race, a huge argument broke out. Let us eavesdrop on the argument. The members of the Board of Elections are Mr. Plurality, Mr. Borda, Mr. Runoff, Mr. Condorcet, and Mr. Socrates.

Mr. P: Mr. B_1 received a *plurality* of the votes; that is, he received the largest number of votes. I admit he did not receive a majority but I still think he must be declared the winner.

Mr. R: But Mr. B_1 received only a little more than one-third of the vote. To elect him would make a mockery of majority rule. I think we should eliminate Mr. B_3 since he received the fewest votes. Then we could hold an election between Mr. B_1 and Mr. B_2. Such a procedure is known as a *runoff election.* In the runoff election someone would get a majority and he would be declared the winner.

Mr. B: No. I have a better plan for selecting a winner. We should give points to each candidate according to how many voters rate him as their first choice, second choice, and so on. The candidate who received the greatest number of points would win.

Mr. P: Too complicated!

Mr. C: In that case you probably won't like my scheme either. I think we should determine if there is one candidate who can beat each of the other candidates in a two-man election. If there is such a person, he would be declared the winner.

Mr. S: It seems we are rather sharply divided on how to decide who should win. I think you will all agree, however, that because of the closeness of the vote, and the fact that no candidate has a majority, we must take unusual action. We should hold another election and, instead of merely asking a voter who is his choice for mayor, we should ask him to rank the candidates in order of his preference. For example, if a voter felt B_2 were his first choice, B_3 his second choice, and B_1 his third choice, he might indicate this as shown on the blackboard (see Figure 10.1).

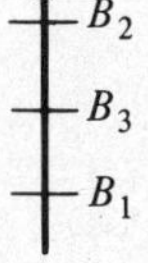

Figure 10.1

Mr. C,
Mr. R, *(in unison):* Good idea!
Mr. P,

Mr. B: That sounds like a good idea but all this does is improve the information we shall have available. It doesn't tell us which of the schemes for deciding the election will be used after we receive the improved information from the new election.

Mr. S: That's true. Let's construct a mathematical model for the process of conducting an election. On the basis of our model, perhaps we may be able to see more clearly how to proceed.

All: Fine. Let's get down to business.

EXERCISES 10.1

1. Suppose there are four candidates for office and 401 voters. Arrange the votes among the candidates so that if the person with the highest number of votes is called the winner, he will have the smallest possible percentage of the total vote. What is this percentage?

2. Suppose the plurality system is being used to decide an election. (A candidate wins with a plurality if he has more votes than any other candidate, though not necessarily a majority of the votes.) Compare the minimal percentage of the vote necessary to win the election if there are
    (a) two candidates,
    (b) three candidates,
    (c) five candidates.
    Generalize.

3.* Discuss the implications of *plurality voting* in primary elections with many candidates. Obtain some information about the size of the field of candidates for presidental primary elections in recent years. (Among the states that conduct presidential primaries are Wisconsin, Oregon, and New Hampshire.)

4.* Has a President of the United States ever been elected without receiving a majority of the popular votes cast? (Recall that the President is not directly elected. A complicated Constitutional procedure is followed that involves the so-called Electoral College.) What percentages of the popular vote did the *plurality* Presidents receive?

5.* Can you think of any recent elections in which there were more than three candidates? Obtain the election results and see if the candidate elected received a majority.

6.* Discuss the implications for majority rule if a strong third party were to achieve national strength and regularly run candidates for the presidency.

7.* Investigate the constitutional procedure for election of the President. Do you think *plurality* voting on the basis of popular vote would be an improvement?

8.* Investigate the election laws in your state and hometown to see if they provide for runoff elections in primaries or regular elections.

9.* The Australian and Irish electoral systems are unusual by United States standards. Investigate how these systems work.

10.2 a model for the board of elections: the candidates and the voter

Mr. P: In order to construct a model for an election, we must first isolate the essential features of the electoral process.

Mr. B: Clearly, we have three considerations in an election. There are candidates, voters, and a procedure for deciding how to select the winner on the basis of the votes cast. Let us call the last an *election procedure*.

Mr. C: Let us begin with the candidates. If there are only two candidates, we have seen that there is no difficulty. Whichever candidate receives the largest number of votes wins, and majority rule is in effect. As soon as there are as many as three candidates, we get into difficulty.

Mr. P: Correct. Let us denote the various candidates by $B_1, B_2, \ldots, B_n$, where n is at least 3. We would then have an election with n candidates.

Mr. S:	I suggest for simplicity that for the moment we limit ourselves to the case where *n* is 3 and there are only the three candidates, B_1, B_2, and B_3.
Mr. B:	We have handled the candidates. Now we must model our voters.
Mr. C:	Certainly all of us except Mr. P agree that the voter has not given us enough information by merely telling us who the candidate of his first choice is. What we need is more precise information from him; that is, we need to know who he thinks is the best candidate, the next best candidate, and so on. Let us use the notation that Mr. S showed us earlier. Thus, a voter might indicate his ranking of candidates as shown in Figure 10.1. This would mean B_2 was his first choice, B_3 his second choice, and B_1 his third choice. Let us call such a diagram a *preference schedule*.
Mr. B:	In that case if there were three candidates, each voter might have one of six possible preference schedules. Here, I'll write them on the board (see Figure 10.2). When a voter comes to the polls, instead of merely voting for his favorite, he would vote for that one of the six preference schedules that represented his views.

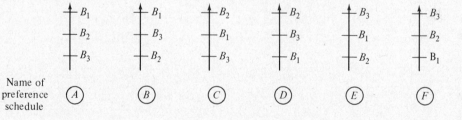

Name of
preference
schedule

(Each candidate is atop precisely two schedules)

Figure 10.2

Mr. P:	I object! The voters are too stupid to get it straight. Besides, it requires a lot more thinking to decide which preference schedule one likes rather than merely to vote for the person one likes best.
Mr. R:	It really does not require too much of the voter. Consider a voter making up his mind whether to vote for B_1, B_2, or B_3.

First he must decide if he prefers B_1 to B_2. Say he likes B_1 better. Then he would have to decide whom he liked better between B_1 and B_3. Say he prefers B_3. Ordinarily, he would now vote for B_3. But surely he is *capable* of deciding between B_1 and B_2 whom he would rank in the second position (B_3 was first). If he selected B_1 over B_2, he then would vote for preference schedule E; if he selected B_2 over B_1, he would vote for schedule F (see Figure 10.2).

Mr. C: But is it not possible that a voter might feel the following way? I prefer B_1 to B_2. I prefer B_2 to B_3. But I prefer B_3 to B_1!

Mr. P: I think we would all agree that if a person ranked candidates this way, then his thinking would not be consistent. I think the typical voter will be consistent in his preferences. But I see another difficulty. Suppose some voter says, "I like B_2, and a plague on both B_3 and B_1." In that case he would be unable to chose one of the six schedules in Figure 10.2.

Mr. B: That seems a reasonable objection. Some voters may be indifferent to two or more candidates. It would mean that we would have to add the schedules I am writing on the blackboard to the six we already have (Figure 10.3). For example, schedule G would be interpreted that the voter preferred B_1 to B_2 and B_3 but was indifferent to both B_2 and B_3. Voting for schedule K would mean that the voter thought B_1 and B_2 were equally good and that both were better than B_3.

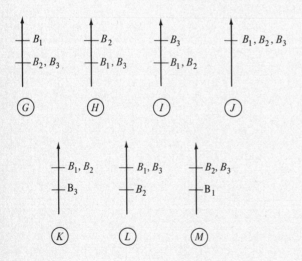

Figure 10.3

Mr. C: Yes. That does seem more accurate.

Mr. S: You are quite right. Technically we should add these additional schedules but, for simplicity, we shall avoid the more complicated situation. Perhaps you will agree to the following:

MODELING ASSUMPTION 1

Given n candidates, each voter can arrange the candidates in a vertical column so that if one candidate B_i lies above another candidate B_j, then B_i is preferred to B_j. Furthermore, the voter will not be indifferent between candidates.

Mr. R: What other assumptions about the voters are necessary?

Mr. S: I think we need one other assumption. Suppose that a voter has chosen a preference schedule and suddenly one candidate dies or withdraws from the race. I think we must assume that the ranking of the remaining candidates remains unchanged and that the voter doesn't alter the relative positions of the remaining candidates. Thus, if a voter likes preference schedule A and B_1 withdraws from the race, the voter still prefers B_2 to B_3 and doesn't suddenly discover with B_1 gone that he prefers B_3 to B_2. Had he preferred them in the order B_1, B_3 (second), B_2 (third), he would have voted for schedule B originally (see Figure 10.2).

MODELING ASSUMPTION 2

If a candidate is removed from consideration, the voters would vote their preference schedules with that name removed and would not reshuffle the relative positions of the remaining candidates.

EXERCISES 10.2

1. If there are four candidates, how many different preference schedules satisfying Modeling Assumption 1 are there?

2. Determine for an arbitrary number n the number of preference schedules that satisfy Modeling Assumption 1. (*Hint:* Use the Principle of Counting, Section 8.3.)

3. If there are four candidates, how many schedules satisfying Modeling Assumption 1 have B_1 in first position? B_1 in second position? B_1 in the last position?

4. Given that there are three candidates, if Modeling Assumption 1 is satisfied, how many schedules have B_1 above B_3? B_1 above B_2? B_2 above B_3?

5. Repeat Exercise 4 using four candidates.

6.* If there are four candidates but indifference between candidates is allowed, how many schedules will there be in addition to those that arise under Modeling Assumption 1.

7.* Suppose that instead of thinking of candidates and voters in elections, we think of alternative bills before a legislative committee whose members must decide which bill to accept. For example, we might have a committee that was trying to decide among alternative school budgets: $B_1 = $ a budget of \$10 million, $B_2 = $ a budget of \$12 million, and $B_3 = $ a budget of \$15 million. Examine the assumptions about candidates and voters to see whether they might apply to this situation of bills and committee decision makers.

8.* Discuss further decision situations where the model we are developing might be of importance. Discuss the validity of the assumptions that have been made for these situations.

9.* Can you think of any additional assumptions that should be made about voter behavior that should be taken into account in our models?

10.* How valid is Modeling Assumption 1?

10.3 a model for the election decision procedure

Mr. P: At long last we reach the election decision procedure. Let me describe my method as it applies to our current framework. Suppose we have counted the number of votes that have been recorded for each of the preference schedules in Figure 10.2. Here, look at this example (Figure 10.4). For convenience I have indicated the schedule name again and below each schedule the number of voters who voted for it. Now we add together the total number of voters who put a given candidate in first place. B_1 gets 28 votes, namely, the 4 people who voted

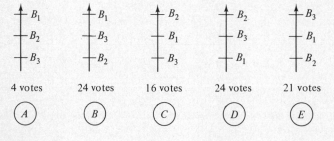

B_1	B_1	B_2	B_2	B_3
B_2	B_3	B_1	B_3	B_1
B_3	B_2	B_3	B_1	B_2
4 votes	24 votes	16 votes	24 votes	21 votes
A	B	C	D	E

Figure 10.4

schedule A rank him first as do the 24 people who voted for schedule B. Candidate B_2 gets $16 + 24 = 40$ votes and B_3 gets 21 votes. Notice that schedule F does not enter into the procedure since no voters chose this schedule. Now to decide the winner, we choose the candidate who received the largest number of first-place votes. In this case B_2 wins with 40 votes.

Mr. B: But we already saw one problem with your system. A candidate may not have a majority and thus, as the number of candidates goes up, the person who wins may have an extremely small percentage of the total first-place votes. For example, in the situation you have just described, the winner has only 45% of the total vote of 89.

Mr. R: I believe my method will be an improvement. Here is what one does. We start off as Mr. P did, determining the number of first-place votes of each of the candidates. Then we hold a runoff election between the two candidates who received the highest number of first-place votes; that is, we eliminate from the running all the lower vote getters except the top two.

Mr. P: But that requires a lot of time and money since a new election will have to be held.

Mr. R: No. That will not be necessary since we have the complete preference schedule for each voter. Furthermore, Modeling Assumption 2 means that we can eliminate the lower vote getters and still determine the winner of the election from the schedules already in our possession. Let us go to the example in Figure 10.4. We have already seen, using Mr. P's analysis, that B_1 received 28 first-place votes, B_2 received 40 first-place votes, and B_3 received 21 first-place votes. Thus we eliminate B_3 and hold a runoff between B_1 and B_2. Now we must compute the number of votes for B_1 and B_2 in the runoff election. Look back at Figure 10.4. Let us compute B_1's vote. Candidate B_1 gets $4 + 24 + 21$ votes.

The 4 votes arise from voters for schedule A since they prefer B_1 to B_2. Twenty-four votes come from the fact that the 24 voters who voted for schedule B prefer B_1 to B_2. The 21 votes come from the fact that the 21 voters who voted for schedule E prefer B_1 to B_2. (Note that we disregard B_3 on schedule E since he was eliminated.) B_2 gets $16 + 24$ votes, from the votes for schedules C and D, respectively. Thus B_2 loses the runoff since B_2 gets 40 votes to B_1's 49. Notice that we have a different winner by my method. B_1 wins rather than B_2!

Mr. P: But B_1 received only 31% of the first-place votes of all the voters. That's even less than what B_2 did. He seems a poor choice of a winner to me. Perhaps someone has a better method.

EXERCISES 10.3

1.   For each of the election results listed in Figure 10.5, determine the winner of the election using plurality voting and the system of a runoff election. Compute the percentage of the first-place vote that the winner receives in each case.

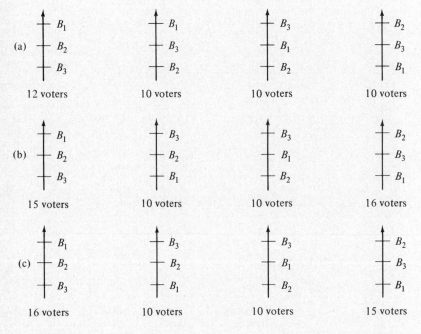

Figure 10.5

2.   If in the example of Figure 10.4 the 21 voters for schedule E had decided to lie about their preferences for B_1 and B_2 and interchange the two, would this have affected the results of the election using plurality voting? the runoff system? (*Note:* Lying about B_1 and B_2 amounts to voting for schedule F instead of for E.)

3.   Suppose there are four candidates for election. Could the following two procedures result in different candidates being elected?
     (a) Hold a runoff between the two candidates with the most first place votes.

(b) Eliminate the candidate with the fewest first place votes. Hold a runoff among the remaining three candidates. Now eliminate the lowest candidate again and hold a runoff between the remaining two candidates.

If you think the answer is yes, construct an example to show this. Discuss which of the procedures (a) or (b) you think is more "democratic."

4.* In a four-candidate election, if no candidate receives a majority, could one decide the election by pairing the winners of first and third place and the second and fourth place in a new election? Can you name some sports where a related procedure is used to decide the winner of a tournament?

10.4 the Borda count

Mr. B: I think I can clear matters up. Let me show you my system. It is based on the idea of giving a candidate "credit" not only for the number of first-place votes he receives but also for the number of his second-, third-, etc. place votes as well. The number of points a candidate receives will be called the *Borda count*. The person with the highest count wins the election.

Suppose we consider a given preference schedule. The number of points a candidate receives will be equal to the number of candidates who lie below him on this schedule. Of course if many voters vote for the same schedule, we multiply the number of points a candidate gets from the schedule by the number of voters who voted for that schedule.

Mr. P: Sounds awfully complicated.

Mr. B: Let me use the example from Figure 10.4 again as an illustration.

Mr. B_1 gets a Borda count of $4(2) + 24(2) + 16(1) + 24(0) + 21(1) = 93$. From schedule A there are 2 people below B_1 so he gets 2 points, which is multiplied by 4 since 4 voters voted for schedule A. From schedule B there are 2 persons below B_1 so he gets 2 points, which is multiplied by 24 since 24 voters selected schedule B_1. From schedule C he gets 1 point since there is one person below him; this is then multiplied by 16 since 16 persons voted for schedule C. He gets no points from schedule D since there are no candidates below B_1. Of course $24 \times 0 = 0$. Finally from schedule E he gets 1 point since there is 1 person below him. This is multiplied by 21, for the 21 persons selecting schedule E. Similarly, we can compute the Borda count of B_2 and B_3.

Mr. B_2's Borda count is $4(1) + 24(0) + 16(2) + 24(2) + 21(0) = 84$.

Mr. B_3's Borda count is $4(0) + 24(1) + 16(0) + 24(1) + 21(2) = 90$.

Under my scheme, the winner of the election would be B_1. I would support Mr. R's conclusion that B_1 should win this election. However, my method will not always yield the same winner as Mr. R's.

EXERCISES 10.4

1. Prove Mr. Borda's assertion that the Borda count and the runoff system do not always yield the same winner by deciding the winner of the elections shown in Figure 10.6 on the basis of the Borda count and the runoff system. Compare these results with those given by the plurality system.

2. Suppose that the system of assigning points in the Borda count for a four-candidate election is altered in the ways described below:
   (a) 3 points for a first place, 2 points for a second place, 1 point for a third place, and 0 points for a fourth place.
   (b) 6 points for a first place, 3 points for a second place, 2 points for a third place, and 1 point for a fourth place.
   (c) 8 points for a first place, 2 points for a second place, —3 points for a third place, and —4 points for a fourth place;
   (d) 4 points for a first place, 5 points for a second place, 6 points for a third place, and 3 points for a fourth place.

   For elections (d) and (e) in Exercise 1, compute the counts of the candidates using the systems just described. Does the final relative position of the candidates differ from system to system? Can you guess what conditions on the points assigned will guarantee that the result will always be the same winner as for the system Mr. Borda used?

3. Suppose points are assigned in a four-candidate election according to the following system: 1 point for a first place, 2 points for a second place, three points for a third place, and 4 points for a fourth place. Suppose the winner is the person with the *lowest point total*. Apply this system to the elections (d) and (e) in Figure 10.6. How does this affect the election? Can you generalize your conclusion to other ways of assigning points where the lowest total yields the winner?

4. Give an example where the Borda count yields the same winner as plurality voting but not the same winner as the runoff system.

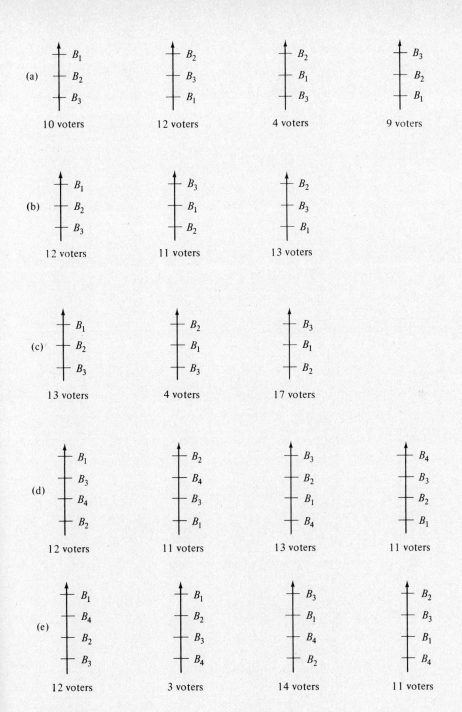

Figure 10.6

407

LISTEN, I CAN LICK HENRY, AND HE CAN LICK YOU...

SO YOU'RE GONNA BE EASY!

YOU IDIOT···DON'T YOU KNOW YOU DON'T HAVE A CHANCE??

RUPE
12-17

5. Give an example using three candidates where plurality voting, the runoff system, and the Borda count give different winners!

10.5 Condorcet's method

Mr. C: I regret that I cannot subscribe to any of the three methods given so far. Let me describe what I consider a far more democratic scheme. Suppose that we can find one candidate who can beat each of the other candidates in a two-man election. Then I believe he should be the winner. Let me illustrate my method for the example in Figure 10.4. First, let us see who is the winner of the election between B_1 and B_2. If we disregard B_3, we discover that B_1 gets $4 + 24 + 21 = 49$ votes. He gets the 4 votes of the voters who voted for schedule A since B_1 lies above B_2 on this schedule. He gets the 24 votes of the voters who voted for schedule B since B_1 lies above B_2 on that schedule, and he gets the 21 votes from the voters for schedule E since after disregarding B_3, B_1 lies above B_2. Since B_2 gets only $16 + 24$ or 40 votes, B_1 can beat B_2. Now let us see how B_1 fares against B_3.

In such an election, B_2 is disregarded and we discover that B_1 gets $4 + 24 + 16$ votes (from schedules A, B, C, respectively) for a total of 44, while B_3 gets $24 + 21 = 45$ votes (from the voters for schedules D and E). Thus B_3 can beat B_1 in a two-way race by the margin of 45 to 44. But, furthermore, B_3 can also beat B_2 in a two-way election. For if we disregard B_1, we discover that B_2 gets $4 + 16 + 24 = 44$ votes (from the persons who voted schedules A, C, and D, respectively), while B_3 gets $24 + 21 = 45$ votes (from the votes for schedules B and E, respectively.) Thus, this two-way race is won by B_3, 45 to 44 over B_2.

* Cartoon "Freddy" by Rupe, courtesy of Publishers-Hall Syndicate.

It seems to me that since B_3 can beat both B_1 and B_2 in a two-way race, majority rule requires that he be declared the winner, despite the fact that he was a loser under the other three systems described!

Mr. P: I can indeed see the logic of Mr. C's claim. I believe his method is best.

Mr. R
and   Mmmmmmm, what do you say Socrates?
Mr. B:

Mr. S: I would have to agree with Mr. C that *when there is* a candidate who can beat every other candidate in a two-way race, he should be the winner of the election. However, Mr. C has pulled the wool over your eyes, gentlemen. There may not always be such a candidate! For example, take the election shown in Figure 10.7. Let us apply Condorcet's method. In the election B_1 versus B_2, B_1 gets $152 + 148$ votes $= 300$ (from schedules A and E), while B_2 gets 151 votes (from schedule D). Thus B_1 beats B_2.

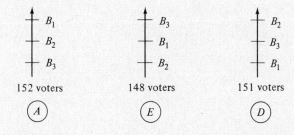

Figure 10.7

In the election between B_2 and B_3, B_2 gets $152 + 151 = 303$ votes (from schedules A and D), while B_3 gets 148 votes (from schedule E). Thus B_2 beats B_3.

But in the contest between B_1 and B_3, B_1 gets 152 votes (from schedule A), while B_3 gets $148 + 151 = 299$ votes (from schedules E and D).

Hence B_3 beats B_1. The Condorcet scheme breaks down and his system is unable to decide a winner. Take a look at the digraph in Figure 10.8; it shows the situation.

Mr. P: But your example shows B_1 beating B_2 (in a two-way race), B_2 beating B_3 in a two-way race, and B_3 beating B_1 in a two-

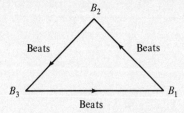

Beats

Beats

Beats

Figure 10.8

way race. Something must be wrong. We ruled out that happening in Modeling Assumption 1.

Mr. S: Unfortunately not. We ruled out this happening for individuals in Modeling Assumption 1 but this example shows, in fact, that Modeling Assumption 1 does not rule out the situation in Figure 10.8 for *groups of voters*. This situation is sometimes called the *voter paradox*.

This voter paradox is not just a curiosity but has some significance in political science. Consider a committee voting on three bills by deciding between pairs of them in some order. For example, the committee might first choose between B_1 and B_2 and then match the winner against B_3. Alternatively, B_3 and B_2 might be compared first and the winner matched against B_1. The voter paradox shows that there will be situations in which the bill that becomes a law will depend on the order in which the votes are taken! Going back to the example just mentioned (Figure 10.7), if we think of B_1, B_2, and B_3 as bills, B_1 becomes law if the order of the voting is B_2 versus B_3 with the winner pitted against B_1.

B_2 becomes law if B_1 and B_3 are voted against one another first, and the winner pitted against B_2.

B_3 becomes law if B_1 and B_2 are voted on first, and then the winner is voted on against B_3.

It seems as if such situations may have actually occurred in the United States Congress (see William Riker, "The Paradox of Voting and Congressional Rules for Voting on Amendments," *The American Political Science Review*, **52**, 1958, pp. 349–356). It turns out that in situations involving bills that are voted on in pairs until a winner is found, the *later* a bill is voted on, the greater the likelihood of its passing. As nice as it would be to avoid the voter paradox, I'm afraid that we are stuck with the possibility that it may occur.

Mr. P: But this seems to put us right back where we were before. How are we going to decide which system to use to decide our election?

EXERCISES 10.5

1. Apply the four different systems of deciding an election on each of the three sets of schedules shown in Figure 10.9.

(a)

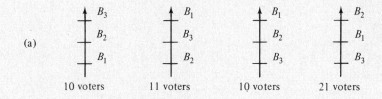

(b)

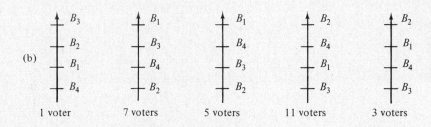

(c)

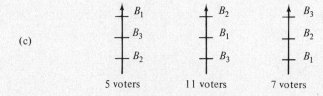

Figure 10.9

2. For each of the following properties, construct a collection of schedules with four candidates having the property noted:
    (a) The Borda count winner is different from the Plurality winner.
    (b) The Condorcet winner is different from the Borda count winner.
    (c) There is no Condorcet winner but all three other methods yield the same winner.
    (d) All four methods yield different winners!
    (e) The Borda count and the plurality vote yield the same winner and the Condorcet and runoff methods yield the same winner.
    (f) The plurality vote yields one winner and the other three methods yield the same winner.

3. Construct a voter paradox collection of schedules for four candidates; i.e., B_1 beats B_2, B_2 beats B_3, B_3 beats B_4, but B_4 beats B_1, where all elections are considered to be 2×2 elections. Can you see how to generalize your example to n candidates?

4. Give an example to show that the following could occur in a four-candidate election. There is some candidate B_1 who can beat the other three in two-way races; but B_2 beats B_3 and B_3 beats B_4, while B_4 beats B_2 in two-way races.

5. A digraph is drawn to show the results of the various two-way elections in a 4-candidate race. (The vertices represent the candidates, two vertices being joined by a directed edge to show who beats whom in a two-way race.) If there is no Condorcet winner, what can you say about this digraph? If there is a Condorcet winner, can the digraph have a directed circuit?

6.* Construct an example to show that a person who prefers A_1 and A_2 first and second in a three-candidate election can improve the chances of A_1 winning by ranking A_2 last, assuming the election will be decided by a Borda count.

7.* Discuss how the Borda count could be used to decide elections where all the schedules in Figure 10.3 as well as those in 10.2 are allowed.

8.* Decide if the following system could be used when there is no Condorcet winner: Count for each candidate the number of persons he can beat in a two-way race. The candidate who gets the largest number wins.

10.6 Arrow's theorem

Mr. S: Perhaps I can shed some further light on the problem of what system to use, by discussing some of the work of Kenneth Arrow, the Nobel prize winning economist.

Up until now we have studied several electoral systems and seen flaws and virtues in all of them. Suppose we take a different tack and attempt to answer the following question: What conditions would we like a *democratic* election decision procedure to obey?

Kenneth Arrow attempted to answer this question in a quite general setting. Imagine we have n voters who have provided a statement of their preferences on three or more alternatives or candidates. In the case of three candidates this means that each voter chooses one of the preference schedules in Figures 10.2 and 10.3. Now, the schedules of these individuals are submitted to a board of elections whose job is to use some *previously defined* system S to choose a ranking of the alternatives (candidates) for society. Note that rather than S selecting a "best" alternative for society, S ranks all the alternatives. This is equivalent to saying that system S chooses one of the 13 schedules in Figures 10.2 and 10.3 as society's ranking of the three alternatives (candidates).

Now Arrow listed four simple rules he wanted the system S to obey:

ADMISSIBILITY OF ALL SCHEDULES: The system S should work (provide a ranking) for any collection of schedules that the voters submitted to the board of elections.

PURE OPTIMALITY: If *every* voter preferred some alternative X to alternative Y, then the system S should choose for society a ranking in which alternative X was preferred to alternative Y.

NONDICTATORSHIP: There is no person P such that in every election the system S chooses as society's ranking of the alternatives the schedule which P submitted. (If this condition is violated, P is a dictator.)

INDEPENDENCE OF IRRELEVANT ALTERNATIVES: The relative position of any alternatives X and Y in society's ranking should depend only on the relative rankings of the two alternatives by the individuals.

The only one of these four conditions whose meaning may not be immediately apparent is the fourth. Here is a simple example illustrating the main idea. Imagine there are four candidates and six voters, whose schedules are shown in Figure 10.10a.

Suppose that when system S is used on these schedules, A_1 is ranked above A_2. (This may or may not seem reasonable but it will illustrate the idea nevertheless.) Now suppose instead of the six schedules in Figure 10.10a, S had to be used on the schedules in Figure 10.10b. When you compare these schedules with the original set (Figure 10.10a), you will see that the *relative* positions of A_1 and A_2 have not been altered, although A_3 and A_4 have been ranked differently. Rule 4 would require A_1 still to be ranked above A_2. Thus A_3 and A_4 were "irrelevant" for the ranking of A_1 and A_2.

All four of these conditions seem extremely reasonable. What decision procedures S satisfy these four conditions? The

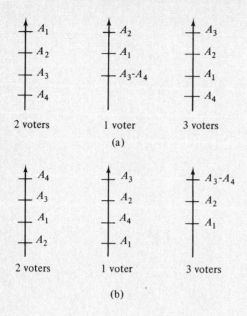

2 voters   1 voter   3 voters

(a)

2 voters   1 voter   3 voters

(b)

Figure 10.10

somewhat startling answer, proved by Arrow in 1953, is that there is no such procedure! In other words, any decision procedure S violates at least one of the four conditions.

Arrow's work has been a jumping off point for much research in an area now known as the *theory of social choice*.

As regards resolving the mayoralty election, the best we can do in the opinion of many experts on social choice is to use the Condorcet procedure applied to the voters' complete preference schedules. In the case that there is no Condorcet winner, resolve the election using the Borda count. In a small town such as ours it is not unreasonable to use a more sophisticated voting method than we could expect to see instituted at the state or national level. Other situations where the Condorcet method or Borda count might be used are in elections for officers of clubs, in student elections, and in school board elections.

EXERCISES 10.6

1. Suppose society must rank the four alternatives $A_1, \ldots, A_4$. The Borda count will be used: Society ranks highest the alternative get-

ting the highest count and society ranks the lowest the alternative getting the lowest count, etc.

(a)   For the set of schedules in Figure 10.11, how would the society rank the alternative if the Borda count were used?

(b)   Show that the fourth Arrow condition is violated by producing a set of schedules where everyone's relative ranking of A_1 and A_2 is as in Figure 10.11 but where the Borda count gives a different ranking than that produced by applying the Borda count to Figure 10.11.

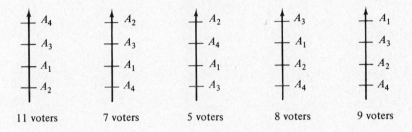

Figure 10.11

(c)   Show that the Borda count will obey conditions 1, 2, and 3.

2.   Suppose the plurality system (first-place votes only) is used to pick first through fourth ranking for society for the schedules in Figure 10.12.

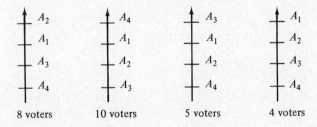

Figure 10.12

(a)   Show that condition 4 is violated by this system.

(b)   Can you produce a set of schedules that show that the plurality system violates condition 2?

(c)   Show that the plurality system satisfies conditions 1 and 3.

3.   Show that if the Condorcet system is used to decide society's first-place winner and the Borda count is used to decide the ranking of the other alternatives, then this "hybrid" system may not satisfy condition 1.

SUGGESTED READING

BLACK, D., *The Theory of Committees and Elections*, Cambridge University Press, London, 1963. A discussion of different methods of conducting elections of various kinds.

FARQUHARSON, R., *Theory of Voting*, Yale University Press, New Haven, Conn., 1969. An attempt to use ideas from game theory in election problems.

LUCE, D. AND H. RAIFFA, *Games and Decisions*, John Wiley & Sons, Inc., New York, N.Y., 1957. Chapter 14 gives an excellent discussion of group decision problems and Arrow's theorem.

11
difference equations and limits to growth

11.1 *simple difference equations*

In Chapter 6 and elsewhere where we have dealt with functions, we have assumed them to be given either as a table of values or in terms of an algebraic expression. As a practical matter, one is often interested in a function for which one does not have a complete description, either as a table of values or an algebraic expression. Under certain circumstances we can find an algebraic description if we are willing to make some assumptions. Here is an example.

Example 1

You have recently purchased a house and are trying to predict your fuel consumption for the coming winter. You have kept careful statistics for the month of October, recording the mean temperature and your burner's oil consumption each day. As it happens, the lowest temperature encountered so far has been 53°F and you would like to be able to predict what the oil consumption would be on days with temperatures 52°, 51°, 50°F, and so on. One way to think about the problem is that we want to fill in the blanks in the table below. Alternatively, we would like to find an algebraic description of the function that relates oil consumption to mean temperature.
One approach to the problem begins by plotting the existing values, as in Figure 11.1, and then asking what points should be plotted for the lower temperature values (indicated with a question mark).

417

Temperature	Fuel Consumption (gal/day)
60°F	2
59	4
58	6
57	8
56	10
55	12
54	14
53	16
52	?
51	?
50	?
49	?

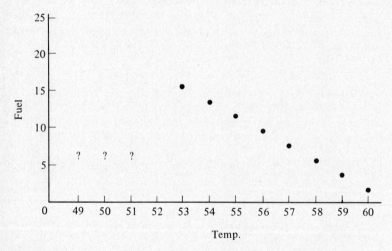

Figure 11.1

Solution 1 (Geometric): The points that have been plotted suggest that the function is linear, that is, that its graph is a straight line. By extending this line with a ruler we can get an idea of the values of the function for lower temperatures. For example, 24 seems a reasonable prediction for the fuel consumption when the temperature is 49°F. This type of analysis is called *linear extrapolation*. It should be clearly understood that it involves the following assumption: Because the function was linear between 60° and 53°F, we can assume that it is linear throughout the whole range of temperatures.

Solution 2 (Algebraic): Let $f(t)$ denote the fuel consumption on a day when the mean temperature is t. Thus, for example, $f(60) = 2$, $f(59) = 4$,

etc. We proceed to calculate the so-called first-order differences $f(60) -$
$f(59)$, $f(59) - f(58)$, $f(58) - f(57)$, etc. These numbers represent the additional fuel needed to take care of a 1° decrease in the temperature from
various base level temperatures.

$$f(60) - f(59) = -2$$
$$f(59) - f(58) = -2$$
$$f(58) - f(57) = -2$$
$$\vdots$$
$$f(54) - f(53) = -2$$

The fact that all the differences are equal down to $f(54) - f(53)$ suggests
that perhaps all the differences are equal, even those for which we have no
data. Of course, this is an assumption that may or may not be true, and it is
the mark of a good model maker to avoid bad assumptions. We can formulate this assumption in the equation

$$f(t) - f(t+1) = +2 \qquad \text{for all values of } t \qquad (11.1)$$

Using this assumption, we can obtain a formula for $f(t)$ by writing
$f(t)$ as a telescoping sum:

$$
\begin{aligned}
f(t) = f(t) - f(t+1) & \qquad \text{i.e., } f(t) = 2 \\
+ f(t+1) - f(t+2) & \qquad\qquad\quad + 2 \\
+ f(t+2) - f(t+3) & \qquad\qquad\quad + 2 \\
& \qquad\qquad\quad + 2 \\
\vdots & \qquad\qquad\quad \vdots \\
+ f(52) - f(53) & \qquad\qquad\quad + 2 \\
+ f(53) & \qquad\qquad\quad + f(53)
\end{aligned}
$$

From the form of the equation on the right, we see that

$$f(t) = (53 - t)(+2) + f(53)$$

because there are $(53 - t)$ differences, each equal to 2, that appear in the
telescoping sum in addition to the term $f(53)$. Replacing $f(53)$ by its value
of 16 and simplifying the equation, we obtain

$$f(t) = -2t + 122 \qquad (11.2)$$

From this equation we can find the fuel consumption for any particular value
of t by substitution. For example, $f(49) = 24$.

Equation (11.1) is called a *difference equation* and Equation (11.2) is the *solution* of the difference equation. Difference equations are usually used when one is dealing with a function f that is defined for positive integer values. In symbols, $f: I \rightarrow R$, where I stands for the set of positive integers and R stands for the collection of all real numbers. In some situations $f(t)$ may be defined for all real numbers. Generally, by a difference equation involving a function f, we mean an equation of the form

$$f(t+1) - f(t) = g(t) \qquad (11.3)$$

which is assumed to hold for all values of t. In our example, (11.1) was a difference equation in which $g(t) = -2$. Note that it is necessary to multiply both sides of (11.1) by -1 in order to get it exactly into the form of (11.3). Other examples of difference equations might be

$$f(t+1) - f(t) = 3t$$

$$f(t+1) - f(t) = t^2 + 1$$

If you reread our solution of Example 1, you will note that it depended heavily on the fact that $g(t)$ was constant and didn't depend on t. Difference equations for which $g(t)$ is a constant always have linear functions as their solutions. In this section we consider only difference equations where $g(t)$ is a fixed number.

In order for a difference equation to be solved, we need to know the value of the function at some particular known value of the variable. This information is known as the *initial condition*. In Example 1 the initial condition was the fact that $f(53) = 16$. The following theorem tells us how to find the solution to a difference equation, provided that we are given an initial condition.

THEOREM 1

Suppose f is a function that satisfies the difference equation and initial condition

$$f(t+1) - f(t) = k$$
$$f(t_0) = a_0 \qquad (11.4)$$

Then for all integral values of t,

$$f(t) = k(t - t_0) + a_0 \qquad (11.5)$$

Proof: First consider the case where $t > t_0$. Then

$$\left.\begin{aligned} f(t) &= f(t) - f(t-1) \\ &+ f(t-1) - f(t-2) \\ &+ f(t-2) - f(t-3) \\ &\quad\cdot \\ &\quad\cdot \\ &\quad\cdot \\ &+ f(t_0+1) - f(t_0) \end{aligned}\right\} \quad t - t_0 \text{ terms each equal to } k$$
$$+ f(t_0)$$

Consequently,

$$f(t) = k(t - t_0) + f(t_0) = k(t - t_0) + a_0$$

Unfortunately, if $t < t_0$, this proof won't work because we can never reach t_0 by starting with t and successively subtracting 1. By starting with t_0 and successively subtracting 1, however, we can reach t. Consequently, we can carry out the previous argument with t and t_0 interchanged, yielding $f(t_0) = k(t_0 - t) + f(t)$. However, this equation is equivalent to $f(t) = k(t - t_0) + f(t_0)$. Since $f(t_0) = a_0$, this is Equation (11.5), which we wished to derive.

In Example 1 our difference equation was inferred from existing data. It sometimes happens that a difference equation can be assumed to exist on theoretical grounds. In the following example, a linear difference equation forms a very simplified model for a country's economic system.

Example 2

Investment is the term used by economists to describe goods such as machine tools, tractors, and other farm equipment, which are used to produce other goods, as opposed to goods which are consumed such as food, safety razors, automobiles, and plastic bags. If one is willing to simplify a bit, one can take it as an economic law that the increase in the gross national product of a country from year t to year $t + 1$ is proportional to the dollar value of investment at year t. If we denote by $P(t)$ the value of the gross national product at year t and by $I(t)$ the value of investment at year t, then an example of this law can be expressed

$$P(t+1) - P(t) = (\tfrac{1}{4})I(t)$$

(We have taken the factor of proportionality to be one-fourth.) Suppose that the country establishes a policy of investing a constant sum of 8 billion dollars per year over the next 20 years. Suppose further that at the start of this plan ($t = 0$) P is 20 billion dollars. We therefore have the following difference equation and initial condition:

$$P(t + 1) - P(t) = 2$$

$$P(0) = 20$$

According to Theorem 1 the formula for P is

$$P(t) = 2t + 20$$

If this model is valid, we can predict that in 6 years after the start of the program the gross national product will be 32 billion dollars, while in 10 years the gross national product will be 40 billion dollars. One method of testing the validity of the assumptions going into the model is to test the actual values of the gross national product against those predicted by the model. It is important to note that even if the model and the actual gross national product agree for a few values of t, there is no assurance that the actual and predicted values will always agree.

   In dealing with the solution of difference equations, it is worthwhile keeping in mind how much they differ from the equations with which you are familiar. Thus, in solving the equation

$$3x - 2 + 4 = 8$$

the solution is the *number* $x = 2$. However, in solving the difference equation

$$f(t + 1) - f(t) = 2$$

we are looking not for a number but a *function*.

EXERCISES 11.1

1. For each of the following problems, find the algebraic expression for $f(t)$.
    (a) $f(t + 1) - f(t) = 2$, $f(7) = 10$
    (b) $f(t + 1) - f(t) = -3$, $f(3) = 0$
    (c) $f(t + 1) - f(t) = 0$, $f(4) = 4$
    (d) $f(t + 1) - f(t) = 1$, $f(-1) = -1$
2. Suppose one has the difference equation $f(t + 1) - f(t) = 2$. Solve this for each of the following initial conditions. Graph each of the func-

tions you obtain on the same set of axes. What do the solutions have in common?

    (a) $f(0) = 0$
    (b) $f(0) = 1$
    (c) $f(1) = 3$

3.   In Example 2 of the text, suppose we consider the following three investment policies:

    (a) $I(t) = 4$ for each t
    (b) $I(t) = 10$ for each t
    (c) $I(t) = 12$ for each t

    Using the initial condition $P(0) = 20$ in all three cases, find $P(t)$ in each of three cases. Plot the graphs of the functions on the same set of axes. Evaluate P at the end of 20 years for each of the three investment policies.

4.   Using the initial condition shown, draw a graph of the function values generated from the difference equation for the given values of t. (The graphs you get will *not* be straight lines.)

    (a) $f(t + 1) = [f(t)]^2 + 4$
           $f(0) = 1;\quad t = 0, 1, 2, 3$
    (b) $f(t + 1) = f(t) - 4t^2$
           $f(0) = 20;\quad t = 0, 1, 2, 3, 4$
    (c) $f(t + 1) = f(t) + 4t$
           $f(0) = 1;\quad t = 0, 1, 2, 3, 4, 5, 6$
    (d) $f(t + 1) = 2f(t) - [f(t)]^2$
           $f(0) = 1;\quad t = 0, 1, 2, \dots, 6$
    (e) $f(t + 1) = 3f(t) - [f(t)]^2$
           $f(0) = 1;\quad t = 0, 1, 2, 3$

5.   A family on a camping trip covers the mileages below in the first few days. Using the type of modeling used in Example 11.1 of the text, determine how far they will have gone at the end of the fiftieth day. (Note that the first few differences are not exactly constant but you may choose a constant that approximates them.) Do you think your solution is likely to be reliable?

Day	Mileage
1	198
2	403
3	600
4	801

6.   A factory normally produces 20 silk purses from 1 lot of 100 sows' ears. From each additional lot of sows' ears, 25 additional purses can

be made. How many silk purses can be made from 10,000 sows' ears?

7.* Suppose $f(t + 2) - f(t) = 3$ for all integer values of t and suppose $f(0) = 0$. Can you find $f(10)$? Can you find $f(11)$?

8.* In Exercise 3 policy (c) involves a yearly investment three times as great as that of policy (a). But policy (c) doesn't lead to a gross national product three times as great as that which arises from policy (a) after 20 years. Is this reasonable? Would the gross national product under policy (c) be three times that under policy (a) if we waited longer, say 100 years? Would it make a difference if we changed the initial condition?

11.2 difference equations and exponential growth

In recent years society has been giving increased attention to the pattern of growth in population, garbage, and gross national product. Many people feel that growth of some of these quantities is so rapid as to spell danger for society. In this section we shall discuss a special type of growth. Exponential growth is what population, epidemics, chain letters, rumors and many other growth phenomena all have in common. Each of these phenomena can be modeled with the following difference equation:

$$f(t + 1) - f(t) = kf(t)$$

where k ($\neq 0$) is some constant factor of proportionality. After algebraic simplification this equation can also be written

$$f(t + 1) = (k + 1)f(t)$$

The following examples show how this equation arises naturally in some of the situations just mentioned.

Example 1

During a political demonstration someone starts a rumor that a group of counter-demonstrators is on its way, seeking a confrontation. We wish to model the spread of the rumor with a view toward predicting how quickly it will reach everyone. For example, will everyone hear the rumor in 15 min if there are 1000 demonstrators? How about 10 min? How about 5 min?

Solution: Our model will be based on the assumption that every person who hears the rumor manages to communicate it in 1 min to three other persons who have not heard it before. In practice, these three people are likely to be contacted at various times during the 1-min interval. For the purposes of the model we assume that they are contacted at the instant the interval ends. Furthermore, an individual does not stop spreading the rumor after 1 min but continues spreading it at the same rate. If we let t denote minutes and let $f(t)$ denote the number of people exposed to the rumor at the end of t minutes, we may assert that

$$f(t+1) - f(t) = 3f(t) \qquad (11.6)$$

Indeed, this equation is merely a notational shorthand for the assertion that the number of new people exposed to the rumor in the period from time t to time $t+1$ [i. e., $f(t+1) - f(t)$] is three times the number of people who had heard the rumor up to time t. This is a direct consequence of our assumption that each person who has heard the rumor will thereafter tell it to three new people in each 1-min time period.

  We assume that at time 0 one person (the creator of the rumor) knows the rumor. This gives the initial condition:

$$f(0) = 1 \qquad (11.7)$$

  To solve the difference equation (11.6) with the given initial condition, we write (11.6) in the form $f(t+1) = 4f(t)$. Now taking t in turn to be 0, 1, 2, etc., we have the following equations:

$f(1) = 4f(0) = 4 \times 1 = 4$ (Use (11.7) for $f(0)$)

$f(2) = 4f(1) = 4 \times 4 = 4^2$ (substituting the value of $f(1)$ from above)

$f(3) = 4f(2) = 4 \times 4^2 = 4^3$ (substituting from the previous line)

$f(4) = 4f(3) = 4 \times 4^3 = 4^4$ (substituting from the previous line)

Clearly we can continue like this forever and at the tth stage we will have the equation

$$f(t) = 4^t$$

  Thus, we know how to calculate how many people have heard the rumor at any given time. After 15 min, 4^{15} people have heard it. The number

4^{15} is an enormous number, somewhat larger than 1 billion (1 billion = 1,000,000,000). As we can see from Figure 11.2, 5 min will be sufficient for everyone to hear the rumor.

$4^0 = 1$

$4^1 = 4$

$4^2 = 16$

$4^3 = 64$

$4^4 = 256$

$4^5 = 1,024$

$4^6 = 4,096$

$4^7 = 16,384$

$4^8 = 64,536$

$4^9 = 262,144$

$4^{10} = 1,048,576$

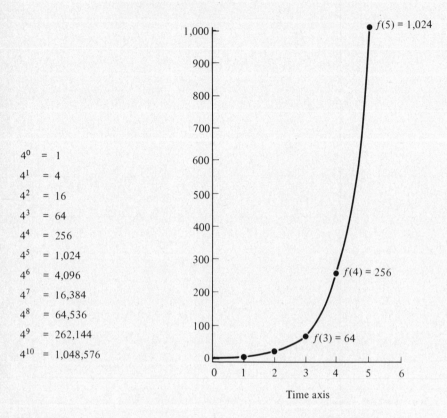

Figure 11.2

The function 4^t in Example 1 is called an *exponential function* because the variable appears in the exponent. Processes that can be modeled by an exponential function a^t, where a is a constant greater than 1, are said to exhibit *exponential growth*.

Example 2

An astronaut returning from Mars brings an extraterrestial virus with him and starts an epidemic. Let us suppose that the disease is communicable by a person for an indefinite period after he has contracted the disease. This is unrealistic but makes a convenient model. Let us assume that the number of healthy people infected by a single carrier has a mean (average) value of one-half a person per day. We wish to find a model that will tell us how many people have had the disease at any time t. If we let t be measured in days and let $f(t)$ denote the number of people who have had the disease at time t, then

$$f(t+1) - f(t) = (\tfrac{1}{2})f(t) \qquad (11.8)$$

Alternatively, $f(t+1) = (1.5)f(t)$. By taking values of t equal to 0, 1, 2, etc., we obtain the following series of equations:

$$f(1) = 1.5f(0)$$

$$f(2) = 1.5f(1)$$

$$f(3) = 1.5f(2)$$

etc.

But at time 0 there is only one person infected, namely, the returning astronaut, so $f(0) = 1$. We can make this substitution in the first equation and get $f(1) = (1.5)(1) = 1.5$. This value of $f(1)$ can then be substituted in the second equation to get

$$f(2) = (1.5)(1.5) = (1.5)^2$$

We can repeat this cycle of calculation and substitution indefinitely and generate the following equations:

$$f(1) = 1.5$$

$$f(2) = (1.5)^2$$

$$f(3) = (1.5)^3$$

$$f(4) = (1.5)^4$$

etc.

From the pattern of these equations we can easily deduce that for every value of t, $f(t) = (1.5)^t$. The graph of this function along with a table of approximate values for the function is shown in Figure 11.3. Notice that it has the same general shape as the function in the last example, getting larger without limit as time increases.

THEOREM 2

The difference equation $f(t+1) - f(t) = kf(t)$, where k is a constant, together with the initial condition $f(0) = a$, has the following solution:

$$f(t) = a(k+1)^t$$

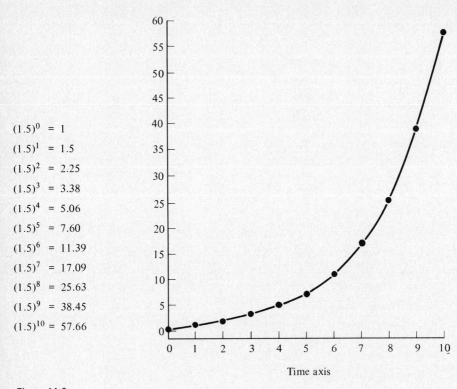

$$(1.5)^0 = 1$$
$$(1.5)^1 = 1.5$$
$$(1.5)^2 = 2.25$$
$$(1.5)^3 = 3.38$$
$$(1.5)^4 = 5.06$$
$$(1.5)^5 = 7.60$$
$$(1.5)^6 = 11.39$$
$$(1.5)^7 = 17.09$$
$$(1.5)^8 = 25.63$$
$$(1.5)^9 = 38.45$$
$$(1.5)^{10} = 57.66$$

Time axis

Figure 11.3

Proof: All we need to do is to substitute this function into the difference equation to see whether it satisfies the difference equation. To do this, note that

$$f(t + 1) = a(k + 1)^{t+1}$$

and so

$$f(t + 1) - f(t) = a(k + 1)^{t+1} - a(k + 1)^t$$

We need to check whether this equals $kf(t)$, which is $ka(k + 1)^t$. But

$$a(k + 1)^{t+1} - a(k + 1)^t = a(k + 1)^t[(k + 1) - 1]$$
$$= a(k + 1)^t(k) = ka(k + 1)^t$$

Thus, we see that $a(k + 1)^t$ satisfies the difference equation and is, therefore, a solution of it.

Having proved this result, we are now in a position to write the solution immediately to any difference equation of the form $f(t + 1) - f(t) = kf(t)$

with the initial condition $f(0) = a$. Then we can, if we wish, make a table of values and a graph of the function f that we find as the solution, as we did in the last two examples. It would be convenient, however, if we could find out something about the function even without graphing it. For example, does it become larger without limit as time increases, or does it level off somewhere and not become larger than a certain number? The next theorem gives some information about this.

THEOREM 3

Let f be the function that satisfies the difference equation $f(t + 1) - f(t) = kf(t)$ with initial condition $f(0) = a \, (a > 0)$. Suppose further that $k > 0$. Then no matter how large a number we choose, $f(t)$ will be larger than that number if t is large enough. On the other hand, if $k = 0$, then $f(t) = a$ for every value of t.

Proof: First we examine the case where $k = 0$. In this case, $f(t + 1) - f(t) = 0$ for all values of t, which means that in each time period there is no change in the value of $f(t)$. This means that $f(t)$ is constant.

Next we consider the case where $k > 0$ and $f(0) > 0$. We begin by substituting $0, 1, 2, \ldots, n - 1$ as values for t in the difference equation.

$$f(1) - f(0) = kf(0)$$
$$f(2) - f(1) = kf(1)$$
$$f(3) - f(2) = kf(2)$$
$$\cdot$$
$$\cdot$$
$$\cdot$$
$$f(n) - f(n - 1) = kf(n - 1)$$

Examining these equations in turn, we can make the following observations.

$k > 0$ and $f(0) > 0$ so $f(1) - f(0) > 0$ so $f(1) > f(0) > 0$

$k > 0$ and $f(1) > 0$ so $f(2) - f(1) > 0$ so $f(2) > f(1) > f(0) > 0$

$k > 0$ and $f(2) > 0$ so $f(3) - f(2) > 0$ so $f(3) > f(2) > f(1) > f(0) > 0$
$$\cdot$$
$$\cdot$$
$$\cdot$$

$k > 0$ and $f(n - 1) > 0$ so $f(n) - f(n - 1) > 0$

so $f(n) > f(n - 1) > \cdots > f(1) > f(0) > 0$

Consequently, for each value of t which is greater than 0, $f(t) > f(0)$. Since $k > 0$, this means $kf(t) > kf(0)$ for each such t. This fact allows us to replace all but the first equation above by inequalities:

$$f(1) - f(0) = kf(0)$$
$$f(2) - f(1) > kf(0)$$
$$f(3) - f(2) > kf(0)$$

$$\cdot$$
$$\cdot$$
$$\cdot$$

$$f(n) - f(n-1) > kf(0)$$

Adding these inequalities gives $f(n) - f(0) > nkf(0)$, which can be rewritten:

$$f(n) > (nk + 1)f(0)$$

The expression on the right is made up of two positive constants k and $f(0)$ and a variable quantity n. Clearly we can make $(nk + 1)$ as large as we like just by taking n to be large enough. Consequently, $f(n)$ can be made as large as we like simply by taking n large enough. Therefore, $f(t)$ becomes larger without limit as t increases.

We shall apply the results of the previous theorems to a simple model for the population explosion.

Example 3

During the twentieth century, up until 1970 the rate of population increase in the United States has been about 0.0146 persons per capita per year, beginning with an initial condition of 76 million in the year 1900. This rate is an amalgam of average birth rate, average death rate, and average immigration rate. It can be interpreted as meaning that in this century for every person residing in the United States at the end of year t, there are 1.0146 persons at the end of the next year $t + 1$. More sensibly, perhaps, for every 10,000 United States residents at time t there will be 10,146 at time $t + 1$. If this rate maintains itself, will the population level off? How many people will be alive in the year 2000?

Solution: Let $f(t)$ denote the population at time t, where we measure t in years beginning with $t = 0$ in 1900. Then

$$f(t + 1) - f(t) = 0.0146f(t)$$

$$f(0) = 76$$

$$(11.9)$$

Using Theorem 2 we immediately write the population function as

$$f(t) = 76(1.0146)^t$$

Thomas Malthus (1766–1834), although not generally regarded as a mathematician, provided the first mathematical theory of population growth. He held that population grows exponentially and has a "natural tendency" to double in about 25 years. He also felt that food supplies could not expand to match this pace and that the inevitable result would be that misery and vice would intervene to lower the population. Malthus' ideas seem oversimplified today, but they have been and still are quite influential. Among other things, he is credited by Charles Darwin and Alfred Wallace, codiscoverers of the Theory of Evolution, with inspiring part of that theory. The essentially pessimistic theories of Malthus were largely a response to the philosophic optimism of writers such as Condorcet whose ideas were much in vogue during and after the French Revolution.
(*Courtesy The Bettmann Archive.*)

Since $k = 0.0146 > 0$, Theorem 3 tells us that the population will increase without limit as t increases. Theorem 3 also tells us that the only way to achieve a limit to the population would be to have k, the rate of increase, be exactly 0 (zero population growth). A table of values and graph for the function $f(t) = 76(1.0146)^t$ is shown in Figure 11.4. The actual population figures are also shown and you can see that our model, based on the function $f(t) = 76(1.0146)^t$, gives a reasonable but not exact fit for the actual figures. A better model would take into account fluctuations in the rate of increase from decade to decade and might try to relate these to social and economic conditions.

We have concerned ourselves in this section with the phenomenon of exponential growth in which a quantity increases at each time step by an amount that is proportional (by a fixed constant of proportionality) to the value of the quantity at the previous time step. Many growth processes in the real world tend to behave this way, and we have given three examples in which the exponential model applies to some extent. As with most models, however, it doesn't apply perfectly. Here are a few specific criticisms of the *goodness of fit*. These criticisms are not mere quibbles but are really quite serious. An applied mathematician would certainly not ignore them but would use them as guides to building a better model.

431

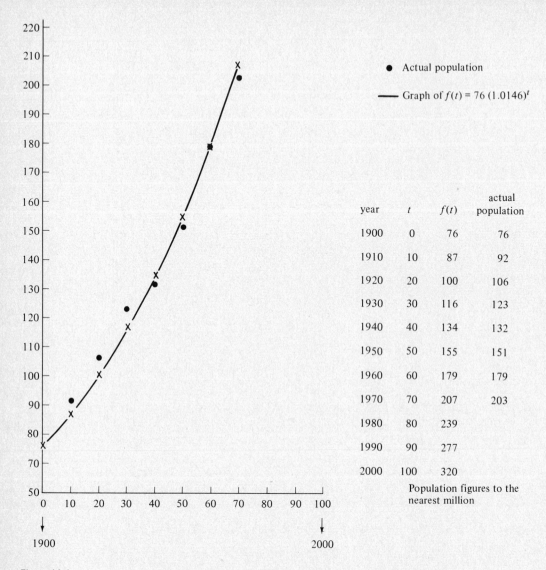

year	t	f(t)	actual population
1900	0	76	76
1910	10	87	92
1920	20	100	106
1930	30	116	123
1940	40	134	132
1950	50	155	151
1960	60	179	179
1970	70	207	203
1980	80	239	
1990	90	277	
2000	100	320	

Population figures to the nearest million

● Actual population

—— Graph of $f(t) = 76 (1.0146)^t$

Figure 11.4

THE RUMOR MODEL

There comes a time when so many people have heard the rumor that a person who wishes to communicate it discovers that all those in his immediate vicinity have heard it. At this point it may become impossible for him to spread the rumor to three people in a minute as assumed by the model.

432

QUESTION 1

A similar type of criticism applies to the epidemic model. Can you state it?

THE POPULATION MODEL

The equation $f(t + 1) - f(t) = 0.0146f(t)$ presents only the current tendency of population growth. If the population continues to grow according to this law, then, as we have seen, it grows larger without limit. There would come a time, however, when pressures of food shortage, overcrowding, etc., would change the equation.

EXERCISES 11.2

1. Solve the following difference equations subject to the initial condition shown. Draw a graph of your solution function.
   (a) $f(t + 1) - f(t) = 2f(t)$
       $f(0) = 4$
   (b) $f(t + 1) - f(t) = 1f(t)$
       $f(0) = 5$
   (c) $f(t + 1) - f(t) = \frac{1}{4}f(t)$
       $f(0) = 800$
   (d) $f(t + 1) - f(t) = \frac{3}{2}f(t)$
       $f(0) = 20$

2. Solve the difference equation

$$f(t + 1) - f(t) = 2f(t)$$

   subject to
   (a) $f(0) = 1$
   (b) $f(0) = 3$
   (c) $f(0) = 6$
   Compare the graphs of the solutions to parts (a), (b), and (c).

3. Suppose one has the difference equation $f(t + 1) - f(t) = (0.5)f(t)$ as in Example 2. Solve this for each of the following initial conditions (changing the initial conditions merely corresponds to changing the number of astronauts who return with the virus).
   (a) $f(0) = 0$
   (b) $f(0) = 3$
   (c) $f(0) = 4$

4. In Example 1 consider each of the following two rates of spread for the rumor.
   (a) One person per minute,
   (b) Two persons per minute.

Using the initial condition $f(0) = 1$, find $f(t)$ in each case. Plot these graphs on the same set of axes. Evaluate f at the end of 5 min in each case.

5. Bacterial cells of a certain type divide into two identical but smaller cells once every hour. In a culture that starts with 150 cells, how many will there be at the end of 10 hr?

6. An important part of a country's capital is its stock of machine tools. These tools are used to produce consumable goods such as bicycles, belt buckles, and candy bars. Some machine tools, however, can be used to make other tools. Suppose that in any 1-year period the number of new machine tools made is equal to 5% of the existing quantity at the beginning of the year. Assuming that machine tools are good forever and don't need to be discarded, write the difference equation governing $f(t)$, the number of machine tools after t years.

7. In Exercise 6, assume now that 3% of all machine tools are discarded in each 1-year period. How does that change the difference equation? What discard rate would be necessary in Exercise 6 in order to avoid having the number of machine tools increase without limit?

8. In the Middle Ages, crafts such as woodworking were learned by apprenticing oneself to a master craftsman. In time one could become a master craftsman and train others. Suppose each master can train two other master craftsman in the space of 10 years, or a mean value of $\frac{1}{5}$ per year. Assume that on the average a master craftsman works for 30 years and then retires or dies. This means that on the average the number of master craftsmen who cease working between times t and $t + 1$ is $\frac{1}{30}f(t)$. Suppose we set $t = 0$ at the beginning of the period that interests us and suppose at that time there are 25 master woodworkers. Write the difference equation governing the function $f(t)$ that gives the number of master woodworkers at time t, and write the initial condition. Solve the difference equation.

9.* Discuss the solution of the difference equation

$$f(t + 1) - f(t) = -2f(t)$$

where $f(0) = 10,000$.
   (a) How does the graph of the solution differ from the graphs of the solutions for Exercise 1?
   (b) Do the function values grow without limit as t (> 0) increases?

10.* Consider the difference equation $f(t + 1) - f(t) = 2f(t)$, where $f(0) = -1$.
   (a) Find the solution function.
   (b) Does the solution function exhibit growth?

(c)   For any given negative constant, can one find a t such that $f(t)$ is less than that constant?

11.* John receives the following chain letter:

> *Dear John,*
>
>   *If you follow the instructions below, you will participate in a scheme to earn yourself and many others $40.*
>
> *(a)   Send $10 to the person from whom you received this letter.*
>
> *(b)   Send an exact copy of this letter to five other people, each of whom will send you $10.*
>
>   *Sincerely,*

Suppose that everyone who receives such a letter complies with it. Let $f(t)$ be the number of people in the chain (i.e. people contacted plus the originator) at time or stage t. Assuming that $f(0) = 1$ (i.e., at time 0 one person started the chain letter), can you write a difference equation for $f(t)$? (Bear in mind that this process differs from the text examples in that once a person writes his five letters, he stops writing letters and no longer contributes to the growth of the number of letters. Thus, the difference equation $f(t + 1) - f(t) = 5f(t)$ is incorrect.) For the equation you determine, find $f(1)$, $f(2)$, $f(3)$, and $f(4)$. Can you explain why chain letters are illegal in most states?

11.3   the limits to growth

In this section we want to discuss some better models for population growth. The underlying notion is, as in Section 11.2, exponential growth but with this novelty: We build into the model the commonsense assumption that growth cannot go on forever before other forces intervene to provide a limit to growth. Thus, if we place a small number of bacteria into a nutrient medium containing no competing organisms, the growth will at first be exponential. There will come a point, however, where the test tube will not support more growth either because the nutrients have been used up or for other reasons. It seems to be a fact that a given environment has a maximum carrying capacity for any organism. We shall now try to incorporate this notion into a difference equation model.

Under optimal growth conditions our organism will have a constant intrinsic rate of natural increase, r (> 0). If this growth rate could continue, we would have a difference equation like this:

$$x(t + 1) - x(t) = rx(t)$$

where $x(t)$ is the function that gives the population of the species at time t. Let us denote by L the largest population of our organism that the environment can support. Then the fraction $[L - x(t)]/L$ can be thought of as a resistance factor to the possibilities for further growth at time t because when $x(t)$ is small compared to L, this fraction is about equal to 1, whereas when $x(t)$ approaches the limit L, the fraction approaches 0. Consequently, we shall use it as a factor to scale down r and write the following difference equation with a variable growth rate as our model:

$$x(t + 1) - x(t) = \frac{r[L - x(t)]}{L} x(t) \qquad (11.10)$$

We note that this can be written, after algebraic simplification, in the form

$$x(t + 1) = \left[r + 1 - \frac{r}{L} x(t) \right] x(t) \qquad (11.11)$$

For future reference we note that we can write (11.11) in the form

$$x(t + 1) = [a - bx(t)]x(t) \qquad (11.12)$$

where a and b are positive constants.

Example 1

Assuming that a bacteria population starts with 29 organisms and that L and r have values of 665 and 1.6, respectively, calculate and plot the function $x(t)$ at these time values for t: 0, 1, 2, . . . , 10.

Solution: The equation governing the growth of the bacteria is obtained by substituting $L = 665$ and $r = 1.6$ in Equation (11.10):

$$x(t + 1) - x(t) = \frac{1.6[665 - x(t)]x(t)}{665} \qquad (11.10')$$

This equation can now be written in the form of (11.12):

$$x(t + 1) = \left[2.6 - \frac{1.6}{665} x(t) \right] x(t) \qquad (11.12')$$

At the start of the process, when $t = 0$, there are 29 organisms so we set $x(0)$ in $(11.12')$ equal to 29 and calculate $x(1) = [2.6 - 2(29)/665](29)$ = 73 (approximately). We can now put $t = 1$ in Equation $(11.12')$, substitute 73 as the value of $x(1)$, and thereby calculate that $x(2) = 178$. With much patience or a calculating machine, we can produce the table of values listed in column (a) of the following table. Column (b) gives empirically observed population levels in a yeast culture that begins with 29 organisms. In Figure 11.5 the theoretical and observed population curves are plotted. As you can

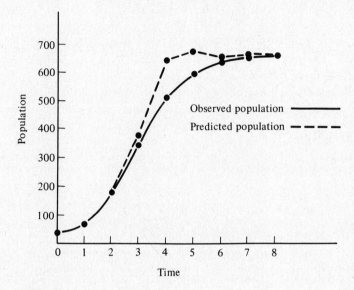

Figure 11.5

Time	Population Predicted by Model	Observed Population
0	29	29
1	73	71
2	178	175
3	386	351
4	645	513
5	676	594
6	658	641
7	669	656
8	663	662
	(a)	(b)

(Data in column (b) adapted from *The Biology of Population Growth*, by Raymond Pearl, Alfred A. Knopf, Inc., New York, N.Y., 1925.)

see, the model fits the reality well at the beginning and end but at $t = 4$ the discrepancy is about 25% of the actual value.

This model for growth has the pleasing feature that it forecasts an equilibrium for the size of the population. By an *equilibrium* we mean a situation in which the population is approximately stable in size for a long period of time. Algebraically, we have equilibrium at time t if

$$x(t + 1) - x(t) = 0$$

In a growth process obeying Equation (11.10) this means

$$\frac{r[L - x(t)]x(t)}{L} = 0$$

One solution to this equation is $x(t) = 0$, that is, if there are no members of species x alive. The other solution occurs when $x(t) = L$, the limiting value of the population size. In Example 1 the nonzero equilibrium occurs at 665. This equilibrium is immediately visually apparent from the graph of $x(t)$ shown in Figure 11.5. Of course, this limiting value of 665 is rarely actually reached and, usually, the population continues to move slowly toward 665. Nevertheless, once one gets close enough to 665, the change is so small that the population is in equilibrium for all practical purposes.

It is interesting to note that difference equations of the form given in (11.12) can serve as a model for other phenomena with no apparent connection with the growth of populations. For example, it has recently been discovered that difference equation (11.12) describes the process by which a technological substitute like synthetic fiber replaces a natural product like wool and cotton fiber. If $x(t)$ denotes the fraction of the total market for fiber that has been captured by the synthetic substitute at time t, then Equation (11.12), with appropriate numbers filled in for a and b, describes the replacement of natural fiber by synthetic fiber.

In the case of population growth we deduced Equation (11.12) on logical grounds. By contrast, it is harder to see why Equation (11.12) should be a good model for technological substitution. The proof of a model is partly in how well it fits the data, however, and Equation (11.12) does fit the data for many replacement processes that have actually taken place: synthetic rubber replacing natural rubber, margarine replacing butter, detergent replacing soap, and other examples.

Now let us return to the dynamics of population growth for a biological organism. We are going to complicate things by imagining that we have not just one species using up environmental resources but two competing species. Now possibilities arise that would not have been conceivable in our earlier situation. There is, for instance, the possibility that one species might be wiped out by the other. To study this situation, let us call our two species X and Y, and let $x(t)$ and $y(t)$ be the functions that give the number of individuals of each of these species at time t. To describe the growth of X we assume again that there is an intrinsic rate of natural increase which must be modified by a resistance factor which depends on the carrying capacity of the environment for the species X and on the amount of X already present. But we further assume that the resistance factor depends on how many individuals of species Y are present. The more Y's present, the harder it is to increase the population of species X. Therefore, we shall take as our difference equation an expression of this form [note the analogy with Equation (11.12)]:

$$x(t + 1) - x(t) = [a - bx(t) - cy(t)]x(t) \qquad (11.13)$$

where a, b, and c are positive constants. Specifically, a is the intrinsic rate of natural increase; b measures the resistance provided by species X to its own further increase; and c measures the resistance provided by species Y to the further increase of species X. Note that when $x(t)$ and $y(t)$ are close enough to 0, the right-hand side is approximately $ax(t)$ and we have simple exponential growth. But as $x(t)$ and $y(t)$ grow larger, the growth rate $a - bx(t) - cy(t)$ becomes small and perhaps even negative. Using analogous assumptions about Y, we obtain a similar equation as our model for species Y:

$$y(t + 1) - y(t) = [k - mx(t) - ny(t)]y(t) \qquad (11.14)$$

where k, m, and n are positive constants.

What does this model tell us about the possibility of the two species reaching an equilibrium? By an equilibrium for two species we mean a situation in which both populations are stable in size, that is, they neither increase nor decrease. In symbols, we have reached an equilibrium at time t provided that

$$x(t + 1) - x(t) = 0$$

and

$$y(t + 1) - y(t) = 0$$

Intuitively, these equations say there has been no change in the number of organisms in each species during one time period. Making these substitutions in (11.13) and (11.14) we have this requirement for equilibrium:

$$[a - bx(t) - cy(t)]x(t) = 0$$

$$[k - mx(t) - ny(t)]y(t) = 0$$

Examining the first equation, we see that either $x(t) = 0$ or $a - bx(t) - cy(t) = 0$. If $x(t) = 0$, species X has been wiped out. This is an equilibrium for species X. Furthermore, the absence of species X allows the other species to seek an equilibrium according to the model we discussed at the outset of this section. We shall reject this possibility, however, and seek an equilibrium achieved under less drastic circumstances. Similarly, in the second equation, we reject the possibility that $y(t) = 0$. This leaves the following conditions for equlibrium:

$$bx(t) + cy(t) = a$$

$$mx(t) + ny(t) = k$$

Now a, b, c, k, m, and n are constants and $x(t)$ and $y(t)$ are variables so what we have here is a pair of linear equations that can be graphed as a pair of straight lines. Perhaps this will seem more familiar if we write, as we often shall, x and y instead of $x(t)$ and $y(t)$. It is also a routine matter to solve these equations simultaneously, as we have shown in Chapter 6.

Example 2

Find the equilibrium values for $x(t)$ and $y(t)$ if these functions satisfy the difference equations

$$x(t + 1) - x(t) = [6.80 - 0.02x(t) - 0.05y(t)]x(t)$$

$$y(t + 1) - y(t) = [8.00 - 0.04x(t) - 0.03y(t)]y(t)$$

Solution: We need to solve the following system simultaneously:

$$[6.80 - 0.02x(t) - 0.05y(t)]x(t) = 0$$

$$[8.00 - 0.04x(t) - 0.03y(t)]y(t) = 0$$

One solution is that $x(t)$ and $y(t)$ are both zero. To find another solution set, take $x(t) = 0$ but assume that $y(t) \neq 0$. Now cancel $y(t)$ and substitute 0 for $x(t)$ in the last equation, finding $8.00 = 0.03y(t)$. This gives $y(t) = 267$ (approximately). Similarly, if we assume that $y(t) = 0$ but $x(t) \neq 0$, we obtain $x(t) = 340$. To find the fourth and final solution, which is also the most interesting one, we assume that both $x(t)$ and $y(t) \neq 0$. After cancellation of $x(t)$ and $y(t)$ we are left with the following system to solve:

$$6.80 = 0.02x + 0.05y$$

$$8.00 = 0.04x + 0.03y$$

The solution is $x = 140$ and $y = 80$. This is shown graphically in Figure 11.6, where the lines correspond to the two equations. It should be noted that it is unfortunately not possible with this analysis to determine when (i.e., at what value of t) any of these equilibrium points will be reached or

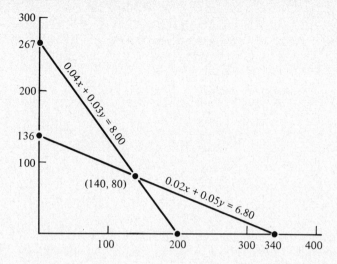

Figure 11.6

even whether any of them will be reached at all. If we had some information about the initial values of x and y and if we did a little more analysis, we could find this information too.

EXERCISES 11.3

1. Find the equilibrium points for the following sets of equations:
   (a) $x(t + 1) - x(t) = [6.2 - 0.03x(t) - 0.04y(t)]x(t)$
       $y(t + 1) - y(t) = [7.6 - 0.06x(t) - 0.02y(t)]y(t)$
   (b) $x(t + 1) - x(t) = [5.0 - 0.002x(t) - 0.0006y(t)]x(t)$
       $y(t + 1) - y(t) = [5.5 - 0.001x(t) - 0.0009y(t)]y(t)$
   (c) $x(t + 1) - x(t) = [8.04 - 0.1x(t) - 0.06y(t)]x(t)$
       $y(t + 1) - y(t) = [4.66 - 0.05x(t) - 0.04y(t)]y(t)$

2.* Suppose there are three species whose growth functions are $x(t)$, $y(t)$, and $z(t)$, which satisfy the following set of difference equations. What are the possibilities for equilibrium?

$$x(t + 1) - x(t) = [3.840 - 0.02x(t) - 0.003y(t) - 0.008z(t)]x(t)$$

$$y(t + 1) - y(t) = [6.465 - 0.0003x(t) - 0.05y(t) - 0.007z(t)]y(t)$$

$$z(t + 1) - z(t) = [5.470 - 0.004x(t) - 0.0006y(t) - 0.08z(t)]z(t)$$

3.* Formulate a difference equation model for two species, X and Y, where Y inhibits the growth of X but X doesn't inhibit or promote the growth

441

of Y and where each species inhibits its own further growth. Can you deduce from your model that an equilibrium exists (with neither species dwindling to zero) if and only if the inhibiting effect of Y on itself is greater than the inhibiting effect of Y on X.

4.* In our examples thus far we have had only positive values for the constants a, b, c, k, m, and n.

    (a) Can you explain why a, b, k, and n would never be negative?

    (b) Can you conceive of a situation in which c or m would be negative? Explain the significance of negative values for c or m or both.

11.4 stable and unstable equilibrium

Wherever the word *equilibrium* is used, it is usually relevant to distinguish between *stable* equilibrium and *unstable* equilibrium. For example, both bottles in Figure 11.7 are in equilibrium but this doesn't tell the whole story.

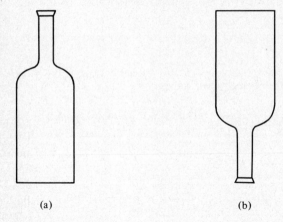

(a)   (b)

Figure 11.7

The bottle standing on its neck is in unstable equilibrium because a slight touch will upset the equilibrium; the other bottle is in stable equilibrium because if it is disturbed slightly, it will return back to its equilibrium position.

We are interested in asking a similar question about the equilibrium of a species or a pair of competing species. First, however, let us begin with the simpler one-species model in Example 1 of the last section:

$$x(t+1) - x(t) = 1.6\frac{[665 - x(t)]x(t)}{665}$$

As we have already observed, the equilibrium values of $x(t)$ occur when $x(t) = 0$ or when $x = 665$. It makes little sense to ask whether $x(t) = 0$ is stable or unstable so let's consider the other case, namely, where $x(t) = 665$. Suppose by some accidental fluctuation the population should fall below 665. Then the rate of increase, $1.6 - [1.6x(t)/665]$, would change from 0 to a positive value. Thus, the population would begin to increase, correcting the initial fluctuation.

If, on the other hand, the initial accidental fluctuation makes $x(t)$ rise above 665, then the rate of increase $1.6 - [1.6x(t)/665]$ becomes negative, and the population decreases toward its equilibrium at 665.

All in all, our species behaves like the bottle in Figure 11.7(a)—small disturbances tend to be corrected rather than magnified. It is possible to show that there is nothing special about the numbers in this particular example that makes the equilibrium stable. If 665 is replaced by any other positive number and 1.6 is replaced by any other growth rate which is not too large, we would discover that the equilibrium at L is stable.

Let us go on to the more interesting case of two competing species. As before, we start with a particular example with particular numbers, namely, Example 2 of Section 11.3.

$$x(t + 1) - x(t) = [6.80 - 0.02x(t) - 0.05y(t)]x(t)$$

$$(11.15)$$

$$y(t + 1) - y(t) = [8.00 - 0.04x(t) - 0.03y(t)]y(t)$$

Recall that the equilibrium in which both species survived was $x = 140$ and $y = 80$ because this is the point where both rates, $6.80 - 0.02x - 0.05y$ and $8.00 - 0.04x - 0.03y$, are zero. Suppose this equilibrium is in existence but at time t_0, say, there is a disturbance in this equilibrium due to some outside force (e.g., an especially dry summer) not represented in this model. To be specific, suppose x increases to 145 while y decreases to 75. What happens now? Will x and y return to their equilibrium values or will one or both species disappear? We can calculate the adjustment made by the species in the next time interval by substituting in (11.15). We obtain

$$x(t_0 + 1) - 145 = [6.80 - 0.02(145) - 0.05(75)]145$$

$$y(t_0 + 1) - 75 = [8.00 - 0.04(145) - 0.03(75)]75$$

Simplifying these expressions, we obtain

$$x(t_0 + 1) = 145 + (1.5)145$$

$$y(t_0 + 1) = 75 - (0.05)75$$

Thus X increases but Y decreases. These changes, however, are in the same direction as the changes in the original disturbance. Therefore, they will not tend to restore the original equilibrium but will instead reinforce the disturbance. The situation is analogous to the bottle standing on its neck that has been given a slight push. The natural dynamics of the situation reinforce the disturbance rather than oppose it. Figure 11.8 shows the state

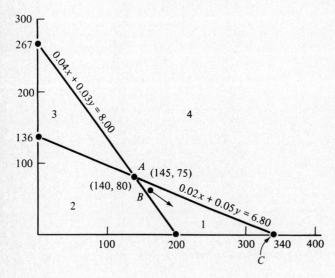

Figure 11.8

of affairs graphically. Point A represents the population levels (140, 80). Point B, (145, 75), shows the population levels after the disturbance. The arrow indicates the direction of change of the populations after the disturbance: X increases and Y decreases.

It can be shown that at any point in Sector 1, the direction of change is the same, namely, down and to the right. Thus, if the equilibrium A is disturbed into this sector, the disturbance will be continued until we reach a point on the X axis where Y has become extinct. Disturbances into the other sectors produce these results:

Sector 2

The disturbance will be opposed and in the next time interval the change will be in the general direction of the equilibrium (although possibly overshooting it).

Sector 3

The disturbance will be continued until X is wiped out.

Sector 4

The disturbance will be opposed and in the next time interval the change will be in the general direction of the equilibrium (although possibly overshooting it).

Assuming the validity of this instability analysis, we can conclude that one of the two species will disappear. The reason is that at the equilibrium A it is only a matter of time before a small disturbance appears that sends the population levels into either Sector 1 or Sector 3.

We shall now try to justify the stability analysis in a theoretical way. Our theorizing in this model will lead us to a principle well-known in ecology, called *Gause's principle*.

For theoretical purposes we replace the particular numbers of our previous discussion with general positive constants:

$$x(t + 1) - x(t) = [a - bx(t) - cy(t)]x(t)$$

$$y(t + 1) - y(t) = [k - mx(t) - ny(t)]y(t)$$

With this much generality, we don't know how to plot the lines so we shall distinguish cases concerning the relationships between the intercepts of the two lines. The x and y intercepts of $a = bx + cy$ are a/b and a/c, respectively: the x and y intercepts of $k = mx + ny$ are k/m and k/n, respectively.

Case I

$a/b > k/m$ and $a/c > k/n$.

Here the lines don't cross in the positive quadrant so there is no equilibrium without the disappearance of one species (see Figure 11.9).

Case II

$a/b < k/m$ and $a/c < k/n$.

This case (see Figure 11.10) is similar to the last. Because the lines don't cross in the positive quadrant, the only equilibrium involves the extinction of one species.

Case III

$a/b > k/m$ and $a/c < k/n$.

The lines do intersect in the positive quadrant (see Figure 11.11) so

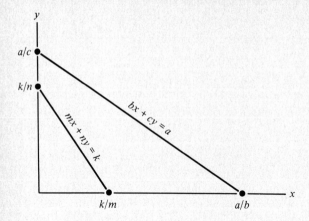

Figure 11.9 : Case I

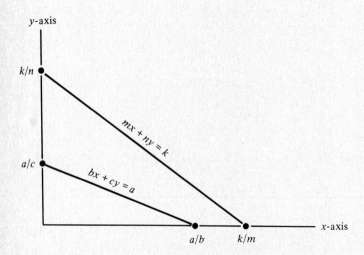

Figure 11.10 : Case II

there is an equilibrium point where both species survive. To determine whether a disturbance will be opposed or continued, we analyze individually the two lines and the half-spaces they determine. All points under the line labeled X satisfy the inequality $a - bx - cy > 0$, while those above line X satisfy the inequality $a - bx - cy < 0$. Similarly, underneath the Y line we have $k - mx - ny > 0$, while above that line the opposite inequality, $k - mx - ny < 0$, holds. Sector 1 consists of points under the X line but above the Y line. Consequently, Sector 1 can be described as the set of points

446

where the following two inequalities hold:

$$a - bx - cy > 0$$

$$k - mx - ny < 0$$

Putting this information into (11.5) we can say that Sector 1 consists precisely of those points where

$$x(t + 1) - x(t) > 0$$

and

$$y(t + 1) - y(t) < 0$$

These inequalities show that species X increases while species Y decreases in Sector 1. The arrow in Sector 1 in Figure 11.11 points in the direction of increasing X and decreasing Y. As in our numerical problem, this indicates an unstable equilibrium. If we took the trouble to work out the state of affairs in the other sectors, we would determine that the proper directions for the arrows are as shown in Figure 11.11.

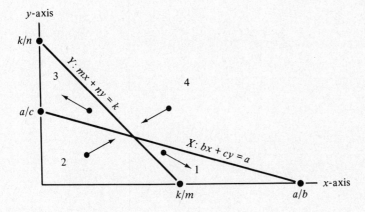

Figure 11.11 : Case III

Case IV

$a/b < k/m$ and $a/c > k/n$.

Again, the lines cross at an equilibrium point in the positive quadrant (see Figure 11.12). Our sector-by-sector analysis, following the pattern laid down in the last case, gives the following results:

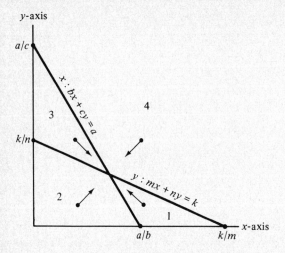

Figure 11.12 : Case IV

Sector 1

$$\left. \begin{array}{l} a - bx - cy < 0 \\ k - mx - ny > 0 \end{array} \right\} \text{ therefore } \left\{ \begin{array}{l} x(t+1) - x(t) < 0 \\ y(t+1) - y(t) > 0 \end{array} \right.$$

Since x decreases and y increases in this sector, the arrow points up and to the left—in other words, back in the direction of the equilibrium. Consequently, a disturbance into this sector will be opposed and in the next time interval the change will be in the general direction of the equilibrium (although possibly overshooting it).

Sector 2

$$\left. \begin{array}{l} a - bx - cy > 0 \\ k - mx - ny > 0 \end{array} \right\} \text{ therefore } \left\{ \begin{array}{l} x(t+1) - x(t) > 0 \\ y(t+1) - y(t) > 0 \end{array} \right.$$

Since x increases and y increases, the arrow points up and to the right. Consequently, a disturbance into this sector will be opposed and in the next time interval the change will be in the general direction of the equilibrium (although possibly overshooting it).

Sector 3

$$\left. \begin{array}{l} a - bx - cy > 0 \\ k - mx - ny < 0 \end{array} \right\} \text{ therefore } \left\{ \begin{array}{l} x(t+1) - x(t) > 0 \\ y(t+1) - y(t) < 0 \end{array} \right.$$

448

Since x increases while y decreases, the arrow points down and to the right. Consequently, a disturbance into this sector will be opposed and in the next time interval the change will be in the general direction of the equilibrium (although possibly overshooting it).

Sector 4

$$\left. \begin{array}{l} a - bx - cy < 0 \\[2mm] k - mx - ny < 0 \end{array} \right\} \text{therefore} \left\{ \begin{array}{l} x(t+1) - x(t) < 0 \\[2mm] y(t+1) - y(t) < 0 \end{array} \right.$$

Since x and y both decrease, the arrow points down and to the left. Consequently, a disturbance into this sector will be opposed and in the next time interval the change will be in the general direction of the equilibrium (although possibly overshooting it).

We have now concluded that the condition that must hold for stable equilibrium is $a/b < k/m$ and $a/c > k/n$. We shall now try to provide a biological interpretation for this. The numbers a and k are environmental carrying capacities for X and Y, respectively; for simplicity we shall assume that they are about equal. This simplifies our inequalities to $c < n$ and $b > m$. These constants have the following significances:

(1) b measures the resistance provided by species X to the increase of species X

(2) m measures the resistance provided by species X to the increase of species Y

(3) c measures the resistance provided by species Y to the increase of species X

(4) n measures the resistance provided by species Y to the increase of species Y

Therefore, $c < n$ and $b > m$ means that each species has a greater inhibiting effect on itself than it has on the other species. This would be the case, for example, if X and Y have overlapping but not identical food requirements. On the other hand, if X and Y have the same food requirements, it is less likely that $c < n$ and $b > m$; therefore the equilibrium is unlikely to be stable. This analysis gives some theoretical justification for an observed phenomenon known as *Gause's principle of competitive exclusion*: no two species with identical *niche* requirements (food, territory, etc.) can coexist.

EXERCISES 11.4

1. Find the equilibrium for the following system.

$$x(t + 1) - x(t) = [0.3 - 0.002x(t) - 0.001y(t)]x(t)$$

$$y(t + 1) - y(t) = [0.25 - 0.0012x(t) - 0.0013y(t)]y(t)$$

Consider disturbances from the equilibrium to each of the following points and determine whether the disturbance will be continued or opposed. Indicate the state of affairs with a diagram and arrows:
(a)   disturbance to (95, 95),
(b)   disturbance to (105, 105),
(c)   disturbance to (95, 105),
(d)   disturbance to (105, 95).

2.* In case IV (where $a/b < k/m$ and $a/c > k/n$) we neglected to analyze the state of affairs when there is a disturbance of the equilibrium point to a point that lies on one of the boundary lines between sectors. Do this analysis.

3.* In addition to the four cases analyzed in the text, one really should consider the possibility that one or more pairs of intercepts are exactly equal. In each of those possible cases, determine whether there is an equilibrium. What can you say about whether disturbances are continued or opposed?

4.* Explain the following remark that appears in the text: "It makes little sense to ask whether $x(t) = 0$ is stable or unstable. . . ."

SUGGESTED READING

GOLDBERG, S., *Introduction to Difference Equations*, John Wiley & Sons, Inc., New York, N.Y., 1958. The mathematical theory of difference equations is developed. In addition, many examples from economics, psychology, and sociology where difference equations can be used as models are given.

SMITH, J. M., *Mathematical Ideas in Biology*, Cambridge University Press, London, 1968. Chapter 2 uses difference equations to model the relationship between predator and prey.

12
appendices

12.1 set theory

DEFINITION 1

A *set* is any collection of objects.

Example 1

Here are some examples of sets:

(1) The set of integer numbers.
(2) The set of real numbers greater than or equal to 4.
(3) The set of English vowels.
(4) The set of primary colors.

The following notation is often useful in defining a set:

$$A = \{x \,|\, x \text{ has property } P\}$$

This is read "*A* is the set of *x* such that *x* has property *P*." The braces stand for "set" and the vertical line stands for "such that." In general, braces around a collection of symbols will denote the set consisting of those symbols.

DEFINITION 2

If a set *A* and a set *B* are such that every element in set *A* is also in set *B*, then *A* is called a *subset* of *B*, and this is denoted $A \subset B$.

Example 2

(1) $\{1, 3\} \subset \{1, 4, 7, 3\}$
(2)   The set of integers is a subset of the set of real numbers.
(3) $\{x \,|\, x \geq 5\} \subset \{x \,|\, x \geq 0\}$

Often it is convenient to generate new sets from old sets.

DEFINITION 3

Let A and B be subsets of set S. The set A', read *the complement of A*, is the set of all those elements in S that are not in A. The set $A \cup B$, read *A union B*, is the set of all those elements in A or in B or in both. The set $A \cap B$, read *A intersect B*, is the set of elements in both A and B.

Example 3

Suppose that $S = \{1, 2, 3, 4\}$ and $A = \{1, 2\}$, $B = \{1, 4\}$. Then

$$A \cup B = \{1, 2, 4\}$$

$$A \cap B = \{1\}$$

$$A' = \{3, 4\}$$

$$B' = \{2, 3\}$$

$$(A \cup B) \cap B' = \{1, 2, 4\} \cap \{2, 3\} = \{2\}$$

Example 4

Let S be the set of cards in an ordinary deck. Let A be the set of spades and B be the set of fours. Then $A \cap B$ is the four of spades and $A \cup B$ the spades together with the fours of clubs, diamonds, and hearts.

In some situations it is useful to have a notation to represent the members of a set. If we wish to state that element a is a member of set A, we shall write $a \in A$.

Example 5

If X is the set of real numbers, we can write $0 \in X$, $-1 \in X$, $\pi \in X$, and $\sqrt{2} \in X$.

12.2   table of square roots

Table 12.2   Square roots[1]

n	$\sqrt{n}$	$\sqrt{10n}$	n	$\sqrt{n}$	$\sqrt{10n}$
1.0	1.00000	3.16228	3.5	1.87083	5.91608
1.1	1.04881	3.31662	3.6	1.89737	6.00000
1.2	1.09545	3.46410	3.7	1.92354	6.08276
1.3	1.14018	3.60555	3.8	1.94936	6.16441
1.4	1.18322	3.74166	3.9	1.97484	6.24500
1.5	1.22474	3.87298	4.0	2.00000	6.32456
1.6	1.26491	4.00000	4.1	2.02485	6.40312
1.7	1.30384	4.12311	4.2	2.04939	6.48074
1.8	1.34164	4.24264	4.3	2.07364	6.55744
1.9	1.37840	4.35890	4.4	2.09762	6.63325
2.0	1.41421	4.47214	4.5	2.12132	6.70820
2.1	1.44914	4.58258	4.6	2.14476	6.78233
2.2	1.48324	4.69042	4.7	2.16795	6.85565
2.3	1.51658	4.79583	4.8	2.19089	6.92820
2.4	1.54919	4.89898	4.9	2.21359	7.00000
2.5	1.58114	5.00000	5.0	2.23607	7.07107
2.6	1.61245	5.09902	5.1	2.25832	7.14143
2.7	1.64317	5.19615	5.2	2.28035	7.21110
2.8	1.67332	5.29150	5.3	2.30217	7.28011
2.9	1.70294	5.38516	5.4	2.32379	7.34847
3.0	1.73205	5.47723	5.5	2.34521	7.41620
3.1	1.76068	5.56776	5.6	2.36643	7.48331
3.2	1.78885	5.65685	5.7	2.38747	7.54983
3.3	1.81659	5.74456	5.8	2.40832	7.61577
3.4	1.84391	5.83095	5.9	2.42899	7.68115

[1]From John Freund, *Statistics: A First Course*, Englewood Cliffs, New Jersey: Prentice-Hall, Inc., 1970, *pp. 317–318*.

Table 12.2 Square roots (cont.)

n	$\sqrt{n}$	$\sqrt{10n}$	n	$\sqrt{n}$	$\sqrt{10n}$
6.0	2.44949	7.74597	**8.0**	2.82843	8.94427
6.1	2.46982	7.81025	**8.1**	2.84605	9.00000
6.2	2.48998	7.87401	**8.2**	2.86356	9.05539
6.3	2.50998	7.93725	**8.3**	2.88097	9.11043
6.4	2.52982	8.00000	**8.4**	2.89828	9.16515
6.5	2.54951	8.06226	**8.5**	2.91548	9.21954
6.6	2.56905	8.12404	**8.6**	2.93258	9.27362
6.7	2.58844	8.18535	**8.7**	2.94958	9.32738
6.8	2.60768	8.24621	**8.8**	2.96648	9.38083
6.9	2.62679	8.30662	**8.9**	2.98329	9.43398
7.0	2.64575	8.36660	**9.0**	3.00000	9.48683
7.1	2.66458	8.42615	**9.1**	3.01662	9.53939
7.2	2.68328	8.48528	**9.2**	3.03315	9.59166
7.3	2.70185	8.54400	**9.3**	3.04959	9.64365
7.4	2.72029	8.60233	**9.4**	3.06594	9.69536
7.5	2.73861	8.66025	**9.5**	3.08221	9.74679
7.6	2.75681	8.71780	**9.6**	3.09839	9.79796
7.7	2.77489	8.77496	**9.7**	3.11448	9.84886
7.8	2.79285	8.83176	**9.8**	3.13050	9.89949
7.9	2.81069	8.88819	**9.9**	3.14643	9.94987

12.3 binomial probability model table

Table 12.3   Binomial model probabilities[2]

N	r	0.05	0.1	0.2	0.3	0.4	0.5	0.6	0.7	0.8	0.9	0.95
2	0	0.902	0.810	0.640	0.490	0.360	0.250	0.160	0.090	0.040	0.010	0.002
	1	0.095	0.180	0.320	0.420	0.480	0.500	0.480	0.420	0.320	0.180	0.095
	2	0.002	0.010	0.040	0.090	0.160	0.250	0.360	0.490	0.640	0.810	0.902
3	0	0.857	0.729	0.512	0.343	0.216	0.125	0.064	0.027	0.008	0.001	
	1	0.135	0.243	0.384	0.441	0.432	0.375	0.288	0.189	0.096	0.027	0.007
	2	0.007	0.027	0.096	0.189	0.288	0.375	0.432	0.441	0.384	0.243	0.135
	3		0.001	0.008	0.027	0.064	0.125	0.216	0.343	0.512	0.729	0.857
4	0	0.815	0.656	0.410	0.240	0.130	0.062	0.026	0.008	0.002		
	1	0.171	0.292	0.410	0.412	0.346	0.250	0.154	0.076	0.026	0.004	
	2	0.014	0.049	0.154	0.265	0.346	0.375	0.346	0.265	0.154	0.049	0.014
	3		0.004	0.026	0.076	0.154	0.250	0.346	0.412	0.410	0.292	0.171
	4			0.002	0.008	0.026	0.062	0.130	0.240	0.410	0.656	0.815
5	0	0.774	0.590	0.328	0.168	0.078	0.031	0.010	0.002			
	1	0.204	0.328	0.410	0.360	0.259	0.156	0.077	0.028	0.006		
	2	0.021	0.073	0.205	0.309	0.346	0.312	0.230	0.132	0.051	0.008	0.001
	3	0.001	0.008	0.051	0.132	0.230	0.312	0.346	0.309	0.205	0.073	0.021
	4			0.006	0.028	0.077	0.156	0.259	0.360	0.410	0.328	0.204
	5			0.002	0.010	0.031	0.078	0.168	0.328	0.590	0.774	
6	0	0.735	0.531	0.262	0.118	0.047	0.016	0.004	0.001			
	1	0.232	0.354	0.393	0.303	0.187	0.094	0.037	0.010	0.002		
	2	0.031	0.098	0.246	0.324	0.311	0.234	0.138	0.060	0.015	0.001	
	3	0.002	0.015	0.082	0.185	0.276	0.312	0.276	0.185	0.082	0.015	0.002
	4		0.001	0.015	0.060	0.138	0.234	0.311	0.324	0.246	0.098	0.031
	5			0.002	0.010	0.037	0.094	0.187	0.303	0.393	0.354	0.232
	6				0.001	0.004	0.016	0.047	0.118	0.262	0.531	0.735
7	0	0.698	0.478	0.210	0.082	0.028	0.008	0.002				
	1	0.257	0.372	0.367	0.247	0.131	0.055	0.017	0.004			
	2	0.041	0.124	0.275	0.318	0.261	0.164	0.077	0.025	0.004		
	3	0.004	0.023	0.115	0.227	0.290	0.273	0.194	0.097	0.029	0.003	
	4		0.003	0.029	0.097	0.194	0.273	0.290	0.227	0.115	0.023	0.004

[2]From John Freund, *Statistics: A First Course*, Englewood Cliffs, New Jersey: Prentice-Hall, Inc., 1970, pp 319–322.

Table 12.3 Binomial model probabilities (cont.)

N	r	0.05	0.1	0.2	0.3	0.4	0.5	0.6	0.7	0.8	0.9	0.95
7	5			0.004	0.025	0.077	0.164	0.261	0.318	0.275	0.124	0.041
	6				0.004	0.017	0.055	0.131	0.247	0.367	0.372	0.257
	7					0.002	0.008	0.028	0.082	0.210	0.478	0.698
8	0	0.663	0.430	0.168	0.058	0.017	0.004	0.001				
	1	0.279	0.383	0.336	0.198	0.090	0.031	0.008	0.001			
	2	0.051	0.149	0.294	0.296	0.209	0.109	0.041	0.010	0.001		
	3	0.005	0.033	0.147	0.254	0.279	0.219	0.124	0.047	0.009		
	4		0.005	0.046	0.136	0.232	0.273	0.232	0.136	0.046	0.005	
	5			0.009	0.047	0.124	0.219	0.279	0.254	0.147	0.033	0.005
	6			0.001	0.010	0.041	0.109	0.209	0.296	0.294	0.149	0.051
	7				0.001	0.008	0.031	0.090	0.198	0.336	0.383	0.279
	8					0.001	0.004	0.017	0.058	0.168	0.430	0.663
9	0	0.630	0.387	0.134	0.040	0.010	0.002					
	1	0.299	0.387	0.302	0.156	0.060	0.018	0.004				
	2	0.063	0.172	0.302	0.267	0.161	0.070	0.021	0.004			
	3	0.008	0.045	0.176	0.267	0.251	0.164	0.074	0.021	0.003		
	4	0.001	0.007	0.066	0.172	0.251	0.246	0.167	0.074	0.017	0.001	
	5		0.001	0.017	0.074	0.167	0.246	0.251	0.172	0.066	0.007	0.001
	6			0.003	0.021	0.074	0.164	0.251	0.267	0.176	0.045	0.008
	7				0.004	0.021	0.070	0.161	0.267	0.302	0.172	0.063
	8					0.004	0.018	0.060	0.156	0.302	0.387	0.299
	9						0.002	0.010	0.040	0.134	0.387	0.630
10	0	0.599	0.349	0.107	0.028	0.006	0.001					
	1	0.315	0.387	0.268	0.121	0.040	0.010	0.002				
	2	0.075	0.194	0.302	0.233	0.121	0.044	0.011	0.001			
	3	0.010	0.057	0.201	0.267	0.215	0.117	0.042	0.009	0.001		
	4	0.001	0.011	0.088	0.200	0.251	0.205	0.111	0.037	0.006		
	5		0.001	0.026	0.103	0.201	0.246	0.201	0.103	0.026	0.001	
	6			0.006	0.037	0.111	0.205	0.251	0.200	0.088	0.011	0.001
	7			0.001	0.009	0.042	0.117	0.215	0.267	0.201	0.057	0.010
	8				0.001	0.011	0.044	0.121	0.233	0.302	0.194	0.075
	9					0.002	0.010	0.040	0.121	0.268	0.387	0.315
	10						0.001	0.006	0.028	0.107	0.349	0.599

Table 12.3   Binomial model probabilities (cont.)

N	r	0.05	0.1	0.2	0.3	0.4	0.5	0.6	0.7	0.8	0.9	0.95
11	0	0.569	0.314	0.086	0.020	0.004						
	1	0.329	0.384	0.236	0.093	0.027	0.005	0.001				
	2	0.087	0.213	0.295	0.200	0.089	0.027	0.005	0.001			
	3	0.014	0.071	0.221	0.257	0.177	0.081	0.023	0.004			
	4	0.001	0.016	0.111	0.220	0.236	0.161	0.070	0.017	0.002		
	5		0.002	0.039	0.132	0.221	0.226	0.147	0.057	0.010		
	6			0.010	0.057	0.147	0.226	0.221	0.132	0.039	0.002	
	7			0.002	0.017	0.070	0.161	0.236	0.220	0.111	0.016	0.001
	8				0.004	0.023	0.081	0.177	0.257	0.221	0.071	0.014
	9				0.001	0.005	0.027	0.089	0.200	0.295	0.213	0.087
	10					0.001	0.005	0.027	0.093	0.236	0.384	0.329
	11							0.004	0.020	0.086	0.314	0.569
12	0	0.540	0.282	0.069	0.014	0.002						
	1	0.341	0.377	0.206	0.071	0.017	0.003					
	2	0.099	0.230	0.283	0.168	0.064	0.016	0.002				
	3	0.017	0.085	0.236	0.240	0.142	0.054	0.012	0.001			
	4	0.002	0.021	0.133	0.231	0.213	0.121	0.042	0.008	0.001		
	5		0.004	0.053	0.158	0.227	0.193	0.101	0.029	0.003		
	6			0.016	0.079	0.177	0.226	0.177	0.079	0.016		
	7			0.003	0.029	0.101	0.193	0.227	0.158	0.053	0.004	
	8			0.001	0.008	0.042	0.121	0.213	0.231	0.133	0.021	0.002
	9				0.001	0.012	0.054	0.142	0.240	0.236	0.085	0.017
	10					0.002	0.016	0.064	0.168	0.283	0.230	0.099
	11						0.003	0.017	0.071	0.206	0.377	0.341
	12							0.002	0.014	0.069	0.282	0.540
13	0	0.513	0.254	0.055	0.010	0.001						
	1	0.351	0.367	0.179	0.054	0.011	0.002					
	2	0.111	0.245	0.268	0.139	0.045	0.010	0.001				
	3	0.021	0.100	0.246	0.218	0.111	0.035	0.006	0.001			
	4	0.003	0.028	0.154	0.234	0.184	0.087	0.024	0.003			
	5		0.006	0.069	0.180	0.221	0.157	0.066	0.014	0.001		
	6		0.001	0.023	0.103	0.197	0.209	0.131	0.044	0.006		
	7			0.006	0.044	0.131	0.209	0.197	0.103	0.023	0.001	

Table 12.3   Binomial model probabilities (cont.)

						p						
N	*r*	0.05	0.1	0.2	0.3	0.4	0.5	0.6	0.7	0.8	0.9	0.95
13	8			0.001	0.014	0.066	0.157	0.221	0.180	0.069	0.006	
	9				0.003	0.024	0.087	0.184	0.234	0.154	0.028	0.003
	10				0.001	0.006	0.035	0.111	0.218	0.246	0.100	0.021
	11					0.001	0.010	0.045	0.139	0.268	0.245	0.111
	12						0.002	0.011	0.054	0.179	0.367	0.351
	13							0.001	0.010	0.055	0.254	0.513
14	0	0.488	0.229	0.044	0.007	0.001						
	1	0.359	0.356	0.154	0.041	0.007	0.001					
	2	0.123	0.257	0.250	0.113	0.032	0.006	0.001				
	3	0.026	0.114	0.250	0.194	0.085	0.022	0.003				
	4	0.004	0.035	0.172	0.229	0.155	0.061	0.014	0.001			
	5		0.008	0.086	0.196	0.207	0.122	0.041	0.007			
	6		0.001	0.032	0.126	0.207	0.183	0.092	0.023	0.002		
	7			0.009	0.062	0.157	0.209	0.157	0.062	0.009		
	8			0.002	0.023	0.092	0.183	0.207	0.126	0.032	0.001	
	9				0.007	0.041	0.122	0.207	0.196	0.086	0.008	
	10				0.001	0.014	0.061	0.155	0.229	0.172	0.035	0.004
	11					0.003	0.022	0.085	0.194	0.250	0.114	0.026
	12					0.001	0.006	0.032	0.113	0.250	0.257	0.123
	13						0.001	0.007	0.041	0.154	0.356	0.359
	14							0.001	0.007	0.044	0.229	0.488
15	0	0.463	0.206	0.035	0.005							
	1	0.366	0.343	0.132	0.031	0.005						
	2	0.135	0.267	0.231	0.092	0.022	0.003					
	3	0.031	0.129	0.250	0.170	0.063	0.014	0.002				
	4	0.005	0.043	0.188	0.219	0.127	0.042	0.007	0.001			
	5	0.001	0.010	0.103	0.206	0.186	0.092	0.024	0.003			
	6		0.002	0.043	0.147	0.207	0.153	0.061	0.012	0.001		
	7			0.014	0.081	0.177	0.196	0.118	0.035	0.003		
	8			0.003	0.035	0.118	0.196	0.177	0.081	0.014		
	9			0.001	0.012	0.061	0.153	0.207	0.147	0.043	0.002	
	10				0.003	0.024	0.092	0.186	0.206	0.103	0.010	0.001
	11				0.001	0.007	0.042	0.127	0.219	0.188	0.043	0.005
	12					0.002	0.014	0.063	0.170	0.250	0.129	0.031
	13						0.003	0.022	0.092	0.231	0.267	0.135
	14							0.005	0.031	0.132	0.343	0.366
	15								0.005	0.035	0.206	0.463

12.4 table of random numbers

On the following pages the digits 0, 1, 2, . . . , 9 are listed in a random order, 50 rows and 35 columns per page. These digits and their arrangement were generated by computer using a method which is designed to produce the same effect as if we had spun a perfectly fair 10 digit spinner (see Figure 12.1) many times and recorded the outcomes.

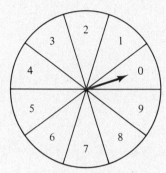

Figure 12.1 : Spinner

If one wishes to find a random sequence of one-digit numbers, this may be done by choosing at random a starting position in the table and reading successive digits from left to right. For example, suppose we begin with the entry 6 which occurs in the eighth column and third row. Reading from left to right produces the sequence

$$65167533681231773095862$$

Perhaps you have noticed that this sequence contains many 6's but not a single 4. Actually, if one chooses a long enough list, each number should occur about equally often. Sometimes one needs a fairly long list for this property of the random number table to manifest itself.

One can also find a random sequence of two-digit numbers by choosing a starting position at random and reading successive two-digit numbers from left to right. This technique is useful for generating a random alternation of two strategies, I_1 and I_2 (see Section 9.4) where each strategy must occur with a specified probability. For example suppose we want I_1 to occur $\frac{1}{4}$ of the time and I_2 to occur $\frac{3}{4}$ of the time. Begin at a randomly chosen position in the table, say the second row and fifth column, and begin reading two-digit numbers from left to right. Under each two-digit number on the list, write a I_1 or I_2 according to this rule: under any of the 25 numbers from 00 to 24 write a I_1 ; under any of the 75 numbers from 25 to 99 write a I_2. This gives :

26, 53, 01, 89, 51, 25, 06, 88, 53, 53, 65, 53, 23, 75, 73, 42, 09

I_2 I_2 I_1 I_2 I_2 I_2 I_1 I_2 I_2 I_2 I_2 I_2 I_1 I_2 I_2 I_2 I_1

Table 12.4   Random numbers[3]

04433	80674	24520	18222	10610	05794	37515
60298	47829	72648	37414	75755	04717	29899
67884	59651	67533	68123	17730	95862	08034
89512	32155	51906	61662	64130	16688	37275
32653	01895	12506	88535	36553	23757	34209
95913	15405	13772	76638	48423	25018	99041
55864	21694	13122	44115	01601	50541	00147
35334	49810	91601	40617	72876	33967	73830
57729	32196	76487	11622	96297	24160	09903
86648	13697	63677	70119	94739	25875	38829
30574	47609	07967	32422	76791	39725	53711
81307	43694	83580	79974	45929	85113	72268
02410	54905	79007	54939	21410	86980	91772
18969	75274	52233	62319	08598	09066	95288
87863	82384	66860	62297	80198	19347	73234
68397	71708	15438	62311	72844	60203	46412
28529	54447	58729	10854	99058	18260	38765
44285	06372	15867	70418	57012	72122	36634
86299	83430	33571	23309	57040	29285	67870
84842	68668	90894	61658	15001	94055	36308
56970	83609	52098	04184	54967	72938	56834
83125	71257	60490	44369	66130	72936	69848
55503	52423	02464	26141	68779	66388	75242
47019	76273	33203	29608	54553	25971	69573
84828	32592	79526	29554	84580	37859	28504
68921	08141	79227	05748	51276	57143	31926
36458	96045	30424	98420	72925	40729	22337
95752	59445	36847	87729	81679	59126	59437
26768	47323	58454	56958	20575	76746	49878
42613	37056	43636	58085	06766	60227	96414
95457	30566	65482	25596	02678	54592	63607
95276	17894	63564	95958	39750	64379	46059
66954	52324	64776	92345	95110	59448	77249
17457	18481	14113	62462	02798	54977	48349
03704	36872	83214	59337	01695	60666	97410
21538	86497	33210	60337	27976	70661	08250
57178	67619	98310	70348	11317	71623	55510
31048	97558	94953	55866	96283	46620	52087
69799	55380	16498	80733	96422	58078	99643
90595	61867	59231	17772	67831	33317	00520
33570	04981	98939	78784	09977	29398	93896
15340	93460	57477	13898	48431	72936	78160
64079	42483	36512	56186	99098	48850	72527
63491	05546	67118	62063	74958	20946	28147
92003	63868	41034	28260	79708	00770	88643
52360	46658	66511	04172	73085	11795	52594
74622	12142	68355	65635	21828	39539	18988
04157	50079	61343	64315	70836	82857	35335
86003	60070	66241	32836	27573	11479	94114
41268	80187	20351	09636	84668	42486	71303

[3]From John Freund, *Modern Elementary Statitics*, Englewood Cliffs, New Jersey: Prentice-Hall, Inc., 1967, *pp. 393–396.*

Table 12.4   Random numbers (cont.)

48611	62866	33963	14045	79451	04934	45576
78812	03509	78673	73181	29973	18664	04555
19472	63971	37271	31445	49019	49405	46925
51266	11569	08697	91120	64156	40365	74297
55806	96275	26130	47949	14877	69594	83041
77527	81360	18180	97421	55541	90275	18213
77680	58788	33016	61173	93049	04694	43534
15404	96554	88265	34537	38526	67924	40474
14045	22917	60718	66487	46346	30949	03173
68376	43918	77653	04127	69930	43283	35766
93385	13421	67957	20384	58731	53396	59723
09858	52104	32014	53115	03727	98624	84616
93307	34116	49516	42148	57740	31198	70336
04794	01534	92058	03157	91758	80611	45357
86265	49096	97021	92582	61422	75890	86442
65943	79232	45702	67055	39024	57383	44424
90038	94209	04055	27393	61517	23002	96560
97283	95943	78363	36498	40662	94188	18202
21913	72958	75637	99936	58715	07943	23748
41161	37341	81838	19389	80336	46346	91895
23777	98392	31417	98547	92058	02277	50315
59973	08144	61070	73094	27059	69181	55623
82690	74099	77885	23813	10054	11900	44653
83854	24715	48866	65745	31131	47636	45137
61980	34997	41825	11623	07320	15003	56774
99915	45821	97702	87125	44488	77613	56823
48293	86847	43186	42951	37804	85129	28993
33225	31280	41232	34750	91097	60752	69783
06846	32828	24425	30249	78801	26977	92074
32671	45587	79620	84831	38156	74211	82752
82096	21913	75544	55228	89796	05694	91552
51666	10433	10945	55306	78562	89630	41230
54044	67942	24145	42294	27427	84875	37022
66738	60184	75679	38120	17640	36242	99357
55064	17427	89180	74018	44865	53197	74810
69599	60264	84549	78007	88450	06488	72274
64756	87759	92354	78694	63638	80939	98644
80817	74533	68407	55862	32476	19326	95558
39847	96884	84657	33697	39578	90197	80532
90401	41700	95510	61166	33757	23279	85523
78227	90110	81378	96659	37008	04050	04228
87240	52716	87697	79433	16336	52862	69149
08486	10951	26832	39763	02485	71688	90936
39338	32169	03713	93510	61244	73774	01245
21188	01850	69689	49426	49128	14660	14143
13287	82531	04388	64693	11934	35051	68576
53609	04001	19648	14053	49623	10840	31915
87900	36194	31567	53506	34304	39910	79630
81641	00496	36058	75899	46620	70024	88753
19512	50277	71508	20116	79520	06269	74173

Table 12.4   Random numbers (cont.)

24418	23508	91507	76455	54941	72711	39406
57404	73678	08272	62941	02349	71389	45605
77644	98489	86268	73652	98210	44546	27174
68366	65614	01443	07607	11826	91326	29664
64472	72294	95432	53555	96810	17100	35066
88205	37913	98633	81009	81060	33449	68055
98455	78685	71250	10329	56135	80647	51404
48977	36794	56054	59243	57361	65304	93258
93077	72941	92779	23581	24548	56415	61927
84533	26564	91583	83411	66504	02036	02922
11338	12903	14514	27585	45068	05520	56321
23853	68500	92274	87026	99717	01542	72090
94096	74920	25822	98026	05394	61840	83089
83160	82362	00350	98536	38155	42661	02363
97425	47335	69709	01386	74319	04318	99387
83951	11954	24317	20345	18134	90062	10761
93085	35203	05740	03206	92012	42710	34650
33762	83193	58045	89880	78101	44392	53767
49665	85397	85137	30496	23469	42846	94810
37541	82627	80051	72521	35342	56119	97190
22145	85304	35348	82854	55846	18076	12415
27153	08662	61078	52433	22184	33998	87436
00301	49425	66682	25442	83668	66236	79655
43815	43272	73778	63469	50083	70696	13558
14689	86482	74157	46012	97765	27552	49617
16680	55936	82453	19532	49988	13176	94219
86938	60429	01137	86168	78257	86249	46134
33944	29219	73161	46061	30946	22210	79302
16045	67736	18608	18198	19468	76358	69203
37044	52523	25627	63107	30806	80857	84383
61471	45322	35340	35132	42163	60332	98851
47422	21296	16785	66393	39249	51463	95963
24133	39719	14484	58613	88717	29289	77360
67253	67064	10748	16006	16767	57345	42285
62382	76941	01635	35829	77516	98468	51686
98011	16503	09201	03523	87192	66483	55649
37366	24386	20654	85117	74078	64120	04643
73587	83993	54176	05221	94119	20108	78101
33583	68291	50547	96085	62180	27453	18567
02878	33223	39199	49536	56199	05993	71201
91498	41673	17195	33175	04994	09879	70337
91127	19815	30219	55591	21725	43827	78862
12997	55013	18662	81724	24305	37661	18956
96098	13651	15393	69995	14762	69734	89150
97627	17837	10472	18983	28387	99781	52977
40064	47981	31484	76603	54088	91095	00010
16239	68743	71374	55863	22672	91609	51514
58354	24913	20435	30965	17453	65623	93058
52567	65085	60220	84641	18273	49604	47418
06236	29052	91392	07551	83532	68130	56970

Table 12.4   Random numbers (cont.)

94620	27963	96478	21559	19246	88097	44926
60947	60775	73181	43264	56895	04232	59604
27499	53523	63110	57106	20865	91683	80688
01603	23156	89223	43429	95353	44662	59433
00815	01552	06392	31437	70385	45863	75971
83844	90942	74857	52419	68723	47830	63010
06626	10042	93629	37609	57215	08409	81906
56760	63348	24949	11859	29793	37457	59377
64416	29934	00755	09418	14230	62887	92683
63569	17906	38076	32135	19096	96970	75917
22693	35089	72994	04252	23791	60249	83010
43413	59744	01275	71326	91382	45114	20245
09224	78530	50566	49965	04851	18280	14039
67625	34683	03142	74733	63558	09665	22610
86874	12549	98699	54952	91579	26023	81076
54548	49505	62515	63903	13193	33905	66936
73236	66167	49728	03581	40699	10396	81827
15220	66319	13543	14071	59148	95154	72852
16151	08029	36954	03891	38313	34016	18671
43635	84249	88984	80993	55431	90793	62603
30193	42776	85611	57635	51362	79907	77364
37430	45246	11400	20986	43996	73122	88474
88312	93047	12088	86937	70794	01041	74867
98995	58159	04700	90443	13168	31553	67891
51734	20849	70198	67906	00880	82899	66065
88698	41755	56216	66852	17748	04963	54859
51865	09836	73966	65711	41699	11732	17173
40300	08852	27528	84648	79589	95295	72895
02760	28625	70476	76410	32988	10194	94917
78450	26245	91763	73117	33047	03577	62599
50252	56911	62693	73817	98693	18728	94741
07929	66728	47761	81472	44806	15592	71357
09030	39605	87507	85446	51257	89555	75520
56670	88445	85799	76200	21795	38894	58070
48140	13583	94911	13318	64741	64336	95103
36764	86132	12463	28385	94242	32063	45233
14351	71381	28133	68269	65145	28152	39087
81276	00835	63835	87174	42446	08882	27067
55524	86088	00069	59254	24654	77371	26409
78852	65889	32719	13758	23937	90740	16866
11861	69032	51915	23510	32050	52052	24004
67699	01009	07050	73324	06732	27510	33761
50064	39500	17450	18030	63124	48061	59412
93126	17700	94400	76075	08317	27324	72723
01657	92602	41043	05686	15650	29970	95877
13800	76690	75133	60456	28491	03845	11507
98135	42870	48578	29036	69876	86563	61729
08313	99293	00990	13595	77457	79969	11339
90974	83965	62732	85161	54330	22406	86253
33273	61993	88407	69399	17301	70975	99129

answers to selected exercises

Chapter 2

2.1   pp. 8–9

1. Mention of falling rock zones or other potentially hazardous areas; location of gasoline and service stations; in each of these cases, the interest to the motorist is obvious.

2. (a)   Transport network—blood system.
   (b)   Pump—heart.
   (c)   Energy storage cell—liver.
   (d)   Computer—brain.
   (e)   Electric wires—nerves.
   (f)   Camera—eye.
   (g)   Baby's bottle—female breast.

4. A globe is a better model of the earth than a flat map when one wants to estimate air distances between widely separated cities. A flat map is a better model than a globe if one wants to carry the map in one's pocket.

5. Model airplanes, model boats, toy trucks, toy cars, doll houses, train sets, toy dishes, and rockets.

7. One reason might be that mice seem willing to run through mazes an endless number of times. On the other hand a monkey might get bored after running through a maze just a few times, so the psychologist would get nothing of value out of the experiment.

2.3   pp. 14–15.

2. Removing one edge between B and C would result in having an odd number of edges (streets) connecting B and C, and therefore there would be no Euler circuit (one of the edges would have to be repeated). Removing two edges between B and C would result in having an even number of edges (streets) connecting B and C, so there would still be an Euler circuit and a shorter distance to travel.

2.4   p. 16.

3. There is no route which duplicates only one intersection. The minimum number of duplications is three.

4. We assumed that the time and distance required to traverse the intersections was negligible. Also, the length of the various streets wasn't taken into consideration. Other neglected factors include the problem of parking the truck while mail is dropped at the special box.

2.5   pp. 18–19.

2. The minimum number is two.

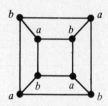

The minimum number is three.

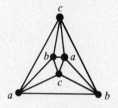

The minimum number is four.

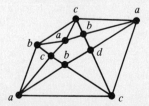

3. The minimum number is four.

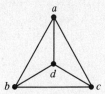

The minimum number is three.

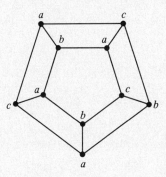

4. The minimum number is three. There can be no solution using three or four colors in which no color is used more than twice because the graph has nine vertices. Such a coloring with five colors is possible.

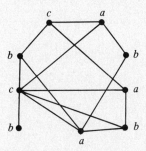

5. Choose any vertex of the graph and look at all the other vertices connected to it by an edge. The worst that can happen is that each of these vertices is connected to all of the others, in which case we must use $k + 1$ colors to color these vertices and the original vertex; otherwise we can clearly still use $\leq k + 1$ colors. Now look at any vertex which hasn't yet been colored. It can be connected to at most k of the vertices which have already been colored, so in coloring this new vertex we can still get away with having $\leq k + 1$ colors. Keep repeating this argument until all vertices are colored.

468

2.6   *pp. 24–30.*

2.   The graph is already drawn. The problem is to find the appropriate coloring for this graph. Such a coloring might be:

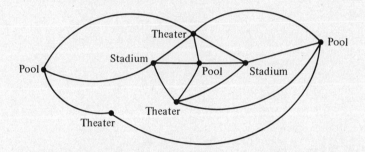

The facilities should be assigned as indicated (other solutions are also possible).

4.   Let vertices represent the students. Connect two vertices provided the students they represent are friendly. This produces the following graph:

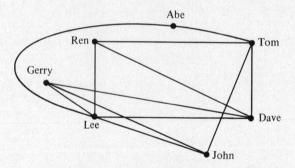

The problem can be converted to finding a Hamilton circuit on this graph. Such a Hamilton circuit might be Abe—Lee—John—Gerry—Dave—Ren—Tom—Abe, so the message can be passed in this order.

6.   Let vertices represent numbers. Connect two vertices if there is a dominoe containing the two numbers. This produces the following graph:

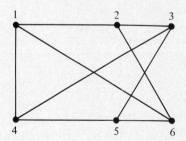

The problem now consists of finding an Euler circuit for this graph. There is no such circuit, since there are an odd number of edges emanating from each vertex, so the dominoes cannot be arranged in the required form.

8. Let vertices represent the cities. Connect two vertices provided the cities they represent are within 80 miles of each other. The problem can be converted to coloring the graph using four colors. Such a coloring might be:

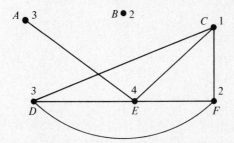

The channels could be assigned in the indicated manner.

9. Let vertices represent the species. Connect two vertices provided the species they represent are incompatible. The problem can be converted to coloring the graph with the minimum number of colors. This minimum number is two (*A* and *B*).

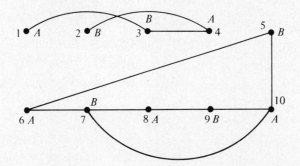

Hence the minimum number of enclosures is two.

Chapter 3

3.1   pp. 42–46.

1. (a) There is an edge connecting a vertex to itself.
   (b) There are infinitely many vertices (and infinitely many edges).
   (c) There is an edge connecting a vertex to itself.
   (d) There is an edge which doesn't culminate with a vertex.
   (e) There are two edges which don't culminate with a vertex.

2.   (a)   There are 15 edges.

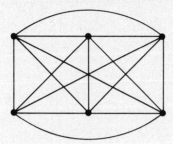

      (c) $\dfrac{n(n-1)}{2}$

3.   In the first graph, $\text{val}(x_1) = 2$, $\text{val}(x_2) = 3$, $\text{val}(x_3) = 4$, $\text{val}(x_4) = 5$, $\text{val}(x_5) = 1$, $\text{val}(x_6) = 4$, $\text{val}(x_7) = 5$, $\text{val}(x_8) = 4$. In the second graph, $\text{val}(x_1) = 2$, $\text{val}(x_2) = 4$, $\text{val}(x_3) = 5$, $\text{val}(x_4) = 3$, $\text{val}(x_5) = 4$, $\text{val}(X_6) = 4$, $\text{val}(x_7) = 4$, $\text{val}(x_8) = 1$, $\text{val}(x_9) = 4$, $\text{val}(x_{10}) = 1$.

4.   (b)   They aren't isomorphic. The first graph has a circuit of length three starting and ending at the top vertex, while the second graph has no circuits of length three.

      (d)   They aren't isomorphic. The first graph has a circuit of length three starting and ending at the top vertex, while the second graph has no circuits of length three.

      (e)   They are isomorphic.

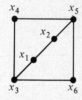 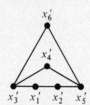

      (g)   They aren't isomorphic. The first graph has five vertices, while the second graph has four.

6.   (a)

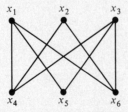

      This is not a planar graph. The fewest number of accidental crossings it can be redrawn with is one:

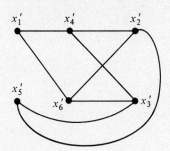

9.

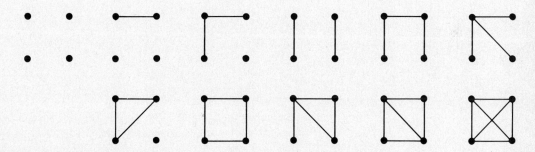

3.2 *pp. 51–52.*

1. (a) $3 \cdot 1 + 2 \cdot 2 + 1 \cdot 3 = 10$, and there exists such a graph:

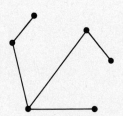

(c) $3 \cdot 1 + 2 \cdot 2 + 2 \cdot 5 = 19$, so there is no such graph.

(e) $2 \cdot 2 + 1 \cdot 6 = 10$, but there is no such graph.

(f) $1 \cdot 2 + 2 \cdot 3 = 8$, and there is such a graph:

(g) $1 \cdot 1 + 1 \cdot 2 + 1 \cdot 3 + 1 \cdot 5 = 11$, so there is no such graph.

3.   The maximum number of edges in a graph with four vertices without multiple edges is six. If we allow multiple edges, we can have an arbitrary number of edges. If a graph has n vertices and no multiple edges then each vertex has valence at most $n - 1$; since there are n vertices, we therefore have at most $n(n - 1)/2$ edges (we must divide by 2 since each edge has been counted twice).

5.   Since G has no multiple edges, x can be connected by an edge to at most all of the remaining vertices of G.

9.   (a) $7 \cdot 4 + 13 \cdot 1 = 41$, so there exists no such hydrocarbon.
      (b) $4 \cdot 4 + 7 \cdot 1 = 23$, so there exists no such hydrocarbon.

10.   (a)   4. (b) 6. (c) 8. (d) 10. If a hydrocarbon has n carbon atoms it can have at most $2n + 2$ hydrogen atoms.

11.   Six.

3.3   pp. 58–60.

2.   (a)   A digraph. (b) A digraph. (c) A graph. (d) A digraph (if A considers B his best friend, B might not necessarily consider A his best friend).

3.   There is an edge directed from team A to team B if A beats B. The digraph is:

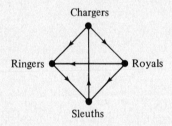

4.   (a)   inval $(x_1) = 0_1$ outval $(x_1) = 1$, inval $(x_2) = 1$, outval $(x_2) = 2$, inval $(x_3) = 1$, outval $(x_3) = 1$, inval $(x_4) = 2$, outval $(x_4) = 0$.
      (b)   inval $(x_1) = 1$, outval $(x_1) = 2$, inval $(x_2) = 1$, outval $(x_2) = 1$, inval $(x_3) = 2$, outval $(x_3) = 1$, inval $(x_4) = 1$, outval $(x_4) = 1$.
      (c)   inval $(x_1) = 2$, outval $(x_1) = 1$, inval $(x_2) = 1$, outval $(x_2) = 2$, inval $(x_3) = 1$, outval $(x_3) = 1$, inval $(x_4) = 2$, outval $(x_4) = 2$.

5.   (a) x_1, x_5.
      (b)   Yes.

3.4 *pp. 66–68.*

2.

Activity	Starting Time	Reason
Bo	0	There is only one path from B to Bo, and it has length 0.
H	0	There is only one path from B to H, and it has length 0.
G	14	There is only one path from B to C, and it has length 14.
P	15	The two paths from B to P have lengths 1 and 15.
S	20	The three paths from B to S have lengths 14, 20, and 6.
M	20	The three paths from B to M have lengths 14, 20, and 6.
E	23	The five paths from B to E have lengths 9, 7, 15, 17, and 23.

The minimum time is 23 days.

4.

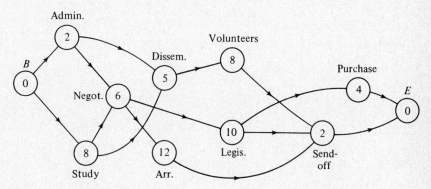

Activity	Starting Time	Reason
Admin.	0	There is one path from B to Admin., and it has length 0.
Study	0	There is one path from B to Study, and it has length 0.
Negot.	8	There are two paths from B to Negot., and they have lengths 2 and 8.
Dissem.	8	There are two paths from B to Dissem., and they have lengths 2 and 8.
Volunteers	13	There are two paths from B to Volunteers, and they have lengths 7 and 13.
Arr.	14	There are two paths from B to Arr., and they have lengths 8 and 14.
Legis.	14	There are two paths from B to Legis., and they have lengths 14 and 8.
Send-off	26	There are six paths from B to Send-off, and they have lengths 15, 24, 21, 20, 18, and 26.
Purchase	24	There are two paths from B to Purchase, and they have lengths 18 and 24.
E	28	There are eight paths from B to E, and they have lengths 22, 28, 17, 23, 26, 20, 22, and 28.

The minimum time is 28. Note that there are two critical paths.

Chapter 4

4.1   pp. 75–80.

1. (a) A path which repeats no edges.
    (b) A path.
    (c) A circuit which repeats no edges.
    (d) Not a path.
    (e) A circuit.
    (f) A path which repeats no edges.
    (g) Not a path.
    (h) A path which repeats no edges.
    (i) A path which repeats no edges.

3. The longest such path has length 7. There are four such paths: $x_{10}, x_9, x_8,$ x_2, x_3, x_7, x_6, x_5; $x_5, x_6, x_7, x_3, x_2, x_8, x_9, x_{10}$; $x_9, x_{10}, x_8, x_2, x_3, x_7, x_6, x_5$; $x_5, x_6, x_7, x_3, x_2, x_8, x_{10}, x_9$. (One might choose to consider the first and second and third and fourth paths identical).

4. (a) One between x_1 and x_3; one between x_2 and x_5; one between x_3 and x_5. In fact there is precisely one path without repeated vertices between any pair of vertices.

5. (a) x_1, x_6, x_3.
   (b) $x_2, x_1, x_6, x_3, x_4, x_5, x_6, x_3, x_2$.
   (c) x_1, x_6, x_3, x_4, x_5.

8. If a path has a repeated edge, then this path must have repeated one of the two vertices joined by that edge.

12. No; there can be no path directed from x to any other vertex of the digraph.

14. (a) 2.
    (b) 3.
    (c) 4.
    (d) 2.
    (e) 1.
    (f) 3.

16. (a)

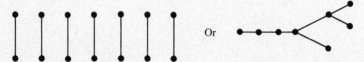

   (b)

18. Acquaintance graphs might work in both cases.

19. Two.

4.2   pp. 89–91.

3. Yes: $x_1, x_2, x_7, x_{11}, x_3, x_7, x_{10}, x_2, x_3, x_4, x_8, x_{13}, x_5, x_8, x_{12}, x_4, x_5, x_6, x_{14},$ $x_{13}, x_{12}, x_{11}, x_{10}, x_9, x_1$.

5. Yes. $x_1, x_2, x_6, x_7, x_2, x_3, x_7, x_5, x_{10}, x_{11}, x_8, x_{10}, x_9, x_8, x_3, x_4, x_5, x_6, x_1$.

4.3 pp. 96–98.

1. There are ten odd-valent vertices, so the theorem says that there are at least five duplications. The minimum number of duplications is seven, which is two more than the estimate of the theorem. We can get these seven duplications by adding the indicated dotted edges:

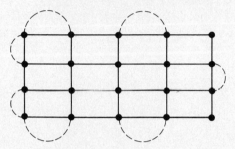

3. (a) Lower estimate supplied by theorem is $\frac{8}{2} = 4$. The minimum number is 4;

    (b) Lower estimate supplied by theorem is $\frac{8}{2} = 4$. The minimum number is actually five;

    (c) Lower estimate supplied by theorem is $\frac{6}{2} = 3$. The minimum number is actually five.

4. If n is even the minimum number is $n/2$. If n is odd, the minimum number is $(n - 1)/2 + r$.

5. If we had an odd number of odd-valent vertices, then the sum of all the valences = sum of all valences of even-valent vertices + sum of all valences of odd-valent vertices = even number + odd number = odd number, which is impossible. See Theorem 2, Section 3.2.

4.4 pp. 103–106.

2. No. Look at the following example:

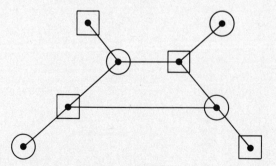

6. The route *ABECDA*, is the cheapest ($399).

9. Probably not, since relative distances would remain about the same, but it is not difficult to find examples where all three solutions are not the same.

11. A knowledge of the traveling salesman problem will help in figuring out the shortest route, thereby spending as little money for gas as possible.

12. $313 (provided by the path *ABECD*).

13. (a) 10 blocks (the route *P, S, B, N, BK, L, P* is such a route).
    (b) No—there is a route *P, L, BK, B, N, S, P* of 10 blocks.

Chapter 5

5.1   p. 111.

1. (a) Keeping records of the date and number of shares of certain stocks or bonds which have been bought or sold.
   (b) Checking tax returns.
   (c) Computing the monthly pension which is due to a retired person.
   (d) Keeping track of the names and number of persons who have registered for a certain section of a certain course.
   (e) Matching compatible personalities.
   (f) Keeping track of heartbeat, respiration rate, etc. of astronauts in space.
   (g) Analysis and tabulation of votes in election districts.

5.3   pp. 123–127.

4. The modification in the list presentation form can be made in the following way: change instruction 5 from "Is $K = 2$"? to "Is $K = 3$"? Exactly the same thing can be done to modify the flow chart.

5.   Modified flow chart:

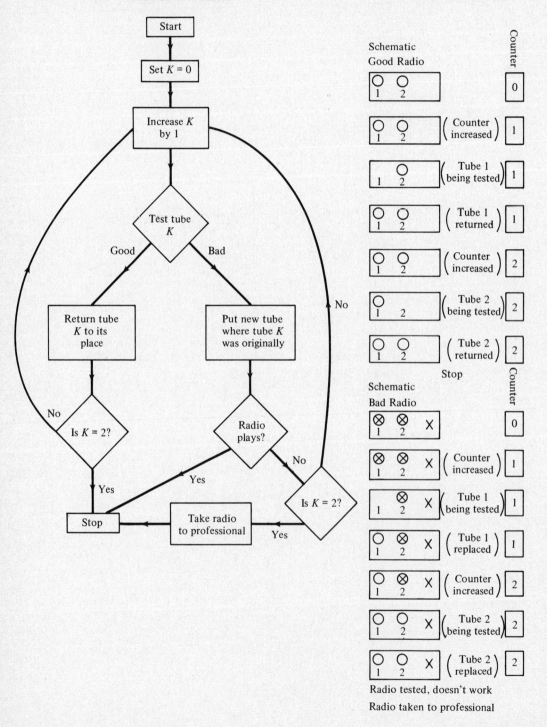

Schematic
Good Radio

Counter

Schematic
Bad Radio

Counter

Radio tested, doesn't work

Radio taken to professional

8. (a)

List	Counter	
1 3 2 4	[0]	
1 3 2 4	[1]	(counter increased by 1)
1 3 2 4	[2]	(counter increased by 1)
1 2 3 4	[2]	(numbers on lines 2 + 3 interchanged)
1 2 3 4	[3]	(counter increased by 1)

Number on line 4 is the answer—algorithm works in this case.

10. No—look at the following list of numbers:

List	Counter	
1 3 2 4	[0]	
1 3 2 4	[1]	(counter increased by 1)
3 1 2 4	[1]	(numbers on lines 1 + 2 interchanged)
3 1 2 4	[2]	(counter increased by 1)
3 2 1 4	[2]	(numbers on lines 2 + 3 interchanged)
3 2 1 4	[3]	(counter increased by 1)

List   *Counter*   (*cont'd*)
3
2
4   ③   (numbers on lines 3 + 4 interchanged)
1

Number on line 1 is not the answer—algorithm doesn't work.

13.   It merely moves the largest one to the last position. Look at the list $\begin{matrix} 4 \\ 3 \\ 2 \\ 1 \end{matrix}$.

15.

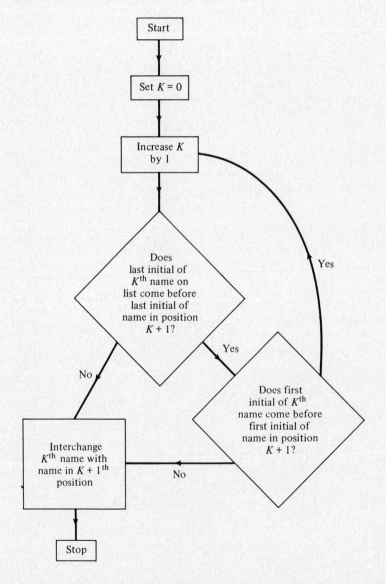

BK

This program doesn't do its job—look at the list *CK*. If two last initials are the same, the program doesn't take first initials into account.

16. Distance from photographer to object being photographed must be taken into account; also, whether to use a flash cube or not.

17. Speed of record being played (e.g. $33\frac{1}{3}$ or 45), degree of loudness, treble and bass levels.

5.4   pp. 133–134.

1. (a)

	HOURS	RATE	PAY
Memory	20	4	
Arithmetic			

1. CLA HOURS

	HOURS	RATE	PAY
	20	4	
		20	

2. MPY RATE

	HOURS	RATE	PAY
	20	4	
		80	

3. STO PAY

	HOURS	RATE	PAY
	20	4	80
		80	

2. (a)

	HOURS	RATE	40-BOX	1.5-BOX	TEMP.	PAY
Memory	70	30	40	1.5		
Arithmetic						

1. CLA HOURS

	HOURS	RATE	40-BOX	1.5-BOX	TEMP.	PAY
	70	30	40	1.5		
	70					

	HOURS	RATE	40-BOX	1.5-BOX	TEMP.	PAY (cont'd)

2. SUB 40-BOX

70	30	40	1.5		
30					

3. MPY 1.5-BOX

70	30	40	1.5		
45					

4. MPY RATE

70	30	40	1.5		
1350					

5. STO TEMP

70	30	40	1.5	1350	
1350					

6. CLA 40-BOX

70	30	40	1.5	1350	
40					

7. MPY RATE

70	30	40	1.5	1350	
1200					

8. ADD TEMP

70	30	40	1.5	1350	
2550					

9. STO PAY

70	30	40	1.5	1350	2550
2550					

3. 1. CLA HOURS
   2. MPY 1.5-BOX
   3. SUB 20-BOX
   4. MPY RATE
   5. STO PAY

This program has just as many instructions as the program for the original formula.

6.   First solution:
1. CLA 9-BOX
2. MPY CENT
3. DIV 5-BOX
4. ADD 32-BOX
5. STO FAHR

       Second Solution:
1. CLA $\frac{9}{5}$-BOX
2. MPY CENT
3. ADD 32-BOX
4. STO FAHR

7. (a)

1. CLA A-BOX
2. ADD B-BOX
3. MPY C-BOX
4. MPY D-BOX
5. STO M-BOX

5.5   *pp. 138–139.*

1. 11—1011, 12—1100, 13—1101, 14—1110, 15—1111, 16—10000, 17—10001, 18—10010, 19—10011, 20—10100.
3. 64, 85, 60, 221.
5. 1048575.
7. 211 is 22, 102 is 11, 222 is 26. 1, 2, 10, 11, 12, 20, 21, 22, 100, 101.
8. 102, 110, 111, 112, 120.
9. 1, 2, 3, 10, 11, 12, 13, 20, 21, 22, 23, 30, 31, 32, 33, 100, 101, 102, 103, 110.

5.6   *pp. 145–146.*

1. (a) 100000. (b) 1000000. (c) 1000101. (d) 110000. (e) 1000001. (f) 1111. (g) 1001110. (h) 100111111.
2. 1111101.
6. (a)   10001110 (187 − 45 = 142).

5.7   pp. 153–155.

1. (a)

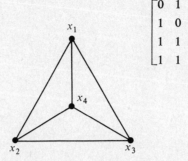

$$\begin{bmatrix} 0 & 1 & 1 & 1 \\ 1 & 0 & 1 & 1 \\ 1 & 1 & 0 & 1 \\ 1 & 1 & 1 & 0 \end{bmatrix}$$

(b)

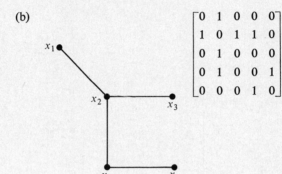

$$\begin{bmatrix} 0 & 1 & 0 & 0 & 0 \\ 1 & 0 & 1 & 1 & 0 \\ 0 & 1 & 0 & 0 & 0 \\ 0 & 1 & 0 & 0 & 1 \\ 0 & 0 & 0 & 1 & 0 \end{bmatrix}$$

(c)

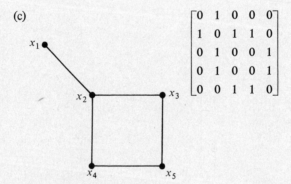

$$\begin{bmatrix} 0 & 1 & 0 & 0 & 0 \\ 1 & 0 & 1 & 1 & 0 \\ 0 & 1 & 0 & 0 & 1 \\ 0 & 1 & 0 & 0 & 1 \\ 0 & 0 & 1 & 1 & 0 \end{bmatrix}$$

2. (a) $\begin{bmatrix} 0 & 0 & 0 & 0 & 0 \\ 1 & 0 & 0 & 0 & 0 \\ 0 & 1 & 0 & 0 & 0 \\ 0 & 0 & 1 & 0 & 0 \\ 0 & 1 & 0 & 1 & 0 \end{bmatrix}$ (b) $\begin{bmatrix} 0 & 1 & 1 & 0 & 0 \\ 0 & 0 & 0 & 0 & 0 \\ 0 & 0 & 0 & 0 & 0 \\ 1 & 0 & 0 & 0 & 0 \\ 0 & 1 & 0 & 1 & 0 \end{bmatrix}$

5. The whole matrix is $\begin{bmatrix} 0 & 1 & 0 \\ 1 & 0 & 1 \\ 0 & 1 & 0 \end{bmatrix}$

6. The valence of x_i is the sum of the entries in the ith row (or i^{th} column).

7. No edge can connect a vertex to itself, by definition of a graph. The same is true for a digraph.

8. $A + B = \begin{bmatrix} 3 & 2 \\ 4 & 8 \end{bmatrix} = B + A, \quad AB = \begin{bmatrix} 8 & 11 \\ 14 & 15 \end{bmatrix}, \quad BA = \begin{bmatrix} 1 & 2 \\ 7 & 25 \end{bmatrix}$

9. $A + B = \begin{bmatrix} 4 & 3 \\ 3 & -1 \end{bmatrix}, \quad AB = \begin{bmatrix} 11 & -7 \\ 9 & -13 \end{bmatrix}, \quad BA = \begin{bmatrix} 2 & 9 \\ 8 & -4 \end{bmatrix}, \quad A^2B = \begin{bmatrix} 29 & -33 \\ 16 & -32 \end{bmatrix}$

13. Let $A = \begin{bmatrix} 1 & -1 \\ -1 & 1 \end{bmatrix}, \quad B = \begin{bmatrix} -1 & -1 \\ -1 & -1 \end{bmatrix}; \quad BA$ is zero.

5.8   *pp. 160–161.*

1. (i) (a) $\begin{bmatrix} 0 & 0 & 1 & 0 \\ 1 & 0 & 0 & 0 \\ 0 & 1 & 0 & 1 \\ 1 & 1 & 0 & 0 \end{bmatrix}$

    (b) $M^2 = \begin{bmatrix} 0 & 1 & 0 & 1 \\ 0 & 0 & 1 & 0 \\ 2 & 1 & 0 & 0 \\ 1 & 0 & 1 & 0 \end{bmatrix}, \quad M^3 = \begin{bmatrix} 2 & 1 & 0 & 0 \\ 0 & 1 & 0 & 1 \\ 1 & 0 & 2 & 0 \\ 0 & 1 & 1 & 1 \end{bmatrix}$

    (c) $\begin{bmatrix} 2 & 2 & 1 & 1 \\ 1 & 1 & 1 & 1 \\ 3 & 2 & 2 & 1 \\ 2 & 1 & 2 & 1 \end{bmatrix}$ (d) Yes.

    (ii) (a) $\begin{bmatrix} 0 & 0 & 1 & 0 \\ 1 & 0 & 1 & 0 \\ 0 & 0 & 0 & 0 \\ 1 & 1 & 1 & 0 \end{bmatrix}$

(b)
$$M^2 = \begin{bmatrix} 0 & 0 & 0 & 0 \\ 0 & 0 & 1 & 0 \\ 0 & 0 & 0 & 0 \\ 1 & 0 & 2 & 0 \end{bmatrix}, \qquad M^3 = \begin{bmatrix} 0 & 0 & 0 & 0 \\ 0 & 0 & 0 & 0 \\ 0 & 0 & 0 & 0 \\ 0 & 0 & 1 & 0 \end{bmatrix}$$

(c)
$$\begin{bmatrix} 0 & 0 & 1 & 0 \\ 1 & 0 & 2 & 0 \\ 0 & 0 & 0 & 0 \\ 2 & 1 & 4 & 0 \end{bmatrix}$$
(d)   No.

(iii)   (a)
$$\begin{bmatrix} 0 & 1 & 1 & 1 \\ 0 & 0 & 1 & 1 \\ 0 & 0 & 0 & 0 \\ 0 & 0 & 1 & 0 \end{bmatrix}$$

(b)
$$M^2 = \begin{bmatrix} 0 & 0 & 2 & 1 \\ 0 & 0 & 1 & 0 \\ 0 & 0 & 0 & 0 \\ 0 & 0 & 0 & 0 \end{bmatrix}, \qquad M^3 = \begin{bmatrix} 0 & 0 & 1 & 0 \\ 0 & 0 & 0 & 0 \\ 0 & 0 & 0 & 0 \\ 0 & 0 & 0 & 0 \end{bmatrix}$$

(c)
$$\begin{bmatrix} 0 & 1 & 4 & 2 \\ 0 & 0 & 2 & 1 \\ 0 & 0 & 0 & 0 \\ 0 & 0 & 1 & 0 \end{bmatrix}$$
(d).   No.

(iv)   (a)
$$\begin{bmatrix} 0 & 1 & 0 & 0 \\ 0 & 0 & 1 & 0 \\ 1 & 0 & 0 & 0 \\ 1 & 1 & 1 & 0 \end{bmatrix}$$

(b)
$$M^2 = \begin{bmatrix} 0 & 0 & 1 & 0 \\ 1 & 0 & 0 & 0 \\ 0 & 1 & 0 & 0 \\ 1 & 1 & 1 & 0 \end{bmatrix}, \qquad M^3 = \begin{bmatrix} 1 & 0 & 0 & 0 \\ 0 & 1 & 0 & 0 \\ 0 & 0 & 1 & 0 \\ 1 & 1 & 1 & 0 \end{bmatrix}$$

(c)
$$\begin{bmatrix} 1 & 1 & 1 & 0 \\ 1 & 1 & 1 & 0 \\ 1 & 1 & 1 & 0 \\ 3 & 3 & 3 & 0 \end{bmatrix}$$
(d)   No.

Chapter 6

6.1   pp. 169–171.

2. 4, 13/2, and 503/2.
4. $f(1) = 2$, $f(2) = 4$, $f(3) = 5$, $f(4) = 1$, $f(5) = 6$, $f(6) = 3$.
6. (a) $p(x, y, z) = 280x + 200y + 400z$.
    (b) $10 \cdot 7 + 15 \cdot 5 + 4 \cdot 10 = 185$.
9. 52¢.
10. $f(1) = 2$, $f(2) = 1$, $f(3) = 4$, $f(4) = 3$, $f(5) = 4$, $f(6) = 5$.
12. (b) $g(x) = 2x + 1$.   (c) $h(x) = 3x - 4$.
    (d) $k(x) = x(x + 1) = x^2 + x$.

6.2   pp. 178–180.

3. $x^2 + 1$ is always a positive number.
4. $\sqrt{x^2 + 3}$ is always a positive number.
6. (a) Yes.   (b) No.   (c) Yes.   (d) No.   (e) No.   (f) Yes.
8. (a)   All real numbers except $x = 7$, $x = 3$.
    (b)   All real numbers except $x = 0$.
    (c)   All real numbers $x \geq 4$.
    (d)   All real numbers $x \geq -5$.
    (e)   All real numbers x except those x with $-3 < x < 1$.
    (f)   All real numbers.
    (g)   All real numbers x except those with $-1 < x < 1$.
9. $f(1) = f(2) = f(3) = f(4) = 0$, yet the function is certainly not identically zero. In general one should look at values other than a, b, c, or d when plotting a function of the form $(x - a)(x - b)(x - c)(x - d)$.

6.3   pp. 182–183.

1. (a)   20, -1, 8.   (b) 20, 52, -20.   (c) -21, -8, -9.
3. $p = \frac{3}{100}L$.
4. $I = 10E$.   (a) 30 amps.   (b) 12 volts.
7. (a) $C = 2\pi r$.   (b) $16\pi \cdot \dfrac{12\pi}{5}$.
8. $t = \frac{1}{3}L - 2$.   246.
9. $I = .06D$.

6.4   pp. 189–191.

1. (a) $-8x + 3y = 7.$ (b) $-7x + 3y = 0.$ (c) $2x + y = 5.$
   (d) $11x + 2y = 62.$
2. (a) $y = 2x + 7.$ (b) $y = -x + 21.$ (c) $y = -5.$
   (d) $y = 9x - 47.$
3. (a) $y = 17x.$ (b) $y = -7x + 5.$ (c) $y = 9.$ (d) $y = 4x - 18.$
5. (a) $212°F.$ (b) $32°F.$ (c) $26\frac{2}{3}°C.$ (d) $-40°.$

6.5   pp. 198–200.

1. (a) The intersection point is $\left(\frac{16}{13}, -\frac{1}{3}\right).$
   (b) No intersection; dual representation.
   (c) No intersection; lines are parallel.
   (d) The intersection point is $(1, 2).$
   (e) The intersection point is $(5, 3).$
   (f) The intersection point is $(8, 1).$
3. Yes; working with the first two equations, we find that $(5, 0)$ is an intersection point of the first two lines. This ordered pair also satisfies the third equation.
4. Let x = number of truck shipments, y = number of train shipments. The data of the problem yield the following equations:

$$x + y = 70$$
$$3x + 5y = 300$$

The solution to these equations is $x = 25$, $y = 45$, so there should be 25 truck shipments and 45 train shipments.
5. Let x = number of packets of deep purple, y = number of packets of light purple. The data of the problem yield the following equations:

$$x + 2y = 5000$$
$$2x + 3y = 7000$$

These equations have the solution $x = -1000$, $y = 3000$. The *no slack* assumption yields a useless model.
6. Let x = number of teams to investigate mergers; y = number of teams to investigate quasimonopolies. The data of the problem yield the following equations:

$$2x + y = 28$$
$$2x + 3y = 48$$

The solution to these equations is $x = 9$, $y = 10$, so there should be 9 teams to investigate mergers and 10 teams to investigate quasimonopolies.

8. Let x = number of fancy tables, y = number of plain tables. The data of the problem yield the following equations:

$$15x + 10y = 600$$

$$24x + 18y = 500$$

These equations have the solution $x = \dfrac{580}{3}$, $y = -230$, so some modification of the model must be made.

9. Let x = number of undershirts, y = number of underpants. The data of the problem yield the following equations;

$$\frac{1}{2}x + \frac{1}{4}y = 200$$

$$18y = 1440$$

The solution to these equations is $x = 370$, $y = 60$, so 370 undershirts and 60 underpants should be made.

6.6   pp. 211–212.

1. (a) $x = y = z = 1$. (b) $x = 1, y = 1, z = 0$.
   (c) $x = 6, y = 1, z = 1$. (d) $x = 1, y = 1, w = 0, z = 0$.
2. (i)(b); (ii)(a); (iii)(b); (iv)(b); (v)(a); (vi)(b).
5. Let x = number of days, y = number of miles, z = number of nights. The data of the problem yield the following equations:

$$120x - y \qquad\qquad = 0$$

$$-10x \qquad + 15z = 10$$

$$10x + 0.1y + 15z = 170$$

The solution to these equations is $x = 5$, $y = 600$, $z = 4$, so the salesman travelled 5 days, 600 miles, and spent 4 nights in a motel.

6.7   p. 221.

1. (a) $-x + 5y < -7$.
   (b) $x > 1$.
   (c) $x < -\frac{1}{7}$.
   (d) $-x - y < 13$.
   (e) $-7x + 8y < 0$.
   (f) $x > 3$.

6.8   *p. 230.*

4.   The resource inequalities are:

$$15x + 10y \leq 600$$
$$24x + 18y \leq 500$$
$$x \geq 0$$
$$y \geq 0$$
$$p = 20x + 15y$$

The corner points are $A = (0, 0)$, $B = \left(\frac{500}{24}, 0\right)$, $C = \left(0, \frac{500}{18}\right)$. The analysis will tell us nothing, since the points where an optimal solution occur do not have integral coordinates.

6.9   *pp. 235–236.*

1.   Let $x =$ number of pounds of G_1 per package,
     $y =$ number of pounds of G_2 per package.
     The resource inequalities are:

$$\left.\begin{array}{l} x + y \leq 200 \\ x + y \leq 200 \end{array}\right\} \quad \text{(These two inequalities imply } x + y = 200.\text{)}$$

$$0 \leq x \leq 80$$
$$y \geq 150$$
$$C = 4x + 6y.$$

The corner points of the feasible region are:

$$A = (50, 150) \text{ and } B = (0, 200)$$

Since at:
$A(50, 150)$ $C = 4(50) + 6(150) = 1100$
$B(0, 200)$ $C = 4(0) + 6(200) = 1200$
The optimal solution is to have 50 pounds of G_1 per package and 150 pounds of G_2 per package.

3.   Let $x =$ number of stoves S, $y =$ number of stoves T. The resource inequalities are:

$$3x + 2y \leq 40$$
$$5x + 4y \leq 40$$
$$x \geq 0$$
$$y \geq 0$$
$$PGS = 50x + 80y$$

The corner points of the feasible region are $A = (0, 0)$ $B = (8, 0)$, $C = (0, 10)$.

At these points, potential gross sales (*PGS*) are:

A (0, 0)   *PGS* = 50(0) + 80(0) = 0
B (8, 0)   *PGS* = 50(8) + 80(0) = 400
C (0, 10)   *PGS* = 50(0) + 80(10) = 800

To maximize potential gross sales, no stoves of type *S* and 10 stoves of type *T* should be produced.

5. Let x = number of dresses, y = number of blouses. The resource inequalities are:

$$4x + 2y \leq 60$$
$$2x + 4y \leq 48$$
$$x \geq 0$$
$$y \geq 0$$
$$p = 8x + 6y$$

The corner points of the feasible region are $A = (0, 0)$, $B = (15, 0)$, $C = (12, 6)$, $D = (0, 12)$.

A (0, 0) $p = 8(0) + 6(0) = 0$
B (15, 0) $p = 8(15) + 6(0) = 120$
C (12, 6) $p = 8(12) + 6(6) = 132$
D (0, 12) $p = 8(0) + 6(12) = 72$

Twelve dresses and six blouses should be made to maximize profit.

Chapter 7

7.1   pp. 247–248.

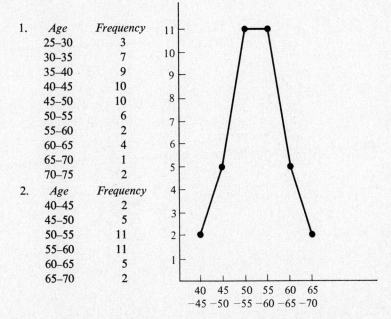

1.	*Age*	*Frequency*
	25–30	3
	30–35	7
	35–40	9
	40–45	10
	45–50	10
	50–55	6
	55–60	2
	60–65	4
	65–70	1
	70–75	2
2.	*Age*	*Frequency*
	40–45	2
	45–50	5
	50–55	11
	55–60	11
	60–65	5
	65–70	2

6.

Score	Frequency
0–10	0
10–20	0
20–30	0
30–40	2
40–50	1
50–60	2
60–70	9
70–80	11
80–90	8
90–100	5

9.

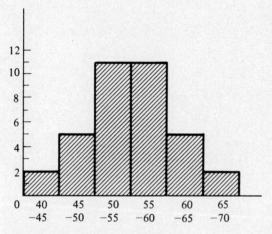

16.

Age	Relative Frequency
40–45	.06
45–50	.14
50–55	.31
55–60	.31
60–65	.14
65–70	.06

22.

Crime	Relative Frequency
A	.19
B	.22
F	.11
R	.30
L	.19

7.2   *pp. 251–252.*

1. 1900
2. Can't be estimated from Table 7.4.
3. Approximately 106 in 1980.

7.3   *pp. 257–258.*

1. Mean = 44.3, median = 42, mode = 46. The mean or median is a better model of the data.
3. Mean = 378.4, median = 308, mode does not exist. The median is a better model of the data.
5. Mean = 73.2, median = 75, mode = 65. The median is a better model of the data.
6. A mean.
7. (a) Mean = 3.33, median = 3.5, mode = 4.
   (b) Mean = 0, median = 0, mode = −2.
   (c) Mean = −.875, median = −1.5, mode does not exist.
   (d) Mean = 66, median = 45, mode = 70.
10. (a) $\dfrac{14 \cdot 30 + 20 \cdot 34}{34} = \; = 32.4$
    (b) $\dfrac{100 + 90 + 100 + 70 + 2(50) + 4(80)}{1 + 1 + 1 + 1 + 2 + 4} = \dfrac{780}{10} = 78.$

7.4   *p. 263.*

1. (a)

   Mean = $4\frac{5}{9}$, median = 5, mode doesn't exist.
   Error Def. I: error = 9 for mean, 7 for median.
   Error Def. II: error = $|(1 - 4\frac{5}{9}) + (2 - 4\frac{5}{9}) + (2 - 4\frac{5}{9}) + (4 - 4\frac{5}{9})$
   $+ (5 - 4\frac{5}{9}) + (5 - 4\frac{5}{9}) + (6 - 4\frac{5}{9}) + (8 - 4\frac{5}{9}) + (8 - 4\frac{5}{9})| = 0$ for
   mean, $|(1 - 5) + (2 - 5) + (2 - 5) + (4 - 5) + (5 - 5)$
   $+ (5 - 5) + (6 - 5) + (8 - 5) + (8 - 5)| = 4$ for median.
   Error Def. III: error = $|1 - 4\frac{5}{9}| + |2 - 4\frac{5}{9}| + |2 - 4\frac{5}{9}| + |4 - 4\frac{5}{9}|$
   $+ |5 - 4\frac{5}{9}| + |5 - 4\frac{5}{9}| + |6 - 4\frac{5}{9}| + |8 - 4\frac{5}{9}| + |8 - 4\frac{5}{9}| = 18\frac{4}{9}$
   for mean, $|1 - 5| + |2 - 5| + |2 - 5| + |4 - 5| + |5 - 5|$
   $+ |5 - 5| + |6 - 5| + |8 - 5| + |8 - 5| = 18$ for median.

7.5   *pp. 268–269.*

1. (a) $\mu = 5, \delta = \sqrt{5.9}.$ (b) $\mu = 1.4, \delta = \sqrt{.05}.$
   (c) $\mu = 3, \delta = \dfrac{\sqrt{8}}{3}.$

2. (a) $\mu = 3\frac{1}{3}, \delta = 1\frac{1}{3}.$ (b) $\mu = 0, \delta = 2.$

    (c) $\mu = -\frac{7}{8}, \delta = \dfrac{\sqrt{3469}}{32}.$ (d) $\mu = 66, \delta = \sqrt{184}.$

4. $\mu = 5, \delta = \sqrt{\dfrac{28}{5}} = \sqrt{5.6} = 2.37.$

6. $1 - 1/h^2 = .99, 1/h^2 = .01, h^2 = 100, h = 10.$ The minimum height should be adjusted to $66 - 2(10) = 46$ inches. The maximum height should be adjusted to $66 + 2(10) = 86$ inches.

8. $4.5 + 1.7h = 7, 1.7h = 2.5, h = 1.47.$
    Percentage $= [1 - 1/(1.47)^2] \times 100\% = 53.7\%$ (approximately).

9. $1 - 1/h^2 = .95, h^2 = 20, h = 4.5. \mu = 9, \delta = 1/\sqrt{2} = .707$ (approximately). Required distance $= 9 + (4.5)(.707) = 12.2$ (approximately).

Chapter 8

8.1   pp. 273–275.

1. $\dfrac{322}{481} = .669.$

3. $\dfrac{23465 - 11564}{23465} = \dfrac{11901}{23465} = .507.$

5. $\dfrac{340}{670} = .507.$

6. (a) $p = \dfrac{1000}{1001}.$

    (b) $p = \frac{5}{6}.$ The stabilized value relative frequency of John's beating Jake at chess is $\frac{5}{6}.$

11. (a) Yes.   (b) Yes.   (c) No.   (d) No.   (e) No.   (f) No.

8.2   pp. 283–285.

1. (a) Yes.
    (b) No—the sum of all the probabilities $\neq 1.$
    (c) No—$P(S) < 0.$
    (d) No—$P(M) > 1.$
    (e) No—the sum of all the probabilities $\neq 1.$
    (f) No—the sum of all the probabilities $\neq 1.$

2. (a) $\dfrac{2}{6} = \dfrac{1}{3}$ (b) $\dfrac{3}{6} = \dfrac{1}{2}$ (c) $\dfrac{2}{6} = \dfrac{1}{3}$

3. $\{A, B, C, D, E, F, G, H, I, J, K, L, M, N, O, P, Q, R, S, T, U, V, W, X, Y, Z\}.$

4. (a) $\dfrac{13}{52} + \dfrac{13}{52} = \dfrac{1}{2}.$ (b) $\dfrac{4}{52} + \dfrac{4}{52} = \dfrac{2}{13}.$ (c) $\dfrac{13}{52} + \dfrac{3}{52} = \dfrac{4}{13}.$

7. $P(B) = \dfrac{10}{30} = \dfrac{1}{3}, P(G) = \dfrac{2}{3}.$

8. (a) $\dfrac{4}{10} = \dfrac{2}{5}$. (b) $\dfrac{2}{5} + \dfrac{2}{10} = \dfrac{3}{5}$. (c) $1 - \dfrac{4}{10} = \dfrac{3}{5}$.

(d) $1 - \dfrac{2}{10} = \dfrac{4}{5}$.

9. (a) $\dfrac{2+3}{9} = \dfrac{5}{9}$. (b) $1 - \dfrac{4}{9} = \dfrac{5}{9}$.

13. No, because the states don't have equal population distributions.

8.3   *pp. 291–293.*

2.

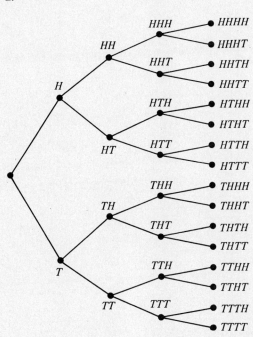

7.

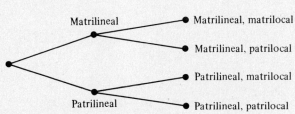

10. $4 \cdot 6 \cdot 5 = 120.$

12.

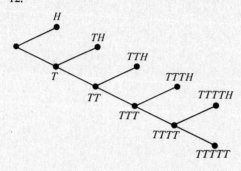

13. $12 \cdot 2 \cdot 2 \cdot 2 \cdot 2 = 192.$

14. $4 \cdot 2 \cdot 4 = 32.$

16. $1 - \dfrac{12 \cdot 11 \cdot 10 \cdot 9 \cdot 8 \cdot 7 \cdot 6}{12^7} = .89.$

18. (a) $1 - \dfrac{12 \cdot 11}{12^2} = .083.$

8.4   *pp. 301–303.*

3. (a) $\dfrac{1}{10} \cdot \dfrac{1}{10} = \dfrac{1}{100}.$ (b) $\dfrac{9}{10} \cdot \dfrac{9}{10} = \dfrac{81}{100}.$ (c) $1 - \dfrac{1}{100} = \dfrac{99}{100}.$

4. $\dfrac{1}{2} \cdot \dfrac{2}{3} = \dfrac{1}{3}.$

6. $\dfrac{1}{5} \cdot \dfrac{1}{5} \cdot \dfrac{1}{5} = \dfrac{1}{125}.$

7. $\dfrac{1}{4} \cdot \dfrac{1}{4} = \dfrac{1}{16}.$

9. $\dfrac{9}{10} \cdot \dfrac{1}{10,000} = \dfrac{9}{100,000}.$

10. $\dfrac{3}{4} \cdot \dfrac{3}{4} = \dfrac{9}{16}.$

11. $\dfrac{365}{365^4} = \dfrac{1}{365^3} = \dfrac{1}{48,590,625}.$

8.5   *pp. 309–310.*

1. $b\left(6, 10, \dfrac{1}{20}\right).$

2. $b\left(9, 20, \dfrac{1}{2}\right).$

3. $b\left(9, 12, \frac{9}{10}\right) + b\left(10, 12, \frac{9}{10}\right) + b\left(11, 12, \frac{9}{10}\right) + b\left(12, 12, \frac{9}{10}\right).$

5. $\left(\frac{1}{5}\right)^5.$

6. $b\left(5, 7, \frac{1}{2}\right) + b\left(6, 7, \frac{1}{2}\right) + b\left(7, 7, \frac{1}{2}\right).$

9. $b(r, N, p)$ = probability of r successes in N attempts = probability of $N - r$ failures in N attempts = $b(N - r, N, 1 - p)$.

10. The sum of these probabilities represents the probability of the whole sample space, which = 1.

8.6   *pp. 316–317.*

1. $\frac{1}{6}(1 + 2 + 3 + 4 + 5 + 6) = \frac{21}{6} = 3\frac{1}{2}.$

3. The possible outcomes are H, TH, TTH, TTT.
   Expected number of kisses = $0 \cdot P(H) + 1 \cdot P(TH) + 2 \cdot P(TTH)$
   $+ 3 \cdot P(TTT) = 1 \cdot \frac{3}{4} \cdot \frac{1}{4} + 2 \cdot \frac{3}{4} \cdot \frac{3}{4} \cdot \frac{1}{4} + 3 \cdot \frac{3}{4} \cdot \frac{3}{4} \cdot \frac{3}{4} = \frac{111}{64}.$
   Expected number of bumps = $1 \cdot P(H) + 1 \cdot P(TH) + 1 \cdot P(TTH)$
   $+ 0 \cdot P(TTT) = 1 \cdot \frac{1}{4} + 1 \cdot \frac{3}{4} \cdot \frac{1}{4} + 1 \cdot \frac{3}{4} \cdot \frac{3}{4} \cdot \frac{1}{4} = \frac{37}{64}.$

5. (a) $\frac{1}{2} \cdot 500 = 250.$ (b) $\frac{1}{3} \cdot 500 = 166\frac{2}{3}.$ (c) $\frac{1}{10} \cdot 500 = 50.$

6. Expected number of heads = $\frac{1}{2} \cdot 10 = 5.$

   $P(5 \text{ heads appear}) = b\left(5, 10, \frac{1}{2}\right) = .246.$

8.7   *pp. 323–324.*

1. $M^2 = \begin{bmatrix} \frac{5}{9} & \frac{4}{9} \\ \frac{4}{9} & \frac{5}{9} \end{bmatrix}, \qquad M^3 = \begin{bmatrix} \frac{13}{27} & \frac{14}{27} \\ \frac{14}{27} & \frac{13}{27} \end{bmatrix}, p_{22}^{(2)} = \frac{5}{9}, p_{12}^{(3)} = \frac{14}{27}, p_{11}^{(3)} = \frac{13}{27}.$

3. $\begin{array}{cc} & E \quad S \end{array}$
   $M = \begin{array}{c} E \\ S \end{array}\begin{bmatrix} \frac{1}{3} & \frac{2}{3} \\ \frac{1}{2} & \frac{1}{2} \end{bmatrix}, \qquad M^4 = \begin{bmatrix} \frac{139}{324} & \frac{185}{324} \\ \frac{185}{432} & \frac{247}{432} \end{bmatrix}, p_{11}^{(4)} = \frac{139}{324}.$

4. $(.9998)(.9998) + (.0002)(.0001).$

5. The entries in the first row of M^3 are $\frac{17}{72}, \frac{385}{864}, \frac{275}{864}$ respectively. Hence,

$$P(\text{being in } A \text{ after 3 transitions}) = \frac{17}{72}$$

$$P(\text{being in } B \text{ after 3 transitions}) = \frac{385}{864}$$

$$P(\text{being in } C \text{ after 3 transitions}) = \frac{275}{864}$$

7.
$$M = \begin{array}{c} \\ E \\ L \end{array}\begin{array}{cc} E & L \\ \left[\begin{array}{cc} \frac{1}{2} & \frac{1}{2} \\ \frac{1}{4} & \frac{3}{4} \end{array}\right] \end{array}, \quad M^4 = \left[\begin{array}{cc} \frac{43}{128} & \frac{85}{128} \\ \frac{85}{256} & \frac{171}{256} \end{array}\right], p_{12}^{(4)} = \frac{85}{128}.$$

9.   The probability that the salesman goes from A to A is $\frac{1}{3}$, which is nonsense.

8.8   *pp. 331–332.*

1. (a) M has no zero entries.

    (b) $M^2 = \left[\begin{array}{cc} \frac{1}{5} & \frac{4}{5} \\ \frac{4}{25} & \frac{21}{25} \end{array}\right]$ no zero entries.

    (c) $M^4 = \left[\begin{array}{ccc} \frac{3}{8} & \frac{3}{8} & \frac{1}{4} \\ \frac{5}{16} & \frac{3}{16} & \frac{1}{2} \\ \frac{5}{16} & \frac{1}{8} & \frac{9}{16} \end{array}\right]$ no zero entries.

    (d) $M^5 = \left[\begin{array}{ccc} \frac{3}{16} & \frac{9}{16} & \frac{1}{4} \\ \frac{1}{16} & \frac{3}{8} & \frac{9}{16} \\ \frac{9}{64} & \frac{31}{64} & \frac{3}{8} \end{array}\right]$ no zero entries.

    (e) $M^2 = \left[\begin{array}{ccc} \frac{7}{24} & \frac{5}{12} & \frac{7}{24} \\ \frac{5}{24} & \frac{11}{24} & \frac{1}{3} \\ \frac{7}{36} & \frac{4}{9} & \frac{13}{36} \end{array}\right]$ no zero entries.

3. (a) M is not regular;
    (b) M is not regular;
    (c) M is regular:

$$[w_1, w_2, w_3]\begin{bmatrix} 0 & \frac{1}{2} & \frac{1}{2} \\ 0 & 0 & 1 \\ \frac{1}{2} & \frac{1}{2} & 0 \end{bmatrix} = [w_1, w_2, w_3]$$

$$\tfrac{1}{2}w_3 = w_1$$

$$\tfrac{1}{2}w_1 + \tfrac{1}{2}w_3 = w_2$$

$$\tfrac{1}{2}w_1 + w_2 = w_3$$

$$w_1 + w_2 + w_3 = 1$$

Hence $w_1 = \frac{2}{9}$, $w_2 = \frac{1}{3}$, $w_3 = \frac{4}{9}$, so the fixed vector is $[\frac{2}{9}, \frac{1}{3}, \frac{4}{9}]$.

(d) M is regular:

$$[w_1, w_2]\begin{bmatrix} .6 & .4 \\ .3 & .7 \end{bmatrix} = [w_1, w_2]$$

$$.6w_1 + .3w_2 = w_1$$

$$.4w_1 + .7w_2 = w_2$$

$$w_1 + w_2 = 1$$

Hence $w_1 = \frac{3}{7}$, $w_2 = \frac{4}{7}$, so the fixed vector is $[\frac{3}{7}, \frac{4}{7}]$.

5.   Less intelligent.

8.9   pp. 341–342.

2.   (a) $u = 1$, $v = w = 0$, $v^2 = 4uw$, so the population is in equilibrium.

      (b) $u = 0$, $v = 1$, $w = 0$, $v^2 \neq 4uw$, so the population is not in equilibrium.

      (c) $u = v = 0$, $w = 1$, $v^2 = 4uw$, so the population is in equilibrium.

      (d) $u = \frac{1}{100}$, $v = \frac{98}{100}$, $w = \frac{1}{100}$, $v^2 \neq 4uw$, so the population is not in equilibrium.

      (e) $u = \frac{16}{25}$, $v = \frac{8}{25}$, $w = \frac{1}{25}$, $v^2 = \frac{64}{(25)^2}$, $4uw = 4 \cdot \frac{16.1}{(25)^2} = \frac{64}{(25)^2}$, $v^2 = 4uw$, so the population is in equilibrium.

      (f) $u = \frac{1}{100}$, $v = \frac{16}{100}$, $w = \frac{83}{100}$, $v^2 = \frac{256}{(100)^2}$, $4uw = 4 \cdot \frac{83.1}{(100)^2} = \frac{252}{(100)^2}$, $v^2 \neq 4uw$, so the population is not in equilibrium.

4.   (a) $u = \frac{150}{300} = \frac{1}{2}$, $v = \frac{120}{300} = \frac{2}{5}$, $w = \frac{30}{300} = \frac{1}{10}$.

Proportion of first generation having $AA = \left(\frac{1}{2} + \frac{1}{2}\left(\frac{2}{5}\right)\right)^2$

$$= \left(\frac{1}{2} + \frac{1}{5}\right)^2 = \left(\frac{7}{10}\right)^2 = \frac{49}{100}.$$

Proportion of first generation having Aa

$$= 2\left(\frac{1}{2} + \frac{1}{2}\left(\frac{2}{5}\right)\right)\left(\frac{1}{10} + \frac{1}{2}\left(\frac{2}{5}\right)\right) = 2\left(\frac{7}{10}\right)\left(\frac{3}{10}\right) = \frac{42}{100}.$$

Proportion of first generation having $aa = \left(\frac{1}{10} + \frac{1}{2}\left(\frac{2}{5}\right)\right)^2$

$$= \left(\frac{3}{10}\right)^2 = \frac{9}{100}.$$

      (b)   No.

6.   (a)   This is possible. Both father and mother can have genes Rr.

      (b)   Not possible. Both parents must have genes rr.

      (c)   This is possible. See (a).

      (d)   This is possible. The father's genes are rr, the mother's genes could be Rr.

      (e)   This is possible. The father's genes are rr, the mother's genes could be Rr (or RR).

Chapter 9

9.1   p. 349.

1. (a)

1	−3

    (b) No row can be eliminated.

(c)

3	1

(d)

2	1	4
3	1	2

(e)

1	−2	3
2	1	1

(f)

1	−1	3
4	2	2

9.2   pp. 354–356.

1. (a) The second row dominates the first, so the first row can be eliminated.

   Expected payoff for second row $= \frac{1}{2}(1) + \frac{1}{2}(2) = \frac{3}{2}$.

   Expected payoff for third row $= \frac{1}{2}(2) + \frac{1}{2}(-3) = -\frac{1}{2}$.

   Hence, the best row is the second.

2. (a) Let p denote the probability on the first column. Then the probability on the second column is $1 - p$. We want $1 \cdot p + 2(1 - p) = 2p - 3(1 - p)$, i.e.

$$p + 2 - 2p = 2p - 3 + 3p,$$

$$5 = 6p, \, p = \frac{5}{6}, \, 1 - p = \frac{1}{6}$$

3. Expected payoff if alarm is sounded $= -20(.05) + -1(.95) = -1.95$.
   Expected payoff if alarm isn't sounded $= -50(.05) + 0(.95) = -2.5$.
   Hence the alarm should always be sounded.

6.

	$\frac{1}{2}$ A	$\frac{1}{2}$ B
Buy Stock	−.01	.08
Buy Bonds	.05	.04

Expected payoff if stock is bought $= \frac{1}{2}(-.01) + \frac{1}{2}(.08) = .035$.

Expected payoff if bonds are bought $= \frac{1}{2}(.05) + \frac{1}{2}(.04) = .045$.

Hence, you should buy bonds.

7.

	$\frac{2}{3}$ Judge Stern	$\frac{1}{3}$ Judge Lenient
Plead Guilty	5	5
Plead Innocent	10	0

Expected fine if you plead guilty $= \frac{2}{3}(5) + \frac{1}{3}(5) = 5 = \frac{15}{3}$

Expected fine if you plead innocent $= \frac{2}{3}(10) + \frac{1}{3}(0) = \frac{20}{3}$

Hence, you should plead guilty.

9.3   pp. 362–366.

2.   Expected profit if he goes on the trip $= \frac{1}{2}[(\frac{3}{4})(4,600) + (\frac{1}{4})(2,600)]$
    $+ \frac{1}{2}[(\frac{3}{4})(1,600) + (\frac{1}{4})(-400)] = 2,600$.
    Expected profit if he doesn't go on the trip $= 3,000$.
    Therefore, he shouldn't go on the trip.

4.

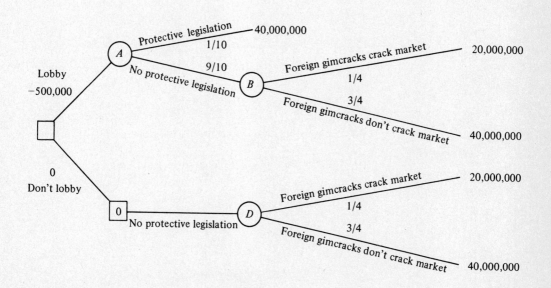

Projected 4-year profit if the industry lobbies $= \frac{1}{10}(40,000,000)$

$$+ \frac{9}{10} \cdot \frac{1}{4}(20,000,000) + \frac{9}{10} \cdot \frac{3}{4}(40,000,000) - 500,000 = 35,000,000.$$

Projected 4-year profit if the industry doesn't lobby $= \frac{1}{4}(20,000,000)$

$$+ \frac{3}{4}(40,000,000) = 35,000,000.$$

Therefore, it doesn't matter what the industry does.

8. Total cost if hurricane is seeded $= 25 + 94.00 = 119.00$ million.
   Total cost if hurricane isn't seeded $= 116$ million.
   Therefore, the hurricane shouldn't be seeded.

9. The cost to Meander is $\frac{1}{6}$ dollar per minute.

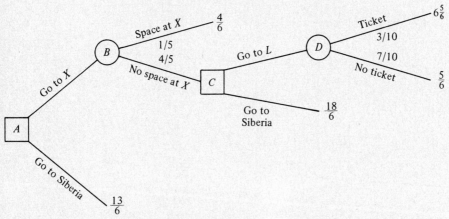

Cost if Meander goes to $L = \frac{3}{10} \cdot \frac{65}{6} + \frac{7}{10} \cdot \frac{5}{6} = \frac{230}{60}$.

Cost if Meander goes to Siberia $= \frac{18}{6} = \frac{180}{60}$ — hence, Meander should go to Siberia if he goes to x and there is no space at x.

Expected cost if Meander goes to $x = \frac{1}{5} \cdot \frac{4}{6} + \frac{4}{5} \cdot \frac{18}{6} = \frac{76}{30}$.

Expected cost if Meander goes to Siberia $= \frac{13}{6} = \frac{65}{30}$.

Therefore, Meander should go to Siberia right away.

9.4   pp. 372–374.

1. (a) $\boxed{2}$ (b) $\boxed{-1}$ (c)

1	−1
−1	1

   (d)

2	−4
−7	3

In (e) and (f), no further simplification is possible.

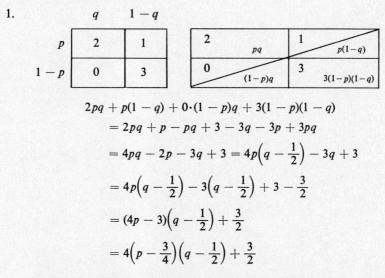

(g)

-1	0
2	-1

2.

	II_1	II_2
I_1	1	-1
I_2	-1	1

No further simplification is possible

5.

	Attack I	Attack II
Defend I	2	-3
Defend II	4	-3

This matrix can be simplified to $\boxed{-3}$, i.e., the situation where II is defended and attacked (resulting in a payoff of $+3$ to the attackers).

8. If it were a zero-sum game, then if A sold 2 widgets in a particular week (for example), B would have to sell -2 widgets that week, which is absurd.

9.5 *pp. 381–382.*

1.

	q	$1-q$
p	2	1
$1-p$	0	3

2 $\quad pq$	1 $\quad p(1-q)$
0 $\quad (1-p)q$	3 $\quad 3(1-p)(1-q)$

$$2pq + p(1 - q) + 0 \cdot (1 - p)q + 3(1 - p)(1 - q)$$
$$= 2pq + p - pq + 3 - 3q - 3p + 3pq$$
$$= 4pq - 2p - 3q + 3 = 4p\left(q - \frac{1}{2}\right) - 3q + 3$$
$$= 4p\left(q - \frac{1}{2}\right) - 3\left(q - \frac{1}{2}\right) + 3 - \frac{3}{2}$$
$$= (4p - 3)\left(q - \frac{1}{2}\right) + \frac{3}{2}$$
$$= 4\left(p - \frac{3}{4}\right)\left(q - \frac{1}{2}\right) + \frac{3}{2}$$

Value of the game $= \frac{3}{2}$. The optimal mixed strategy for the row player is to play the first row $\frac{3}{4}$ of the time, the second row $\frac{1}{4}$ of the time. The optimal mixed strategy for the column player is to play the first column $\frac{1}{2}$ the time, the second column $\frac{1}{2}$ the time.

4.   The analysis will not work. For example, consider the matrix

3	2
1	−1

The first row dominates the second and the second column dominates the first. The value of the game is 2. Applying mixed strategy analysis to the matrix, we get $3pq + 2p(1-q) + (1-p)q - (1-p)(1-q) = 3pq + 2p - 2pq + q - pq\,1 + p + q - pq = -pq + 3p + 2q - 1 = -p(q-3) + 2q - 1 = -p(q-3) + 2(q-3) + 6 - 1 = -(p-2)(q-3) + 5,$ which implies that the value of the game is 5 (also, the p and q values are impossible).

9.6   pp. 387–388.

1.   (a)   The matrix can be simplified to $\boxed{3,1}$. There is no cooperative improvement.

(b)   The matrix can be simplified to $\boxed{3,2}$. $\boxed{4,3}$ is a cooperative improvement.

(c)   The matrix can be simplified to $\boxed{1,2}$. $\boxed{4,3}$ is a cooperative improvement.

9.7   pp. 391–392.

1.   (a)   0, 0 is the only equilibrium pair. 10, 9 is a cooperative improvement over this pair.

(b)   4, −1 is the only equilibrium pair. There is no cooperative improvement over this pair.

(c)   0, 0 is the only equilibrium pair. There is no cooperative improvement over this pair.

Chapter 10

10.1   pp. 397–398.

1.   The winner gets 101 votes; the other three candidates get 100 votes apiece. The winner gets 25.3 percent of the vote.

10.2   *pp. 401–402.*

1. 24
2. $n(n - 1)(n - 2)\ldots 3\cdot 2\cdot 1 = n!$
3. 6; 6; 6.
5. 12; 12; 12.
6. 51. Together with the 24 schedules satisfying Modeling Assumption 1, there are 75 schedules.

10.3   *pp. 404–405.*

1. (b) B_3 is the plurality winner; B_2 is the run-off winner.
    (c) B_3 is the plurality winner; B_3 is the run-off winner.
2. No; yes. B_2 would win the run-off instead of B_1

10.4   *pp. 406–408.*

2. For the election (d) in Exercise 1, system (b) gives B_1 120 points, B_2 139 points, B_3 169 points, and B_4 136 points, while system (d) gives B_1 192 points, B_2 211 points, B_3 233 points, and B_4 210 points.
4. Use election (c) in Figure 10.9.
5. Use an election with 10 votes for schedule B (see Figure 10.2), 9 votes for schedules D, and 8 votes for schedule F.

10.5   *pp. 411–412.*

1. For the election (*b*) the plurality winner is B_2, the run-off winner is B_2, the Borda Count winner is B_1 and the Condorcet winner is B_2. For the election (*c*), the plurality winner is B_2, the run-off winner is B_3, the Borda Count winner is B_2, and there is no Condorcet winner
2. (a)   Use the election in Figure 10.9 (b).
    (b)   Modify the schedules in Figure 10.9 (*a*), by placing B_4 in the fourth position on each.
    (c)   Refer to Figure 10.2. Use 7 votes for schedule A, 3 votes for schedule B, 6 votes for schedule D, and 5 votes for schedule E, modified by placing B_4 at the bottom of each of these schedules.
5. The invalence of every vertex of the digraph is greater than zero. No.
7. The number of points a candidate receives from a preference schedule equals the number of candidates below him on that schedule.

8. Provided the system has a means of breaking a tie, when two or more candidates can beat equal numbers of opponents in two-way races, then this system could be used. Whether it would represent a superior system to the Borda Count is unclear.

Chapter 11

11.1   pp. 422–424.

1. (a) $f(t) = 2(t - 7) + 10$.
   (b) $f(t) = -3(t - 3)$.
   (c) $f(t) = 4$.
   (d) $f(t) = t$.
3. (a) $p(t) = t + 20$; when $t = 20$, $p(t) = 40$.
   (c) $p(t) = 3t + 20$; when $t = 20$, $p(t) = 80$.

4. (a)

t	$f(t)$
0	1
1	5
2	29
3	845

(b)

t	$f(t)$
0	20
1	16
2	0
3	−36
4	−100

(c)

t	$f(t)$
0	1
1	5
2	13
3	25
4	41
5	61
6	85

(d)

t	$f(t)$
0	1
1	1
2	1
3	1
4	1
5	1
6	1

(e)

t	$f(t)$
0	1
1	2
2	2
3	2

5. Let $M(t)$ be the mileage gone after t days. The table suggests that $M(t + 1) - M(t) = 200$ and $M(1) = 200$ (approximately). Thus $M(t) = 200t$. When $t = 50$, $M(t) = 10,000$. This will not be reliable if the 200 mi/day average doesn't persist. If the terrain changes it may not persist.

7.* We can find $f(10)$ this way:

$$f(10) - f(8) = 3$$
$$f(8) - f(6) = 3$$
$$f(6) - f(4) = 3$$
$$f(4) - f(2) = 3$$
$$f(2) - f(0) = 3$$

Adding gives $f(10) - f(0) = 15$. Since $f(0) = 0$, $f(10) = 15$. We can't find $f(11)$, but we could if we knew $f(t)$ for some odd integer value of t.

11.2 pp. 433–435.

1. (a) $f(t) = 4(3)^t$

t	$f(t)$
0	4
1	12
2	36

(b) $f(t) = 5(2)^t$

t	$f(t)$
0	5
1	10
2	20

(c) $f(t) = 800(1.25)^t$

t	$f(t)$
0	800
1	1000
2	1250

(d) $f(t) = 20(2.5)^t$

t	$f(t)$
0	20
1	50
2	125

3.   (a) $f(t) = 0.$
     (b) $f(t) = 3(1.5)^t.$
     (c) $f(t) = 4(1.5)^t.$
5. $f(t) = 150(2)^t.$
     $f(10) = (150)(1024) = 153,600.$
6. $f(t + 1) - f(t) = .05\, f(t).$
7. $f(t + 1) - f(t) = .02\, f(t).$
     5% discard rate prevents growth.
9.* $f(t) = 10,000(-1)^t$

t	$f(t)$
0	10,000
1	−10,000
2	10,000
3	−10,000

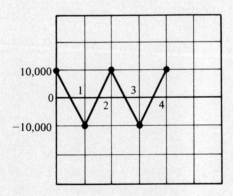

(a)   The graph oscillates.
(b)   The only values ever assumed by $f(t)$ are +10,000 and −10,000. Thus it does not grow without limit.

11.* $$f(t + 2) - f(t + 1) = 5[f(t + 1) - f(t)]$$
     i.e. $$f(t + 2) = 6f(t + 1) - 5f(t).$$

This equation allows us to compute $f(2)$ (take $t = 0$), $f(3)$, ... using $f(0)$

and $f(1)$. $f(0)$ is given as 1 by the initial condition. $f(1)$ cannot be determined from our equation but clearly equals 6.

t	$f(t)$
0	1
1	6
2	31
3	156
4	781

$f(t)$ grows larger without limit, but since there are only a finite number of people available to participate, something has got to give. What usually happens is that after a few stages the letter writers have difficulty finding any *new* participants and so they can't collect any money.

11.3 pp. 441–442.

1. (a) $x = 100, y = 80$.
   (b) $x = 1000, y = 5000$.
   (c) $x = 42, y = 64$.
2. $x = 150, y = 120, z = 60$.

11.4 p. 450.

1. Equilibrium is at $x = 100, y = 100$.
   (a) Let $x(t) = 95, y(t) = 95$.

   $$x(t + 1) - 95 = (.3 - .190 - .095)95 = 1.425$$
   $$y(t + 1) - 95 = (.25 - .114 - .1235)95 = 1.1875$$

   (b) Let $x(t) = 105, y(t) = 105$.

   $$x(t + 1) - 105 = (.300 - .210 - .105)105 = -1.575$$
   $$y(t + 1) - 105 = (.2500 - .1260 - .1365)105 = -1.3125$$

   (c) Let $x(t) = 95, y(t) = 105$.

   $$x(t + 1) - 95 = (.300 - .190 - .105)95 = .475$$
   $$y(t + 1) - 105 = (.2500 - .114 - .1365)105 = -.0525$$

(d) Let $x(t) = 105$, $y(t) = 95$.

$$x(t + 1) - 105 = (.300 - .210 - .095)105 = -.525$$
$$y(t + 1) - 95 = (.2500 - .1260 - .1235)95 = .0475$$

4.* If $x(t) = 0$, it is impossible for there to be a "disturbance" making $x(t) > 0$ (an extinct species will not reappear spontaneously) or $x(t) < 0$ (when counting members of a species you can't have less than nothing).

index